Theoretical and Computational Research in Various Scheduling Models

Theoretical and Computational Research in Various Scheduling Models

Editors

Chin-Chia Wu
Win-Chin Lin

MDPI • Basel • Beijing • Wuhan • Barcelona • Belgrade • Manchester • Tokyo • Cluj • Tianjin

Editors
Chin-Chia Wu
Feng Chia University
Taiwan

Win-Chin Lin
Feng Chia University
Taiwan

Editorial Office
MDPI
St. Alban-Anlage 66
4052 Basel, Switzerland

This is a reprint of articles from the Special Issue published online in the open access journal *Mathematics* (ISSN 2227-7390) (available at: https://www.mdpi.com/journal/mathematics/special_issues/Theor_Comput_Res_Var_Sched_Model).

For citation purposes, cite each article independently as indicated on the article page online and as indicated below:

LastName, A.A.; LastName, B.B.; LastName, C.C. Article Title. *Journal Name* **Year**, *Volume Number*, Page Range.

ISBN 978-3-0365-3018-5 (Hbk)
ISBN 978-3-0365-3019-2 (PDF)

© 2022 by the authors. Articles in this book are Open Access and distributed under the Creative Commons Attribution (CC BY) license, which allows users to download, copy and build upon published articles, as long as the author and publisher are properly credited, which ensures maximum dissemination and a wider impact of our publications.

The book as a whole is distributed by MDPI under the terms and conditions of the Creative Commons license CC BY-NC-ND.

Contents

About the Editors . . . . **vii**

Preface to "Theoretical and Computational Research in Various Scheduling Models" . . . . . **ix**

Ruyan He and Jinjiang Yuan
Two-Agent Preemptive Pareto-Scheduling to Minimize Late Work and Other Criteria
Reprinted from: *Mathematics* **2020**, *8*, 1517, doi:10.3390/math8091517 . . . . **1**

Yuan Zhang, Zhichao Geng and Jinjiang Yuan
Two-Agent Pareto-Scheduling of Minimizing Total Weighted Completion Time and Total Weighted Late Work
Reprinted from: *Mathematics* **2020**, *8*, 2070, doi:10.3390/math8112070 . . . . **19**

Dung-Ying Lin and Tzu-Yun Huang
A Hybrid Metaheuristic for the Unrelated Parallel Machine Scheduling Problem
Reprinted from: *Mathematics* **2021**, *9*, 768, doi:10.3390/math9070768 . . . . **37**

Chen-Yang Cheng, Shih-Wei Lin, Pourya Pourhejazy, Kuo-Ching Ying and Yu-Zhe Lin
No-Idle Flowshop Scheduling for Energy-Efficient Production: An Improved Optimization Framework
Reprinted from: *Mathematics* **2021**, *9*, 1335, doi:10.3390/math9121335 . . . . **57**

Wen-Tso Huang, Cheng-Chang Lu and Jr-Fong Dang
Improving the Return Loading Rate Problem in Northwest China Based on the Theory of Constraints
Reprinted from: *Mathematics* **2021**, *9*, 1397, doi:10.3390/math9121397 . . . . **75**

Alessio Angius, András Horváth and Marcello Urgo
A Kronecker Algebra Formulation for Markov Activity Networks with
Phase-Type Distributions
Reprinted from: *Mathematics* **2021**, *9*, 1404, doi:10.3390/math9121404 . . . . **91**

Ting-Chun Lo and Bertrand M. T. Lin
Relocation Scheduling in a Two-Machine Flow Shop with Resource Recycling Operations
Reprinted from: *Mathematics* **2021**, *9*, 1527, doi:10.3390/math9131527 . . . . **113**

Anna Antonova, Konstantin Aksyonov and Olga Aksyonova
An Imitation and Heuristic Method for Scheduling with Subcontracted Resources
Reprinted from: *Mathematics* **2021**, *9*, 2098, doi:10.3390/math9172098 . . . . **149**

Shu-Shun Liu, Agung Budiwirawan and Muhammad Faizal Ardhiansyah Arifin
Non-Sequential Linear Construction Project Scheduling Model for Minimizing Idle Equipment Using Constraint Programming (CP)
Reprinted from: *Mathematics* **2021**, *9*, 2492, doi:10.3390/math9192492 . . . . **171**

About the Editors

Chin-Chia Wu is a Professor in the Department of Statistics, Feng Chia University, Taiwan. He received his Doctoral Degree from the Graduate Institute of Management, School of Management, at the National Taiwan University of Science and Technology, Taiwan, in 1997. His teaching and research interests include applied statistics and operations research. He has published more than 180 papers in SCI/SSCI journals.

Win-Chin Lin is an associate professor in the Department of Statistics, Feng Chia University, Taiwan. He received his doctoral degree from the Department of Statistics, Iowa State University, USA, in 1998. His teaching and research interests include applied statistics, experimental designs, and scheduling.

Preface to "Theoretical and Computational Research in Various Scheduling Models"

The long-standing field of research with great practical value in operation research involves designing effective methods to find the best solution to perform certain jobs or policies with or without certain constraints. The literature shows that numerous papers have been written on scheduling theory and its applications over a long time. To solve various scheduling problems in different real-life environments, the literature also assesses many studies on developing exact methods or approximate algorithms, through deriving computational complexity, or evaluating their performance by using simulation results.

However, there are a lot of challenging scheduling problems in new application domains which have yet to be further explored. Thus, scheduling issues have always been a popular field of research, with many potential real-life applications including assignment, manufacturing, and logistics.

This Special Issue aims to provide a bridge to facilitate the interaction between the researcher and the practitioner in scheduling questions. Although discrete mathematics is a common method to solve scheduling problems, the further development of this method is limited due to the lack of general principles, which poses a major challenge to this research field. Papers that have made significant contributions to methodological progress or created model innovations to solve major and well-documented scheduling problems are welcome. The studies can be theoretical, methodological, computational, or application-oriented. In addition, relevant statistical applications in social systems are also welcome. Potential topics include but are not limited to the following:

- Scheduling in flow shops, open shops, or job shops settings
- Scheduling on parallel machines or in assembly flow shop
- Scheduling in green manufacturing environment
- Scheduling with multiple competing agents
- Scheduling in intelligent logistics
- Scheduling in time-dependent processing times
- Statistical methods application to engineering or relevant disciplinary

Chin-Chia Wu, Win-Chin Lin
Editors

Article

Two-Agent Preemptive Pareto-Scheduling to Minimize Late Work and Other Criteria

Ruyan He * and Jinjiang Yuan

School of Mathematics and Statistics, Zhengzhou University, Zhengzhou 450001, China; yuanjj@zzu.edu.cn
* Correspondence: heruyan219@163.com or heruyan219@gs.zzu.edu.cn

Received: 12 August 2020; Accepted: 28 August 2020; Published: 5 September 2020

Abstract: In this paper, we consider three preemptive Pareto-scheduling problems with two competing agents on a single machine. In each problem, the objective function of agent A is the total completion time, the maximum lateness, or the total late work while the objective function of agent B is the total late work. For each problem, we provide a polynomial-time algorithm to characterize the trade-off curve of all Pareto-optimal points.

Keywords: two-agent; Pareto-scheduling; late work; trade-off curve; polynomial time

1. Introduction

In recent decades, scheduling with two competing agents and scheduling with late-work criterion have been two hot topics in scheduling research. However, research for the combination of the two topics has not been studied extensively. One reason for this phenomenon stems from the fact that the single-machine scheduling problem for minimizing the total late work is already NP-hard when preemption is not allowed. Given the polynomial solvability of the preemptive scheduling problem for minimizing the total late work, we study the single-machine two-agent preemptive Pareto-scheduling problems with the total late work being one of the criteria.

Problem Formulation: Consider two competing agents A and B. For each agent $X \in \{A, B\}$, let $\mathcal{J}^X = \{J_1^X, J_2^X, \ldots, J_{n_X}^X\}$ be the set of jobs of agent X, where $\mathcal{J}^A \cap \mathcal{J}^B = \emptyset$ and the jobs in $\mathcal{J}^X$ are called the X-jobs. Each job J_j^X has a processing time $p_j^X > 0$ and a due date $d_j^X \geq 0$ which are integrally valued. The $n = n_A + n_B$ independent jobs in $\mathcal{J}^A \cup \mathcal{J}^B$ need to be preemptively processed on a single machine. Let $P_X = \sum_{j=1}^{n_X} p_j^X$ and $P = P_A + P_B$. All jobs considered in this paper are available at time zero. Our problems allow us to assume that the maximum due date of all jobs is at most P.

Let σ be a feasible schedule which assigns the jobs for processing in pieces in the interval $[0, +\infty)$. To enhance the flexibility of analysis, we allow the existence of idle times in a feasible schedule. The *completion time* of job J_j^X in σ is denoted as $C_j^X(\sigma)$. The *late work* of J_j^X in σ, denoted $Y_j^X(\sigma)$, is the amount of processing time of J_j^X after its due date d_j^X in σ. If $Y_j^X(\sigma) = 0$, J_j^X is called *early* in σ. If $0 < Y_j^X(\sigma) < p_j^X$, J_j^X is called *partially early* in σ. If $Y_j^X(\sigma) = p_j^X$, J_j^X is called *late* in σ.

The scheduling criteria related to our research are given by $\sum C_j^A = \sum_{j=1}^{n_A} C_j^A(\sigma)$ (the *total completion time* of the A-jobs under schedule σ), $L_{\max}^A = \max\{C_j^A(\sigma) - d_j^A : 1 \leq j \leq n_A\}$ (the *maximum lateness* of the A-jobs in schedule σ), and $\sum_j Y_j^X = \sum_{j=1}^{n_X} Y_j^X(\sigma)$ (the *total late work* of the X-jobs under schedule σ). Then the three *Pareto-scheduling problems* studied in this paper are given by

$$\begin{cases} 1|\text{pmtn}|^\#(\sum C_j^A, \sum Y_j^B), \\ 1|\text{pmtn}|^\#(L_{\max}^A, \sum Y_j^B), \\ 1|\text{pmtn}|^\#(\sum Y_j^A, \sum Y_j^B). \end{cases}$$

For each of the above three problems, we aim to find all the Pareto-optimal points of the problem and for each point a corresponding Pareto-optimal schedule. The formal definitions of Pareto-optimal point and Pareto-optimal schedule can be found in T'kindt and Billaut [1]. In this paper, the set of Pareto-optimal points forms a curve. Then we use the term *trade-off curve* to describe the set of Pareto-optimal points.

Literature Review: There is a huge amount of literature in scheduling with two competing agents and scheduling with late work criterion. Limited to the space of this paper, we only review some most related results.

Two-agent scheduling was first introduced by Agnetis et al. [2]. The authors proposed two important models for single-machine scheduling with two competing agents: constrained optimization model and Pareto optimization model. They studied nine problems arising from the different criteria combinations for the two agents and stated time complexity results for most of the resulting cases, where the scheduling criteria include the maximum regular cost function, the total (weighted) completion time and the number of tardy jobs. Early research on two-agent scheduling problems can also be found in Cheng et al. [3,4], Lee et al. [5], Leung et al. [6], and Ng et al. [7]. Agnetis et al. [8] applied the concept of price of fairness in resource allocation to two-agent single-machine scheduling problems, in which one agent aims at minimizing the total completion time, while the other agent wants to minimize the maximum tardiness with respect to a common due date. They further discussed the problem in which both agents wish to minimize the total completion time of their own jobs. Zhang et al. [9] studied the price of fairness in a two-agent single-machine scheduling problem in which both agents A and B want to minimize their own total completion time, and agent B has exactly two jobs.

There are few results on the two-agent scheduling problems with precedence constraints. Agnetis et al. [2] considered the two-agent problem with precedence constraints on single machine, i.e., $1|prec|f_{max}^A : f_{max}^B$ and solved this problem in polynomial time. Mor and Mosheiov [10] extended a classical single-machine scheduling problem, where the objective is to minimize maximum cost, given general job-dependent cost functions and general precedence constraints which was solved by the well-known Lawler's Algorithm. First, they allowed the option of job rejection. Then, they studied the more general setting of two competing agents, where job rejection is allowed either for one agent or for both. They showed that both extensions can be solved in polynomial time. Gao and Yuan [11] considered the Pareto-scheduling with two agents A and B for minimizing the total completion time of A-jobs and a maximum cost of B-jobs with precedence constraints. They showed that the problem can be solved in polynomial time. More recent results on two-agent scheduling problems can be found in Agnetis et al. [12], Liu et al. [13], Oron et al. [14], Perez-Gonzalez and Framinan [15], and Yuan [16,17].

Scheduling related to late-work criterion was first studied by Blazewicz and Finke [18]. Since then, several research groups have focused on this performance measure, obtaining a set of interesting results. Potts and Van Wassenhove [19] considered a single-machine scheduling problem where the goal is to minimize the total amount of late work. They showed that the problem is NP-hard and presented a pseudo-polynomial-time algorithm. Potts and Van Wassenhove [20] further proposed a branch-bound algorithm for the same problem and presented two fully polynomial-time approximation schemes with running times $O(\frac{n^2}{\epsilon})$ and $O(\frac{n^3}{\epsilon})$, respectively. Hariri et al. [21] considered a single-machine problem of minimizing the total weighted late work. They presented an $O(n \log n)$ algorithm for the preemptive total weighted late work problem. For papers that consider scheduling problem of minimizing other total late-work criteria, the reader may refer to the survey paper of Sterna [22].

Up to now, people have done little research about the combination of two-agent scheduling with late-work criterion. Wang et al. [23] addressed a two-agent scheduling problem where the objective is to minimize the total late work of the first agent, with the restriction that the maximum lateness of the second agent cannot exceed a given value. For small-scale problem instances, they established two pseudo-polynomial dynamic programming algorithms. For medium- to large-scale problem instances, they presented a branch-and-bound algorithm. Zhang and Wang [24] presented a two-agent scheduling problem where the objective is to minimize the total weighted late work of agent A, while keeping the maximum cost of agent B cannot exceed a given bound U. They addressed the complexity of those problems, and presented the optimal polynomial-time algorithms or pseudo-polynomial-time algorithm to solve the scheduling problems, respectively. Zhang and Yuan [25] considered the same problem as above and further studied the three versions of the problem.

Our research also uses some results in the single-machine preemptive scheduling with forbidden intervals (or maintenance activities), i.e., $1, h_m|\text{pmtn}|f$, where "$1, h_m$" means that there are m forbidden intervals on the single machine and "f" is the objective function to be minimized. Without reviewing this scheduling topic in detail, we only state two known results used in our discussion. Lee [26] showed that problem $1, h_m|\text{pmtn}|\sum C_j$ can be solved by the preemptive SPT rule in $O(m + n \log n)$ time and problem $1, h_m|\text{pmtn}|L_{\max}$ can be solved by the preemptive EDD rule in $O(m + n \log n)$ time.

Our Contributions: In Section 2, we introduce some notations and definitions and present several important lemmas. In Section 3, we show that the trade-off curve of problem $1|\text{pmtn}|^\#(\sum C_j^A, \sum Y_j^B)$ can be determined in $O(nn_A n_B)$ time. In Section 4, we show that the trade-off curve of problem $1|\text{pmtn}|^\#(L_{\max}^A, \sum Y_j^B)$ can be determined in $O(nn_A n_B)$ time. In Section 5, we show that the trade-off curve of problem $1|\text{pmtn}|^\#(\sum Y_j^A, \sum Y_j^B)$ can be determined in $O(n \log n)$ time. Finally, some concluding remarks are given in Section 6.

2. Preliminaries

Let $\mathcal{J}^A \cup \mathcal{J}^B$ be the job instance to be preemptively scheduled on a single machine. The preemption assumption allows us to schedule each job in pieces. For a piece $J_{jj'}^X$ of job J_j^X, we use $p_{jj'}^X$ to denote the length (processing time) of $J_{jj'}^X$ and use $C_{jj'}^X(\sigma)$ to denote the completion time of $J_{jj'}^X$ in schedule σ.

In the following, we consider the Pareto-scheduling problem $1|\text{pmtn}|^\#(f^A, \sum Y_j^B)$ on instance $\mathcal{J}^A \cup \mathcal{J}^B$, where f^A is a regular objective function of the A-jobs or $f^A = \sum Y_j^A$. We use $\Omega(\mathcal{J}^A, \mathcal{J}^B)$ to denote the set of all Pareto-optimal points of this problem.

In a schedule σ of $\mathcal{J}^A \cup \mathcal{J}^B$, each B-job J_j^B is partitioned into two parts: the *early part* $J_j^{BE}(\sigma)$ and the *late part* $J_j^{BY}(\sigma)$, where $J_j^{BE}(\sigma)$ is processed before time d_j^B in σ and $J_j^{BY}(\sigma)$ is processed after time d_j^B in σ. Moreover, $p_j^{BE}(\sigma) = p_j^B - Y_j^B(\sigma)$ and $p_j^{BY}(\sigma) = Y_j^B(\sigma)$ are used to denote the lengths (processing times) of $J_j^{BE}(\sigma)$ and $J_j^{BY}(\sigma)$, respectively. A part of length 0 is called a *trivial part*. We allow the existence of trivial parts to enhance flexibility in analysis. Then we have

$$\sum_{j=1}^{n_B} Y_j^B(\sigma) = \sum_{j=1}^{n_B} p_j^{BY}(\sigma) = P_B - \sum_{j=1}^{n_B} p_j^{BE}(\sigma). \tag{1}$$

For convenience, we renumber the B-jobs such that

$$d_1^B \leq d_2^B \leq \cdots \leq d_{n_B}^B \leq P \tag{2}$$

and keep this numbering throughout this paper. Let $\sigma_0^B = (J_1^B, J_2^B, \ldots, J_{n_B}^B)$ which schedules the B-jobs in the EDD order described in (2). From Potts and Van Wassenhove [19], the optimal value of problem $1|\text{pmtn}|\sum Y_j$ on instance $\mathcal{J}^B$ is given by $T_{\max}(\sigma_0^B)$. Then we have the following lemma.

Lemma 1. *For each point $(f^*, Y^*) \in \Omega(\mathcal{J}^A, \mathcal{J}^B)$, we have $T_{\max}(\sigma_0^B) \leq Y^* \leq P_B$.*

To make our analysis operational, we now consider an integer $Y^* \in [T_{\max}(\sigma_0^B), P_B]$ and present a procedure to schedule the B-jobs preemptively with a particular structure. This procedure imitates the algorithm in Hariri et al. [21] for solving problem $1|\text{pmtn}|\sum w_j Y_j$.

Algorithm 1: For scheduling the B-jobs according to the value of Y^*.

Input: The B-jobs $\mathcal{J}^B$ with the EDD order in (2) and an integer $Y^* \in [T_{\max}(\sigma_0^B), P_B]$.

Step 1: Determine the minimum index $j^* \in \{1, 2, \ldots, n_B\}$ such that $p_1^B + p_2^B + \cdots + p_{j^*}^B \geq Y^*$. Then $p_1^B + p_2^B + \cdots + p_{j^*-1}^B < Y^*$. Please note that if $Y^* = 0$, then we have $j^* = 1$. We call $J_{j^*}^B$ the *critical B-job* corresponding to Y^*.

Step 2: Decompose the critical B-job $J_{j^*}^B$ into two parts $J_{j^*}^{BE}$ and $J_{j^*}^{BY}$ such that

$$p_{j^*}^{BY} = Y^* - \sum_{j=1}^{j^*-1} p_j^B \text{ and } p_{j^*}^{BE} = p_{j^*}^B - p_{j^*}^{BY} = \sum_{j=1}^{j^*} p_j^B - Y^*.$$

We call $J_{j^*}^{BE}$ and $J_{j^*}^{BY}$ the *early part* and the *late part* of $J_{j^*}^B$, respectively, corresponding to Y^*. Set $\mathcal{J}^{BE}(Y^*) = \{J_{j^*}^{BE}, J_{j^*+1}^B, J_{j^*+2}^B, \ldots, J_{n_B}^B\}$ and $\mathcal{J}^{BY}(Y^*) = \{J_1^B, J_2^B, \ldots, J_{j^*-1}^B, J_{j^*}^{BY}\}$.

Step 3: Generate a schedule $\sigma^{B(Y^*)}$ of the B-jobs $\mathcal{J}^B = \mathcal{J}^{BE}(Y^*) \cup \mathcal{J}^{BY}(Y^*)$ in the following way:

(3.1) From time $P^* := P + 1$, schedule the jobs (or pieces) in $\mathcal{J}^{BY}(Y^*)$ consecutively in the order $J_1^B, J_2^B, \ldots, J_{j^*-1}^B, J_{j^*}^{BY}$.

(3.2) Schedule the jobs (or pieces) in $\mathcal{J}^{BE}(Y^*)$ by using the algorithm in Hariri et al. [21] for solving problem $1|pmtn|\sum Y_j$ on instance $\mathcal{J}^{BE}(Y^*)$:

Beginning from time $d_{n_B}^B$, schedule the jobs (or pieces) in $\mathcal{J}^{BE}(Y^*)$ backwards in the order $J_{n_B}^B, J_{n_B-1}^B, \ldots, J_{j^*+1}^B, J_{j^*}^{BE}$ such that each job (or piece) in $\mathcal{J}^{BE}(Y^*)$ is scheduled as late as possible subject to its due date.

Output: The schedule $\sigma^{B(Y^*)}$ of the B-jobs.

It can be observed that Procedure(Y^*) runs in $O(n)$ time. An objective function of the A-jobs, denoted f^A, is called *regular* if f^A is nondecreasing in the completion times of the A-jobs. Please note that $\sum C_j^A$ and $L_{\max}^A$ are regular, but $\sum Y_j^A$ is not regular since the preemptive assumption. The following lemma is critical in our discussion.

Lemma 2. *Consider problem $1|pmtn|^\#(f^A, \sum Y_j^B)$ on instance $\mathcal{J}^A \cup \mathcal{J}^B$, where either $f^A = \sum Y_j^A$ or f^A is a regular objective function of the A-jobs. Assume that $(f^*, Y^*) \in \Omega(\mathcal{J}^A, \mathcal{J}^B)$ and let $\sigma^{B(Y^*)}$ be the schedule of $\mathcal{J}^B$ generated by Procedure(Y^*). Then there exists a Pareto-optimal schedule π corresponding to (f^*, Y^*) in which the B-jobs are scheduled in the same manner as that in $\sigma^{B(Y^*)}$. Such a Pareto-optimal schedule π is called a Y^*-standard schedule in the sequel.*

Proof. Let σ be a Pareto-optimal schedule corresponding to (f^*, Y^*) such that $C_{\max}(\sigma)$ is as small as possible. Then no idle exists in σ, and so, $C_{\max}(\sigma) = P$.

The late parts of B-jobs can be scheduled arbitrarily late without affecting the objective values f^A and $\sum Y_j^B$. Thus, by shifting the late parts of B-jobs in σ, we obtain a new Pareto-optimal schedule σ_1 corresponding to (f^*, Y^*) such that the following property (P1) holds for σ_1.

(P1) The late parts of B-jobs are scheduled consecutively in the interval $[P^*, P^* + Y^*]$ in an arbitrary order without idle time.

We next generate a Pareto-optimal schedule σ_2 corresponding to (f^*, Y^*) such that following property (P2) holds for σ_2.

(P2) For every nontrivial early part $J_j^{BE}(\sigma_2)$ and every nontrivial late part $J_k^{BY}(\sigma_2)$ among the B-jobs, we have $j \geq k$.

If σ_1 has the property (P2), we just set $\sigma_2 = \sigma_1$. Otherwise, there are a nontrivial early part $J_j^{BE}(\sigma_1)$ and a nontrivial late part $J_k^{BY}(\sigma_1)$ among the B-jobs, such that $j < k$. From (2), we have $d_j^B \leq d_k^B$. Let $\delta = \min\{p_j^{BE}(\sigma_1), p_k^{BY}(\sigma_1)\}$. By exchanging an amount of length δ between $J_j^{BE}(\sigma_1)$ and $J_k^{BY}(\sigma_1)$ in σ_1, we obtain a new schedule without changing the objective values but with improving in the direction we need. Repeating this procedure, we eventually obtain a new Pareto-optimal schedule σ_2 corresponding to (f^*, Y^*) such that both properties (P1) and (P2) hold for σ_2.

Let σ_3 be the schedule obtained from σ_2 by rescheduling (if necessary) the early parts of B-jobs in the order $J_1^{BE}(\sigma) \prec J_2^{BE}(\sigma) \prec \cdots \prec J_{n_B}^{BE}(\sigma)$. Since the EDD property described in (2), the early parts of B-jobs in σ_2 are also early in σ_3. Then σ_3 is a Pareto-optimal schedule corresponding to (f^*, Y^*) such that properties (P1) and (P2), and additionally, the following property (P3), hold for σ_3.

(P3) The early parts of B-jobs are scheduled in the order $J_1^{BE}(\sigma_3) \prec J_2^{BE}(\sigma_3) \prec \cdots \prec J_{n_B}^{BE}(\sigma_3)$.

Please note that $\sum_{j=1}^{n_B} Y_j^B(\sigma_3) = \sum_{j=1}^{n_B} Y_j^B(\sigma^{B(Y^*)}) = Y^*$. Since σ_3 has the three properties (P1)–(P3), from Procedure (Y^*) for generating $\sigma^{B(Y^*)}$, we know that σ_3 and $\sigma^{B(Y^*)}$ have the same early parts and late parts of B-jobs. Then, in both schedules, the early parts of B-jobs are given by $\mathcal{J}^{BE}(Y^*) = \{J_{j^*}^{BE}, J_{j^*+1}^B, J_{j^*+2}^B, \ldots, J_{n_B}^B\}$ and the late parts of B-jobs are given by $\mathcal{J}^{BY}(Y^*) = \{J_1^B, J_2^B, \ldots, J_{j^*-1}^B, J_{j^*}^{BY}\}$, as defined in Step 2 of Procedure(Y^*).

Now let π be the schedule obtained from σ_3 by the following two actions: (i) from time P^*, reschedule the late parts in $\mathcal{J}^{BY}(Y^*)$ consecutively in the order $J_1^B, J_2^B, \ldots, J_{j^*-1}^B, J_{j^*}^{BY}$, and (ii) without changing the processing order of A-jobs and the processing order of the early parts of B-jobs, reschedule them such that the early parts of B-jobs are scheduled as late as possible subject to their due dates, and then, the A-jobs are scheduled as early as possible.

Clearly, in schedule π, the B-jobs are scheduled in the same manner as that in $\sigma^{B(Y^*)}$. Then we have $\sum_{j=1}^{n_B} Y_j^B(\pi) = Y^*$. From the construction of π, we have $f^A(\pi) \leq f^A(\sigma_3) = f^*$. The Pareto-optimality of (f^*, Y^*) further implies that $(f^A(\pi), \sum_{j=1}^{n_B} Y_j^B(\pi)) = (f^*, Y^*)$. Consequently, π is a required Pareto-optimal schedule corresponding to (f^*, Y^*). The lemma follows. □

From Lemmas 1 and 2, the Pareto-scheduling problem $1|\text{pmtn}|^\#(f^A, \sum Y_j^B)$ on instance $\mathcal{J}^A \cup \mathcal{J}^B$ can be solved by the following general approach:

For each value $Y^* \in [T_{\max}(\sigma_0^B), P_B]$, run Procedure$(Y^*)$ to obtain the schedule $\sigma^{B(Y^*)}$ of the B-jobs. Determine the intervals occupied by the B-jobs in $\sigma^{B(Y^*)}$ and regards these intervals as forbidden intervals. The intervals which are not occupied by the B-jobs in $\sigma^{B(Y^*)}$ is called the free-time intervals. Then solve problem $1, h_m|\text{pmtn}|f^A$ on instance $\mathcal{J}^A$ to obtain a Y^*-standard schedule.

The above approach cannot be implemented in polynomial time since it enumerates all the possible choices of Y^*. Therefore, in the next three sections, for $f^A \in \{\sum C_j^A, L_{\max}^A, \sum Y_j^A\}$, we will present polynomial-time algorithms, respectively, to characterize the trade-off curves.

To this end, we set $Y^{(0)} = T_{\max}(\sigma_0^B)$, and run Procedure$(Y^{(0)})$ to obtain the schedule $\sigma^{B(Y^{(0)})}$. Assume that the intervals occupied by the B-jobs are given by $h_1, h_2, \ldots, h_m$, where $h_i = [\tau_1^{(i)}, \tau_2^{(i)}]$ is the i-th interval, $i = 1, 2, \ldots, m$, such that

$$0 \leq \tau_1^{(1)} < \tau_2^{(1)} < \tau_1^{(2)} < \tau_2^{(2)} < \cdots < \tau_1^{(m)} < \tau_2^{(m)}. \tag{3}$$

From the implementation of Procedure$(Y^{(0)})$, we have

$$\tau_1^{(m)} = P^* \text{ and } \tau_2^{(m)} = P^* + Y^{(0)}. \tag{4}$$

For each $Y^* \in [Y^{(0)}, P_B]$, we define i^* to be the maximum index in $\{1, 2, \ldots, m-1\}$ such that $Y^* - Y^{(0)} \geq \sum_{i=1}^{i^*-1}(\tau_2^{(i)} - \tau_1^{(i)})$ and let $\tau^* \in [\tau_1^{(i^*)}, \tau_2^{(i^*)})$ such that $Y^* - Y^{(0)} = \sum_{i=1}^{i^*-1}(\tau_2^{(i)} - \tau_1^{(i)}) + (\tau^* - \tau_1^{(i^*)})$.

From the implementation of Procedure(Y^*) again, the set of time intervals occupied by the B-jobs in schedule $\sigma^{B(Y^*)}$, denoted by $\mathcal{I}(\sigma^{B(Y^*)})$, is given by

$$\mathcal{I}(\sigma^{B(Y^*)}) = \{[\tau^*, \tau_2^{(i^*)}], [\tau_1^{(i^*+1)}, \tau_2^{(i^*+1)}], \ldots, [\tau_1^{(m-1)}, \tau_2^{(m-1)}], [P^*, P^* + Y^*]\}. \quad (5)$$

We will write $h_{i^*} = [\tau^*, \tau_2^{(i^*)}]$, $h_i = [\tau_1^{(i)}, \tau_2^{(i)}]$ for $i = i^* + 1, i^* + 2, \ldots, m - 1$, and $h_m = [P^*, P^* + Y^*]$.
The above discussion will help us to construct the trade-off curves easily.

3. The First Problem

In this section, we consider problem $1|\text{pmtn}|^\#(\sum C_j^A, \sum Y_j^B)$ on instance $\mathcal{J}^A \cup \mathcal{J}^B$. By the job-exchanging argument, we can verify that the A-jobs must be scheduled in the SPT order in every Pareto-optimal schedule. Thus, in this section, we renumber the A-jobs by the SPT order such that $p_1^A \leq p_2^A \leq \cdots \leq p_{n_A}^A$. Then we only consider the schedules in which the A-jobs are scheduled in the order $J_1^A \prec J_2^A \prec \cdots \prec J_{n_A}^A$.

Given a point $(C^*, Y^*) \in \Omega(\mathcal{J}^A, \mathcal{J}^B)$, let σ be the Y^*-standard schedule of $\mathcal{J}^A \cup \mathcal{J}^B$. Then the set of forbidden intervals (occupied by the B-jobs) is given by (5) and the A-jobs are preemptively scheduled in the order $J_1^A \prec J_2^A \prec \cdots \prec J_{n_A}^A$ from time 0 in the free-time intervals as early as possible. Thus, there are $m - i^* + 1$ forbidden intervals and the first forbidden interval in σ is given by $h_{i^*} = [\tau^*, \tau_2^{(i^*)}]$.

If $P_A \leq \tau^*$, then all the A-jobs are scheduled before the first forbidden interval h_{i^*} in σ. In this case, we have no further action.

In general, suppose that $P_A > \tau^*$. Then at least one A-job completes after h_{i^*} in σ. Let $J_{k(\sigma)}^A$ be the first A-job which completes after h_{i^*} in σ. Then, there are totally $n_A - k(\sigma) + 1$ A-jobs completing after h_{i^*} in σ.

For each index $j \in \{k(\sigma), k(\sigma) + 1, \ldots, n_A\}$, we define $i[j]$ to be the interval index such that J_j^A completes after interval $h_{i[j]}$ and before interval $h_{i[j]+1}$ in σ, or equivalently, $\tau_2^{(i[j])} < C_j^A(\sigma) \leq \tau_1^{(i[j]+1)}$. We further define

$$e(\sigma) = \min\{C_j^A(\sigma) - \tau_2^{(i[j])} : j = k(\sigma), k(\sigma) + 1, \ldots, n_A\}. \quad (6)$$

An A-job J_j^A with $j \in \{k(\sigma), k(\sigma) + 1, \ldots, n_A\}$ is called a *crucial A-job* in σ if $C_j^A(\sigma) - \tau_2^{(i[j])} = e(\sigma)$. Please note that if J_j^A is a crucial A-job in σ, then the interval $[\tau_2^{(i[j])}, C_j^A(\sigma)]$ is fully occupied by job J_j^A in σ, implying that J_j^A is the first A-job completing after interval $h_{i[j]}$ in σ. In this case, we call interval $h_{i[j]}$ the *nearest* forbidden interval corresponding to crucial A-job J_j^A in σ.

Set $\delta_l(\sigma) = \tau_2^{(l)} - \tau_1^{(l)}(\sigma)$, $i^* \leq l \leq m$, to be the length of the forbidden interval h_l in σ. In particular, if $l = i^*$, then $\tau_1^{(l)}(\sigma) = \tau^*$. Let

$$\theta(\sigma) = \min\{e(\sigma), \delta_{i^*}(\sigma)\}. \quad (7)$$

Please note that when the schedule σ is given, the A-job index $k(\sigma)$ can be determined in $O(n_A)$ time, the interval indices $i[j]$ for $j \in \{k(\sigma), k(\sigma) + 1, \ldots, n_A\}$ can be determined in $O(n_A)$ time. After that, the value $e(\sigma)$ defined in (6) can be determined in $O(n_A)$ time. Finally, the value $\theta(\sigma)$ can be determined by its definition in (7) in constant time. Then we have the following lemma.

Lemma 3. *Given the Y^*-standard schedule σ in advance, the values $k(\sigma)$ and $\theta(\sigma)$ can be determined in $O(n_A)$ time.*

For each $Y \in [Y^*, Y^* + \theta(\sigma)]$, let σ' be the Y-standard schedule. Then σ' is obtained from σ by shifting the first $Y - Y^*$ units of h_{i^*} to the last forbidden interval and then moving the A-jobs in $\{J_{k(\sigma)}^A, J_{k(\sigma)+1}^A, \ldots, J_{n_A}\}$ left to eliminate the idle times accordingly. This means that $C_j^A(\sigma') \leq$

$C_j^A(\sigma) - (Y - Y^*)$ for $j \in \{k(\sigma), k(\sigma)+1, \ldots, n_A\}$. Assume that the total completion time of A-jobs in σ' is C. According to Lemma 2, (C, Y) is a Pareto-optimal point. In the following, we consider the trade-off curve between (C^*, Y^*) and (C, Y). For convenience, point (C, Y) is simply called point Y.

We will show that the trade-off curve for $Y \in [Y^*, Y^* + \theta(\sigma))$ is a line segment. However, the point $Y^* + \theta(\sigma)$ may have the singularity.

Lemma 4. *For each point $(C, Y) \in \Omega(\mathcal{J}^A, \mathcal{J}^B)$ with $Y \in [Y^*, Y^* + \theta(\sigma))$, we have $\frac{C-C^*}{Y-Y^*} = -(n_A - k(\sigma) + 1)$.*

Proof. For each $Y \in [Y^*, Y^* + \theta(\sigma))$, we have $Y - Y^* < \theta(\sigma) \le e(\sigma)$. When we change σ to σ', no crucial A-jobs are moved left across their corresponding nearest forbidden intervals in σ'. As a result, compared with σ, each of the completion times of the $n_A - k(\sigma) + 1$ A-jobs in $\{J_{k(\sigma)}^A, J_{k(\sigma)+1}^A, \ldots, J_{n_A}^A\}$ has decreased $Y - Y^*$ units in σ'. Thus, we have $C - C^* = -(n_A - k(\sigma) + 1)(Y - Y^*)$, as required. □

Let $\mathcal{J}_c^A(\sigma)$ be the set of crucial A-jobs in σ. We use $\Delta(\sigma)$ to denote the total length of all the nearest forbidden intervals corresponding to the $\kappa(\sigma)$ crucial A-jobs in σ, i.e.,

$$\Delta(\sigma) = \sum \{\delta_{i[j]}(\sigma) : J_j^A \in \mathcal{J}_c^A(\sigma)\}.$$

The following lemma is only used to display the singularity of point $Y^* + \theta(\sigma)$.

Lemma 5. *For the point $(C, Y) \in \Omega(\mathcal{J}^A, \mathcal{J}^B)$ with $Y = Y^* + \theta(\sigma)$, we have the following three statements.*

(i) *If $e(\sigma) > \delta_{i^*}(\sigma)$, then $C = C^* - (n_A - k(\sigma) + 1)\theta(\sigma)$.*
(ii) *If $e(\sigma) \le \delta_{i^*}(\sigma)$ and the first crucial A-job is nearest to h_{i^*} in σ, then $C = C^* - \Delta(\sigma) - (n_A - k(\sigma))\theta(\sigma)$.*
(iii) *If $e(\sigma) \le \delta_{i^*}(\sigma)$ and no crucial A-job is nearest to h_{i^*} in σ, then $C = C^* - \Delta(\sigma) - (n_A - k(\sigma) + 1)\theta(\sigma)$.*

Proof. When σ changes to σ', each of the $n_A - k(\sigma) + 1$ A-jobs in $\{J_{k(\sigma)}^A, J_{k(\sigma)+1}^A, \ldots, J_{n_A}^A\}$ is moved left $\theta(\sigma)$ units, and in the case that $e(\sigma) = \theta(\sigma)$, the crucial A-jobs are also moved left across their corresponding nearest forbidden intervals in σ'. Thus, we have

$$C = C^* - \Delta^*(\sigma') - (n_A - k(\sigma) + 1)\theta(\sigma), \tag{8}$$

where $\Delta^*(\sigma') = 0$ if $e(\sigma) > \theta(\sigma)$ and $\Delta^*(\sigma') = \sum \{\delta_{i[j]}(\sigma') : J_j^A \in \mathcal{J}_c^A(\sigma)\}$ if $e(\sigma) = \theta(\sigma)$. The key point is that $\delta_{i^*}(\sigma') = \delta_{i^*}(\sigma) - \theta(\sigma)$ and $\delta_l(\sigma') = \delta_l(\sigma)$ for $l = i^*+1, i^*+2, \ldots, m$.

Under the assumption of (i), we have $\Delta^*(\sigma') = 0$. From (8), we have $C = C^* - (n_A - k(\sigma) + 1)\theta(\sigma)$.

Under the assumption of (ii), we have $\Delta^*(\sigma') = \Delta(\sigma) - \theta(\sigma)$. From (8), we have $C = C^* - \Delta(\sigma) - (n_A - k(\sigma))\theta(\sigma)$.

Under the assumption of (iii), we have $\Delta^*(\sigma') = \Delta(\sigma)$. From (8), we have $C = C^* - \Delta(\sigma) - (n_A - k(\sigma) + 1)\theta(\sigma)$. The lemma follows. □

Theorem 1. *Algorithm 2 generates the trade-off curve of $1|pmtn|^\#(\sum C_j^A, \sum Y_j^B)$ in $O(nn_A n_B)$ time.*

Proof. The correctness of Algorithm 2 is guaranteed by Lemmas 2 and 4. We estimate the time complexity of the algorithm in the following.

The preprocessing procedure runs in $O(n_A \log n_A + n_B \log n_B)$ time. Each of Steps (1.1) and (1.2) runs in $O(n_B)$ time, so Step 1 runs in $O(n_B)$ time.

After Step 1, the algorithm has K iterations. In each iteration, either one forbidden interval is eliminated or at least one A-job is moved left across its corresponding nearest forbidden interval. Since $m \le n_B$, we have $K = O(n_A n_B)$.

At each iteration, Step 2 runs in $O(n_A + m) = O(n)$ time. From Lemma 3, Step (3.1) runs in $O(n_A)$ time. Thus, each iteration runs in $O(n)$ time.

The above discussion establishes the $O(nn_An_B)$-time complexity of Algorithm 2. □

Algorithm 2: Trade-off curve of problem $1|\text{pmtn}|^{\#}(\sum C_j^A, \sum Y_j^B)$.

Input: Instance $\mathcal{J}^A \cup \mathcal{J}^B$.

Preprocessing: Renumber the A-jobs such that $p_1^A \leq p_2^A \leq \cdots \leq p_{n_A}^A$ and renumber the B-jobs such that $d_1^B \leq d_2^B \leq \cdots \leq d_{n_B}^B$.

Step 1: Do the following:

(1.1) Generate schedule σ_0^B which schedules the B-jobs in the order $J_1^B \prec J_2^B \prec \cdots \prec J_{n_B}^B$ in the interval $[0, P_B]$ without idle times. Then calculate the value $Y^{(0)} = T_{\max}(\sigma_0^B)$.

(1.2) Run Procedure($Y^{(0)}$) to obtain the schedule $\sigma^B(Y^{(0)})$ of the B-jobs. Determine the intervals occupied by the B-jobs in $\sigma^B(Y^{(0)})$, say $h_1, h_2, \ldots, h_m$, where $h_i = [\tau_1^{(i)}, \tau_2^{(i)}]$ is the i-th interval, $i = 1, 2, \ldots, m$, as described in (3). Then regard $h_1, h_2, \ldots, h_m$ as forbidden intervals which will be updated in the implementation of the algorithm. We take the convention that the forbidden intervals are just occupied by the B-jobs.

(1.3) Set $t_0 := 1$ and set $i := 0$.

Step 2: Do the following:

(2.1) Generate the $Y^{(i)}$-standard schedule σ_i of $\mathcal{J}^A \cup \mathcal{J}^B$ in which $h_{t_i}, h_{t_i+1}, \ldots, h_m$ are the forbidden intervals and the A-jobs are preemptively scheduled in the order $J_1^A \prec J_2^A \prec \cdots \prec J_{n_A}^A$ as early as possible. Determine the value $C^{(i)} = \sum_{j=1}^{n_A} C_j^A(\sigma_i)$.

(2.2) If $J_{n_A}^A$ (and so, every A-job) is scheduled before the first forbidden interval h_{t_i} in σ_i, then set $K = i$ and go to Step 4. (In this case, we have obtained the whole trade-off curve.)
If $J_{n_A}^A$ completes after h_{t_i} in σ_i, then go to Step 3. (In this case, we use $J_{k(\sigma_i)}$ to denote the first A-job completing after h_{t_i} in σ_i.)

Step 3: Do the following:

(3.1) Calculate the values $k(\sigma_i)$ and $\theta(\sigma_i)$.

(3.2) Define a left closed right open segment $\mathcal{L}_i$ in the interval $[Y^{(i)}, Y^{(i)} + \theta(\sigma_i))$ by the following way:

$$\mathcal{L}_i : \frac{C - C^{(i)}}{Y - Y^{(i)}} = -(n_A - k(\sigma_i) + 1), \quad Y \in [Y^{(i)}, Y^{(i)} + \theta(\sigma_i)).$$

(3.3) Set $Y^{(i+1)} := Y^{(i)} + \theta(\sigma_i)$ and $h_m := [P^*, P^* + Y^{(i+1)}]$. Moreover, if $\theta(\sigma_i) = \delta_{t_i}(\sigma_i)$, then set $t_{i+1} := t_i + 1$; and if $\theta(\sigma_i) < \delta_{t_i}(\sigma_i)$, then set $t_{i+1} := t_i$ and $h_{t_{i+1}} := [\tau_1^{(t_i)} + \theta(\sigma_i), \tau_2^{(t_i)}]$ (which is obtained from interval h_{t_i} by deleting the first $\theta(\sigma_i)$ units.)

(3.4) Set $i := i + 1$. Go to Step 2.

Step 4: Output the trade-off curve $\Omega(\mathcal{J}^A, \mathcal{J}^B) = \mathcal{L}_0 \cup \mathcal{L}_1 \cup \cdots \cup \mathcal{L}_{K-1} \cup \{(C^{(K)}, Y^{(K)})\}$.

It can be observed that the total interruption time (i.e., the number of interruptions) of all the jobs in each schedule generated by Algorithm 2 is upper bounded by $1 + m \leq 1 + n_B$.

Let us consider the job instance I_1 displayed in Table 1. The trade-off curve of problem $1|\text{pmtn}|^{\#}(\sum C_j^A, \sum Y_j^B)$ on instance I_1 is shown in Figure 1.

Table 1. The job instance I_1.

J_i^X	J_1^A	J_2^A	J_3^A	J_4^A	J_1^B	J_2^B	J_3^B
p_i^X	1	1	2	3	2	4	2
d_i^X	0	0	0	0	3	4	8

Please note that $P^* = P + 1 = 16$. Let $\Omega = \Omega(\mathcal{J}^A, \mathcal{J}^B)$. The key steps in applying Algorithm 2 to solve the instance I_1 are as follows:

(i) Generate the schedule $\sigma_0^B = (J_1^B, J_2^B, J_3^B)$ and calculate the Y-value $Y^{(0)} = T_{\max}(\sigma_0^B) = 2$. Then generate the schedule $\sigma^{B(Y^{(0)})}$, and the forbidden interval set $\mathcal{I}^{B(Y^{(0)})} = \{h_1, h_2, h_3\}$ is determined, where $h_1 = [0,4]$, $h_2 = [6,8]$, and $h_3 = [16,18]$. Then, for each $y \in (Y^{(0)}, P_B] = (2,8]$, $\sigma^{B(y)}$ and $\mathcal{I}^{B(y)}$ can be easily generated.

(ii) Generate the $Y^{(0)}$-standard schedule σ_0 of $\mathcal{J}^A \cup \mathcal{J}^B$ in which h_1, h_2, h_3 are the forbidden intervals and the A-jobs are preemptively scheduled in the order $J_1^A \prec J_2^A \prec J_3^A \prec J_4^A$ as early as possible. Determine the value $C^{(0)} = \sum_{j=1}^{n_A} C_j^A(\sigma_0) = 34$. Then σ_0 is a Pareto-optimal schedule corresponding to $(34, 2) \in \Omega$.

(iii) Calculate the values $k(\sigma_0) = 1$ and $\theta(\sigma_0) = 1$. Then line segment $\mathcal{L}_0$, which is the trade-off curve in the interval $[2,3)$, satisfies $C = -4Y + 42$. Calculate $Y^{(1)} = Y^{(0)} + \theta(\sigma_0) = 3$ and the forbidden interval set $\mathcal{I}^{B(Y^{(1)})} = \{h_1, h_2, h_3\}$, where $h_1 = [1,4]$, $h_2 = [6,8]$, and $h_3 = [16,19]$. Generate σ_1, and calculate $C^{(1)} = \sum_{j=1}^{n_A} C_j^A(\sigma_1) = 27$. Then the intermediate point $(27, 3) \in \Omega$ is a jump discontinuity point and σ_1 is a Pareto-optimal schedule corresponding to $(27, 3)$.

(iv) Calculate the values $k(\sigma_1) = 2$ and $\theta(\sigma_1) = 1$. Then the line segment $\mathcal{L}_1$, which is the trade-off curve in the interval $[3,4)$, satisfies $C = -3Y + 36$. Calculate $Y^{(2)} = Y^{(1)} + \theta(\sigma_1) = 4$ and the forbidden interval set $\mathcal{I}^{B(Y^{(2)})} = \{h_1, h_2, h_3\}$, where $h_1 = [2,4]$, $h_2 = [6,8]$, and $h_3 = [16,20]$. Generate σ_2, and calculate $C^{(2)} = \sum_{j=1}^{n_A} C_j^A(\sigma_2) = 20$. Then the intermediate point $(20, 4) \in \Omega$ is a jump discontinuity point and σ_2 is a Pareto-optimal schedule corresponding to $(20, 4)$.

(v) Calculate the values $k(\sigma_2) = 3$ and $\theta(\sigma_2) = 2$. Then the line segment $\mathcal{L}_2$, which is the trade-off curve in the interval $[4,6)$, satisfies $C = -2Y + 28$. Calculate $Y^{(3)} = Y^{(2)} + \theta(\sigma_2) = 6$ and the forbidden interval set $\mathcal{I}^{B(Y^{(3)})} = \{h_2, h_3\}$, where $h_2 = [6,8]$ and $h_3 = [16,22]$. Generate σ_3, and calculate $C^{(3)} = \sum_{j=1}^{n_A} C_j^A(\sigma_3) = 16$. Then the intermediate point $(16, 6) \in \Omega$ is a break point and σ_3 is a Pareto-optimal schedule corresponding to $(16, 6)$.

(vi) Calculate the values $k(\sigma_3) = 4$ and $\theta(\sigma_3) = 1$. Then the line segment $\mathcal{L}_3$, which is the trade-off curve in the interval $[6,7)$, satisfies $C = -Y + 22$. Calculate $Y^{(4)} = Y^{(3)} + \theta(\sigma_3) = 7$ and the forbidden interval set $\mathcal{I}^{B(Y^{(4)})} = \{h_2, h_3\}$, where $h_2 = [7,8]$ and $h_3 = [16,23]$. Generate σ_4, and calculate $C^{(4)} = \sum_{j=1}^{n_A} C_j^A(\sigma_4) = 14$. Then the intermediate point $(14, 7) \in \Omega$ is a jump discontinuity point and σ_4 is a Pareto-optimal schedule corresponding to $(14, 7)$.

(vii) Finally, we conclude that $\Omega = \{\mathcal{L}_0 \cup \mathcal{L}_1 \cup \mathcal{L}_2 \cup \mathcal{L}_3 \cup \{(14,7)\}\}$ as displayed in Figure 1.

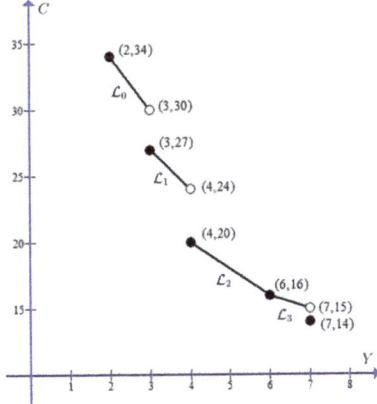

Figure 1. Trade-off curve.

4. The Second Problem

In this section, we consider problem $1|\text{pmtn}|^{\#}(L^A_{\max}, \sum Y^B_j)$ on instance $\mathcal{J}^A \cup \mathcal{J}^B$. By the job-shifting argument, we can verify that for each Pareto-optimal point, there is a corresponding Pareto-optimal schedule in which the A-jobs are scheduled in the EDD order. Thus, in this section, we renumber the A-jobs by the EDD order such that $d^A_1 \leq d^A_2 \leq \cdots \leq d^A_{n_A}$. Then we only consider the schedules in which the A-jobs are scheduled in the order $J^A_1 \prec J^A_2 \prec \cdots \prec J^A_{n_A}$.

For a point $(L^*, Y^*) \in \Omega(\mathcal{J}^A, \mathcal{J}^B)$, let σ be a Y^*-standard schedule of $\mathcal{J}^A \cup \mathcal{J}^B$. Then the set of forbidden intervals (occupied by the B-jobs) is given by (5) and the A-jobs are preemptively scheduled in the order $J^A_1 \prec J^A_2 \prec \cdots \prec J^A_{n_A}$ from time 0 in the free-time intervals. Thus, there are $m - i^* + 1$ forbidden intervals and the first forbidden interval in σ is given by $h_{i^*} = [\tau^*, \tau^{(i^*)}_2]$.

If $P_A \leq \tau^*$, then all the A-jobs are scheduled before the first forbidden interval h_{i^*} in σ. In this case, we have no further action.

In general, suppose that $P_A > \tau^*$. Then at least one A-job completing after h_{i^*} in σ. Let $J^A_{k(\sigma)}$ be the first A-job which completes after h_{i^*} in σ. Then, there are totally $n_A - k(\sigma) + 1$ A-jobs completing after h_{i^*} in σ.

Let $L^* = L^A_{\max}(\sigma)$ be the maximum lateness of A-jobs in σ. An A-job J^A_j with $k(\sigma) \leq j \leq n_A$ is called *critical* in σ if $L^A_j(\sigma) = L^A_{\max}(\sigma)$. Again, we use $\mathcal{J}^A_c(\sigma)$ to denote the set of all critical A-jobs. Then $\mathcal{J}^A_c(\sigma) = \{J^A_j : L^A_j(\sigma) = L^*, k(\sigma) \leq j \leq n_A\}$.

For each critical A-job J^A_j, we define $i[j]$ to be the interval index such that J^A_j completes after interval $h_{i[j]}$ and before interval $h_{i[j]+1}$ in σ, or equivalently, $\tau^{(i[j])}_2 < C^A_j(\sigma) \leq \tau^{(i[j]+1)}_1$. Define

$$\lambda(\sigma) = \max\{C^A_j(\sigma) - \tau^{(i[j])}_2 : J^A_j \in \mathcal{J}^A_c(\sigma)\}. \tag{9}$$

A critical A-job J^A_j is called a *desired A-job* in σ if $C^A_j(\sigma) - \tau^{(i[j])}_2 = \lambda(\sigma)$. In this case, interval $h_{i[j]}$ is called the nearest forbidden interval corresponding to desired A-job J^A_j in σ. We further define

$$BL(\sigma) = \max\{L^A_j(\sigma) : C^A_j(\sigma) \leq \tau^*\}, \tag{10}$$

and

$$\Delta(\sigma) = L^* - BL(\sigma). \tag{11}$$

In the case that no A-job completes before interval h_{i^*}, we define $BL(\sigma) = -\infty$ and $\Delta(\sigma) = +\infty$. Moreover, we define

$$\vartheta(\sigma) = \min\{\min\{\lambda(\sigma), \delta_{i^*}(\sigma)\}, \Delta(\sigma)\}. \tag{12}$$

Please note that when the schedule σ is given, the values $C^A_j(\sigma)$ and $L^A_j(\sigma)$, $j = 1, 2, \ldots, n_A$, and L^* can be determined in $O(n_A)$ time. Then the set $\mathcal{J}^A_c(\sigma)$ and the interval indices $i[j]$ for $J^A_j \in \mathcal{J}^A_c(\sigma)$ can be determined in $O(n_A)$ time. After that, the value $\lambda(\sigma)$, $BL(\sigma)$ and $\Delta(\sigma)$ can be determined in $O(n_A)$ time. Finally, the value $\vartheta(\sigma)$ can be determined by its definition in (12) in constant time. Then we have the following lemma.

Lemma 6. *Given the Y^*-standard schedule σ in advance, the values $\Delta(\sigma)$ and $\vartheta(\sigma)$ can be determined in $O(n_A)$ time.*

If $\Delta(\sigma) = 0$, then $BL(\sigma) = L^*$. In this case, (L^*, Y^*) is the last Pareto-optimal point.

Suppose that $\Delta(\sigma) > 0$. Then $\vartheta(\sigma) > 0$. For each $Y \in [Y^*, Y^* + \vartheta(\sigma)]$, let σ' be the Y-standard schedule. Then σ' is obtained from σ by shifting $Y - Y^*$ units of h_{i^*} to the last forbidden interval and then moving the A-jobs in $\{J^A_{k(\sigma)}, J^A_{k(\sigma)+1}, \ldots, J_{n_A}\}$ left to eliminate the idle times accordingly. This means that $L^A_j(\sigma') \leq L^A_j(\sigma) - (Y - Y^*)$ for $j \in \{k(\sigma), k(\sigma) + 1, \ldots, n_A\}$. Assume that the maximum lateness of A-jobs in σ' is L. According to Lemma 2, (L, Y) is a Pareto-optimal point. In the following, we consider the trade-off curve between (L^*, Y^*) and (L, Y). For convenience, point (L, Y) is simply called point Y.

We will show that the trade-off curve for $Y \in [Y^*, Y^* + \vartheta(\sigma))$ is a line segment. Discussion for the singularity of the point $Y^* + \vartheta(\sigma)$ will be omitted. This does not affect the characterization of the trade-off curve.

Lemma 7. *Suppose that $\Delta(\sigma) > 0$. For each point $(L, Y) \in \Omega(\mathcal{J}^A, \mathcal{J}^B)$ with $Y \in [Y^*, Y^* + \vartheta(\sigma))$, we have $\frac{L-L^*}{Y-Y^*} = -1$.*

Proof. For each $Y \in [Y^*, Y^* + \vartheta(\sigma))$, we have $Y - Y^* < \vartheta(\sigma)$. When we change σ to σ', no desired A-jobs are moved left across their corresponding nearest forbidden intervals in σ'. As a result, compared with σ, each desired job must move forward $Y - Y^*$ units and the other A-jobs move forward at least $Y - Y^*$ units in σ'. Thus, the desired A-jobs in σ are also critical A-jobs in σ'. Then we have $\frac{L-L^*}{Y-Y^*} = -1$, as required. □

Theorem 2. *Algorithm 3 generates the trade-off curve of $1|pmtn|^{\#}(L_{max}^A, \sum Y_j^B)$ in $O(nn_A n_B)$ time.*

Algorithm 3: Trade-off curve of problem $1|pmtn|^{\#}(L_{max}^A, \sum Y_j^B)$.

Input: Instance $\mathcal{J}^A \cup \mathcal{J}^B$.
Preprocessing: Renumber the A-jobs such that $d_1^A \leq d_2^A \leq \cdots \leq d_{n_A}^A$ and renumber the B-jobs such that $d_1^B \leq d_2^B \leq \cdots \leq d_{n_B}^B$.
Step 1: Do the following:
(1.1) Generate schedule σ_0^B which schedules the B-jobs in the order $J_1^B \prec J_2^B \prec \cdots \prec J_{n_B}^B$ in the interval $[0, P_B]$ without idle times. Then calculate the value $Y^{(0)} = T_{max}(\sigma_0^B)$.
(1.2) Run Procedure($Y^{(0)}$) to obtain the schedule $\sigma^{B(Y^{(0)})}$ of the B-jobs. Determine the intervals occupied by the B-jobs in $\sigma^{B(Y^{(0)})}$, say $h_1, h_2, \ldots, h_m$, where $h_i = [\tau_1^{(i)}, \tau_2^{(i)}]$ is the i-th interval, $i = 1, 2, \ldots, m$, as described in (3). Then regard $h_1, h_2, \ldots, h_m$ as forbidden intervals which will be updated in the implementation of the algorithm. We take the convention that the forbidden intervals are just occupied by the B-jobs.
(1.3) Set $t_0 := 1$ and set $i := 0$.
Step 2: Do the following:
(2.1) Generate the $Y^{(i)}$-standard schedule σ_i of $\mathcal{J}^A \cup \mathcal{J}^B$ in which $h_{t_i}, h_{t_i+1}, \ldots, h_m$ are the forbidden intervals and the A-jobs are preemptively scheduled in the order $J_1^A \prec J_2^A \prec \cdots \prec J_{n_A}^A$ as early as possible. Determine the values $L^{(i)} = L_{max}^A(\sigma_i)$.
(2.2) Determine the value $\Delta(\sigma_i)$. Moreover, if $\Delta(\sigma_i) > 0$, determine the value $\vartheta(\sigma_i)$.
(2.3) If $\Delta(\sigma_i) = 0$, then set $K = i$ and go to Step 4. (In this case, we have obtained the whole trade-off curve.)
If $\Delta(\sigma_i) > 0$, then go to Step 3. (In this case, we have $\vartheta(\sigma_i) > 0$.)
Step 3: Define a left closed right open segment $\mathcal{L}_i$ in the interval $[Y^{(i)}, Y^{(i)} + \vartheta(\sigma_i))$ by the following way:
$$\mathcal{L}_i : \frac{L - L^{(i)}}{Y - Y^{(i)}} = -1, \ Y \in [Y^{(i)}, Y^{(i)} + \vartheta(\sigma_i)).$$
(3.3) Set $Y^{(i+1)} := Y^{(i)} + \vartheta(\sigma_i)$ and $h_m := [P^*, P^* + Y^{(i+1)}]$. Moreover, if $\vartheta(\sigma_i) = \delta_{t_i}(\sigma_i)$, then set $t_{i+1} := t_i + 1$; and if $\vartheta(\sigma_i) < \delta_{t_i}(\sigma_i)$, then set $t_{i+1} := t_i$ and $h_{t_i+1} := [\tau_1^{(t_i)} + \vartheta(\sigma_i), \tau_2^{(t_i)}]$ (which is obtained from interval h_{t_i} by deleting the first $\vartheta(\sigma_i)$ units.)
(3.4) Set $i := i + 1$. Go to Step 2.
Step 4: Output the trade-off curve $\Omega(\mathcal{J}^A, \mathcal{J}^B) = \mathcal{L}_0 \cup \mathcal{L}_1 \cup \cdots \cup \mathcal{L}_{K-1} \cup \{(L^{(K)}, Y^{(K)})\}$.

Proof. Correctness of Algorithm 3 is guaranteed by Lemmas 2 and 7. The time complexity can be estimated by the similar way of Theorem 1 by putting Lemma 6 in discussion. □

It can be observed that the total interruption time (i.e., the number of interruptions) of all the jobs in each schedule generated by Algorithm 3 is upper bounded by $1 + m \leq 1 + n_B$.

Let us consider the job instance I_2 displayed in Table 2. The trade-off curve of problem $1|\text{pmtn}|^\#(L_{\max}^A, \sum Y_j^B)$ on instance I_2 is shown in Figure 2.

Table 2. The instance I_2.

J_i^X	J_1^A	J_2^A	J_3^A	J_1^B	J_2^B	J_3^B
p_i^X	4	2	3	5	2	2
d_i^X	4	7	11	3	7	12

Please note that $P^* = P + 1 = 19$. Let $\Omega = \Omega(\mathcal{J}^A, \mathcal{J}^B)$. The key steps in applying Algorithm 3 to solve the instance I_2 are as follows:

(i) Generate the schedule $\sigma_0^B = (J_1^B, J_2^B, J_3^B)$ and calculate the Y-value $Y^{(0)} = T_{\max}(\sigma_0^B) = 2$. Then generate the schedule $\sigma^{B(Y^{(0)})}$, and the forbidden interval set $\mathcal{I}^{B(Y^{(0)})} = \{h_1, h_2, h_3, h_4\}$ is determined, where $h_1 = [0,3]$, $h_2 = [5,7]$, $h_3 = [10,12]$, and $h_4 = [19,21]$. Then, for each $y \in (Y^{(0)}, P_B] = (2,9]$, $\sigma^{B(y)}$ and $\mathcal{I}^{B(y)}$ can be easily generated.

(ii) Generate the $Y^{(0)}$-standard schedule σ_0 of $\mathcal{J}^A \cup \mathcal{J}^B$ in which h_1, h_2, h_3, h_4 are the forbidden intervals and the A-jobs are preemptively scheduled in the order $J_1^A \prec J_2^A \prec J_3^A$ as early as possible. Determine the value $L^{(0)} = L_{\max}^A(\sigma_0) = 6$. Then σ_0 is a Pareto-optimal schedule corresponding to $(6,2) \in \Omega$.

(iii) Calculate the values $\Delta(\sigma_0) = 6 > 0$ and $\vartheta(\sigma_0) = 1$. Then the line segment $\mathcal{L}_0$, which is the trade-off curve in the interval $[2,3]$, satisfies $L = -Y + 8$. Calculate $Y^{(1)} = Y^{(0)} + \vartheta(\sigma_0) = 3$ and the forbidden interval set $\mathcal{I}^{B(Y^{(1)})} = \{h_1, h_2, h_3, h_4\}$, where $h_1 = [1,3]$, $h_2 = [5,7]$, $h_3 = [10,12]$, and $h_4 = [19,22]$. Generate σ_1, and calculate $L^{(1)} = L_{\max}^A(\sigma_1) = 4$. Then the intermediate point $(4,3) \in \Omega$ is a jump discontinuity point and σ_1 is a Pareto-optimal schedule corresponding to $(4,3)$.

(iv) Calculate the values $\Delta(\sigma_1) = 4 > 0$ and $\vartheta(\sigma_1) = 2$. Then the line segment $\mathcal{L}_1$, which is the trade-off curve in the interval $[3,5]$, satisfies $L = -Y + 7$. Calculate $Y^{(2)} = Y^{(1)} + \vartheta(\sigma_1) = 5$ and the forbidden interval set $\mathcal{I}^{B(Y^{(2)})} = \{h_2, h_3, h_4\}$, where $h_2 = [5,7]$, $h_3 = [10,12]$, and $h_4 = [19,24]$. Generate σ_2, and calculate $L^{(2)} = L_{\max}^A(\sigma_2) = 2$. Then the intermediate point $(2,5) \in \Omega$ is a continuity point and σ_2 is a Pareto-optimal schedule corresponding to $(2,5)$.

(v) Calculate the values $\Delta(\sigma_2) = 2 > 0$ and $\vartheta(\sigma_2) = 1$. Then the line segment $\mathcal{L}_2$, which is the trade-off curve in the interval $[5,6]$, satisfies $L = -Y + 7$. Calculate $Y^{(3)} = Y^{(2)} + \vartheta(\sigma_2) = 6$ and the forbidden interval set $\mathcal{I}^{B(Y^{(2)})} = \{h_2, h_3, h_4\}$, where $h_2 = [6,7]$, $h_3 = [10,12]$, and $h_4 = [19,25]$. Generate σ_3, and calculate $L^{(3)} = L_{\max}^A(\sigma_3) = 0$. Then the intermediate point $(0,6) \in \Omega$ is a jump discontinuity point and σ_3 is a Pareto-optimal schedule corresponding to $(0,6)$.

(vi) Calculate the value $\Delta(\sigma_3) = 0$. Then, we conclude that $\Omega = \{\mathcal{L}_0 \cup \mathcal{L}_1 \cup \mathcal{L}_2 \cup \{(0,6)\}\}$ as displayed in Figure 2.

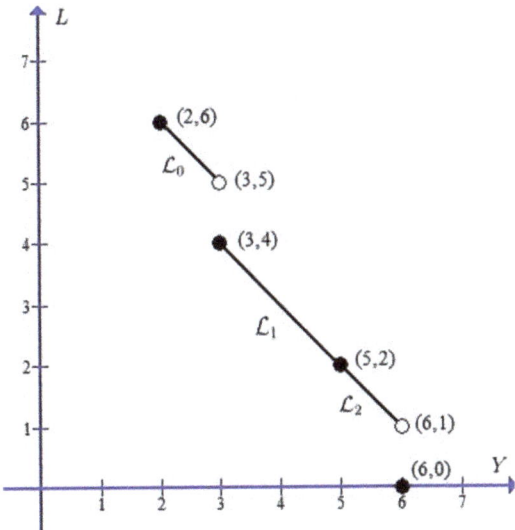

Figure 2. Trade-off curve.

5. The Third Problem

In this section, we consider problem $1|\text{pmtn}|^{\#}(\sum Y_j^A, \sum Y_j^B)$ on instance $\mathcal{J}^A \cup \mathcal{J}^B$. We renumber the A-jobs by the EDD order such that $d_1^A \leq d_2^A \leq \cdots \leq d_{n_A}^A$.

For a point $(Y^A, Y^B) \in \Omega(\mathcal{J}^A, \mathcal{J}^B)$, a Y^B-standard schedule of $\mathcal{J}^A \cup \mathcal{J}^B$ corresponding to (Y^A, Y^B) can be obtained in the following way in $O(n)$ time: (i) Run Procedure(Y^B) to obtain the schedule $\sigma^{B(Y^B)}$ of the B-jobs, and (ii) run the algorithm in Hariri et al. [21] for solving problem $1|\text{pmtn}|\sum Y_j$ to schedule the A-jobs in the free-time intervals not occupied by the B-jobs in $\sigma^{B(Y^B)}$. Thus, we only need to consider the trade-off curve of problem $1|\text{pmtn}|^{\#}(\sum Y_j^A, \sum Y_j^B)$ on instance $\mathcal{J}^A \cup \mathcal{J}^B$. We first establish a nice property for problem $1|\text{pmtn}|\sum Y_j$ in the following lemma.

Lemma 8. *Let $\mathcal{J} = \{J_1, J_2, \ldots, J_n\}$ be a job instance of problem $1|\text{pmtn}|\sum Y_j$. Let $\mathcal{U}$ be a subset of $\mathcal{J}$ such that there is a schedule of instance $\mathcal{J}$ such that all the jobs in $\mathcal{U}$ are early. Then there is an optimal schedule of problem $1|\text{pmtn}|\sum Y_j$ on instance $\mathcal{J}$ such that all the jobs in $\mathcal{U}$ are early.*

Proof. We first prove the result for problem $1|\text{pmtn}|\sum Y_j$ without maintenance intervals by induction on $|\mathcal{U}|$. The result holds trivially if $|\mathcal{U}| = 0$.

Inductively, suppose that $|\mathcal{U}| = k \geq 1$, $\mathcal{U} = \{J_{j_1}, J_{j_2}, \ldots, J_{j_k}\}$, $d_{j_1} \leq d_{j_2} \leq \cdots \leq d_{j_k}$, and there is a feasible schedule of instance $\mathcal{J}$ such that all the jobs in $\mathcal{U}$ are early. Moreover, the result holds for every proper subset of $\mathcal{U}$ (the induction hypothesis).

Since $\mathcal{U} \setminus \{J_{j_k}\}$ is a proper subset of $\mathcal{U}$, from the induction hypothesis, there is an optimal schedule π of problem $1|\text{pmtn}|\sum Y_j$ on instance $\mathcal{J}$ such that all the $k-1$ jobs $J_{j_1}, J_{j_2}, \ldots, J_{j_{k-1}}$ are early in π. Since all the jobs in $\mathcal{U}$ are early in some feasible schedule, we have $p_{j_1} + p_{j_2} + \cdots + p_{j_k} \leq \max\{d_{j_1}, d_{j_2}, \ldots, d_{j_k}\} = d_{j_k}$. This implies that all the $k-1$ jobs $J_{j_1}, J_{j_2}, \ldots, J_{j_{k-1}}$ are completed by time d_{j_k} in π and at least p_{j_k} units of time in the interval $[0, d_{j_k}]$ are not occupied by the $k-1$ jobs $J_{j_1}, J_{j_2}, \ldots, J_{j_{k-1}}$.

If $C_{j_k}(\pi) \leq d_{j_k}$, i.e., J_{j_k} is early in π, then π is a required optimal schedule.

Suppose in the following that $C_{j_k}(\pi) > d_{j_k}$. Then there is a certain index $i \in \{0, 1, \ldots, p_{j_k}\}$ such that for job J_{j_k}, the first i unit pieces $J_{j_k,1}(\pi), J_{j_k,2}(\pi), \ldots, J_{j_k,i}(\pi)$ are early in π and the last $p_{j_k} - i$ unit pieces $J_{j_k,i+1}(\pi), \ldots, J_{j_k,p_{j_k}}(\pi)$ are late in π. Let $\mathcal{S}$ be the time space which consists of the last $p_{j_k} - i$ units of time in the interval $[0, d_{j_k}]$ that are not occupied by the $k-1$ jobs $J_{j_1}, J_{j_2}, \ldots, J_{j_{k-1}}$ and the i unit pieces $J_{j_k,1}(\pi), J_{j_k,2}(\pi), \ldots, J_{j_k,i}(\pi)$ of J_{j_k}. Let $\mathcal{T}$ be the time space which consists of the $p_{j_k} - i$

13

units of time that are occupied by the $p_{j_k} - i$ unit pieces $J_{j_k,i+1}(\pi),\ldots,J_{j_k,p_{j_k}}(\pi)$ of J_{j_k}. Let σ be the schedule of $\mathcal{J}$ obtained from π by exchanging the subschedules in $\mathcal{S}$ and in $\mathcal{T}$. Then J_{j_k} is early in σ. Moreover, $\sum Y_j(\sigma) \leq \sum Y_j(\pi) - |\mathcal{T}| + |\mathcal{S}| = \sum Y_j(\pi)$, implying that σ is also optimal. Now all the jobs in $\mathcal{U} = \{J_{j_1}, J_{j_2}, \ldots, J_{j_k}\}$ are early in σ. Consequently, σ is an optimal schedule of problem $1|pmtn|\sum Y_j$ on instance $\mathcal{J}$ such that all the jobs in $\mathcal{U}$ are early. The result follows by the induction principle. □

We next use Lemma 8 to prove the following useful lemma.

Lemma 9. *Let $\mathcal{J} = \{J_1, J_2, \ldots, J_n\}$ be a job instance of problem $1|pmtn|\sum Y_j$. Let π be a schedule of the jobs in $\mathcal{J}$. Then there is an optimal schedule σ of problem $1|pmtn|\sum Y_j$ on instance $\mathcal{J}$ such that $Y_j(\sigma) \leq Y_j(\pi)$ for $j = 1, 2, \ldots, n$.*

Proof. For each $j \in \{1, 2, \ldots, n\}$, we partition J_j into two parts $J_{j'}$ and $J_{j''}$ such that $p_{j'} = p_j - Y_j(\pi)$ is the early work of J_j in π, $p_{j''} = Y_j(\pi)$ is the late work of J_j in π, and $d_{j'} = d_{j''} = d_j$. Let $\mathcal{J}' = \{J_{j'}, J_{j''} : j = 1, 2, \ldots, n\}$. Let π' be the schedule of $\mathcal{J}'$ which is obtained from π by just regarding the early part of J_j in π as job $J_{j'}$ and regarding the late part of J_j in π as job $J_{j''}$. Then all the jobs in $\{J_{j'} : j = 1, 2, \ldots, n\}$ are early in π'. According to Lemma 8, there is an optimal schedule σ of problem $1|pmtn|\sum Y_j$ on instance $\mathcal{J}'$ such that all the jobs in $\{J_{j'} : j = 1, 2, \ldots, n\}$ are early. Since the preemptive assumption, the two instances $\mathcal{J}$ and $\mathcal{J}'$ have no essential difference for problem $1|pmtn|\sum Y_j$, σ is an optimal schedule of problem $1|pmtn|\sum Y_j$ on instance $\mathcal{J}$. The result follows by noting that $Y_j(\sigma) = Y_{j''}(\sigma) \leq p_{j''} = Y_j(\pi)$ for $j = 1, 2, \ldots, n$. □

Let Y^{AB} be the optimal value of problem $1|pmtn|\sum Y_j$ on instance $\mathcal{J}^A \cup \mathcal{J}^B$. We have the following lemma.

Lemma 10. *For each point $(Y^A, Y^B) \in \Omega(\mathcal{J}^A, \mathcal{J}^B)$, we have $Y^A + Y^B = Y^{AB}$.*

Proof. Let σ be a Pareto-optimal schedule of problem $1|pmtn|^{\#}(\sum Y_j^A, \sum Y_j^B)$ on instance $\mathcal{J}^A \cup \mathcal{J}^B$ such that $\sum Y_j^A(\sigma) = Y^A$ and $\sum Y_j^B(\sigma) = Y^B$. From Lemma 9, there is an optimal schedule σ' of $1|pmtn|\sum Y_j$ on instance $\mathcal{J}^A \cup \mathcal{J}^B$ such that $\sum Y_j^A(\sigma') \leq \sum Y_j^A(\sigma)$ and $\sum Y_j^B(\sigma') \leq \sum Y_j^B(\sigma)$. The optimality of σ' implies that $\sum Y_j(\sigma') = Y^{AB}$. From the property of Pareto-optimal point, we can obtain that $\sum Y_j^A(\sigma') = \sum Y_j^A(\sigma)$ and $\sum Y_j^B(\sigma') = \sum Y_j^B(\sigma)$. Thus, we have $Y^A + Y^B = Y^{AB}$. The result follows. □

Theorem 3. *The trade-off curve of problem $1|pmtn|^{\#}(\sum Y_j^A, \sum Y_j^B)$ on instance $\mathcal{J}^A \cup \mathcal{J}^B$ can be determined in $O(n \log n)$ time.*

Proof. Let $Y_A^{(0)}$ be the optimal value of problem $1|pmtn|Y_j^A$ on instance $\mathcal{J}^A$. Let $Y_B^{(0)}$ be the optimal value of problem $1|pmtn|Y_j^B$ on instance $\mathcal{J}^B$. Recall that Y^{AB} is the optimal value of problem $1|pmtn|\sum Y_j$ on instance $\mathcal{J}^A \cup \mathcal{J}^B$. From Hariri et al. [21], $Y_A^{(0)}$, $Y_B^{(0)}$, and Y^{AB} can be determined in $O(n_A \log n_A)$ time, $O(n_B \log n_B)$ time, and $O(n \log n)$ time, respectively.

Please note that $Y_A^{(0)}$ is the minimum total late work of A-jobs among all Pareto-optimal points and $Y_B^{(0)}$ is the minimum total late work of B-jobs among all Pareto-optimal points. Thus, from Lemma 10, the trade-off curve is the line segment

$$Y^A + Y^B = Y^{AB}, \quad Y_B^{(0)} \leq Y^B \leq Y^{AB} - Y_A^{(0)},$$

connecting point $(Y^{AB} - Y_B^{(0)}, Y_B^{(0)})$ to point $(Y_A^{(0)}, Y^{AB} - Y_A^{(0)})$. So, the overall complexity to obtain the trade-off curve is given by $O(n \log n)$. □

It can be observed that the total interruption time (i.e., the number of interruptions) of all the jobs in each Pareto-optimal schedule is upper bounded by $\max\{n_A, n_B\} + 2$.

Let us consider the job instance I_3 displayed in Table 3. The trade-off curve of problem $1|pmtn|^{\#}(\sum Y_j^A, \sum Y_j^B)$ on instance I_3 is shown in Figure 3.

Table 3. The instance I_3.

J_i^X	J_1^A	J_2^A	J_3^A	J_4^A	J_5^A	J_1^B	J_2^B	J_3^B
p_i^X	4	2	3	2	2	3	3	3
d_i^X	3	4	7	11	14	2	8	13

Please note that $P^* = P + 1 = 23$. Let $\Omega = \Omega(\mathcal{J}^A, \mathcal{J}^B)$. The key steps to solve the instance I_3 are as follows:

(i) Generate the schedule $\sigma_0^B = (J_1^B, J_2^B, J_3^B)$ and calculate the Y^B-value $Y^{(0)} = T_{\max}(\sigma_0^B) = 1$. Then generate the schedule $\sigma^{B(Y^{(0)})}$, and the forbidden interval set $\mathcal{I}^{B(Y^{(0)})} = \{h_1, h_2, h_3, h_4\}$ is determined, where $h_1 = [0, 2]$, $h_2 = [5, 8]$, $h_3 = [10, 13]$, and $h_4 = [23, 24]$. Then, for each $y \in (Y^{(0)}, P_B] = (1, 9]$, $\sigma^{B(y)}$ and $\mathcal{I}^{B(y)}$ can be easily generated.

(ii) Generate the $Y^{(0)}$-standard schedule σ_0 of $\mathcal{J}^A \cup \mathcal{J}^B$ in which h_1, h_2, h_3, h_4 are the forbidden intervals and the A-jobs are preemptively scheduled by running the algorithm in Hariri et al. [21] for solving problem $1|pmtn|\sum Y_j$ in the free-time intervals not occupied by the B-jobs in $\sigma^{B(Y^0)}$. Determine the value $Y^A(\sigma_0) = \sum Y_j^A(\sigma_0) = 7$. Then σ_0 is a Pareto-optimal schedule corresponding to $(7, 1) \in \Omega$.

(iii) By using the same method as (i) and (ii), we schedule A-jobs first. Generate the schedule $\sigma^{A(Y^{(0)})}$, where $Y^{(0)} = T_{\max}(\sigma_0^A) = 2$. Generate the schedule σ_1 of $\mathcal{J}^A \cup \mathcal{J}^B$ in which the B-jobs are preemptively scheduled by running the algorithm in Hariri et al. [21] for solving problem $1|pmtn|\sum Y_j$ in the free-time intervals not occupied by the A-jobs in $\sigma^{A(Y^{(0)})}$. Then σ_1 is a Pareto-optimal schedule corresponding to $(2, 6) \in \Omega$. From Lemma 10, σ_1 is the final Pareto-optimal schedule. Then the trade-off curve is just the line segment in the interval $[1, 6]$, which satisfies $Y^A = -Y^B + 8$ as displayed in Figure 3.

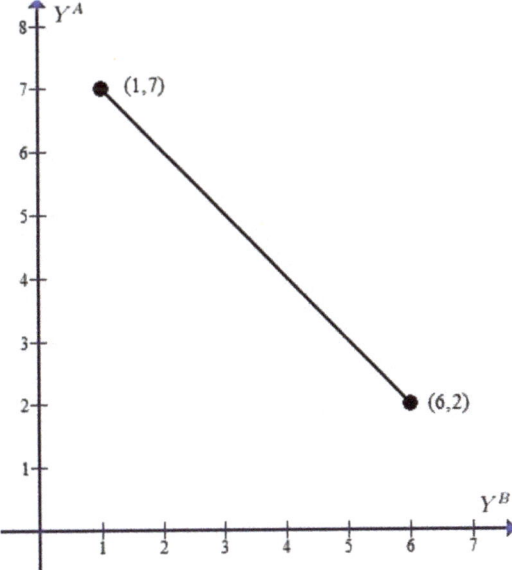

Figure 3. Trade-off curve.

6. Conclusions

This paper considers three preemptive Pareto-scheduling problems with two competing agents on a single machine. Two agents compete to perform their respective jobs on a common single machine and each agent has his own criterion to optimize. In each problem, the goal of agent A is to minimize the total completion time, the maximum lateness, or the total late work while agent B wants to minimize the total late work. For each problem, we provide a polynomial-time algorithm to characterize the trade-off curve of all Pareto-optimal points.

Late-work criterion can be met in all cases where the penalty imposed on a solution depends on the number of tardy units of jobs performed in a system. For example, in production planning where the manufacturer is concerned with minimizing any order delays which cause financial loss, in control systems where the accuracy of control procedures depends on the amount of information provided as their input, in agriculture where performance measures based on due dates, and so on. In the case where two criteria need to be minimized, the trade-off curve results an ideal solution. Once the trade-off curve is characterized, decision makers can make decisions as needed.

For the future research, the trade-off curve of the problem $1|\text{pmtn}|^{\#}(\sum C_j^A, \sum w_j Y_j^B)$ or $1|\text{pmtn}|^{\#}(L_{\max}^A, \sum w_j Y_j^B)$ is worthy of study. Since the existence of precedence constraints on scheduling problems reflects real-life problems, it is also worthy to study the two-agent problems with precedence constraints. Another interesting future research direction is to investigate fairness issues when the total late work is one of the criteria in two-agent scheduling problems.

Author Contributions: Conceptualization, methodology, and writing—original manuscript: R.H.; project management, supervision, and writing—review: J.Y. All authors have read and agreed to the published version of the manuscript.

Funding: This research was funded by the National Natural Science Foundation of China under grant numbers 11671368 and 11771406.

Acknowledgments: The authors would like to thank the Associate Editor and two anonymous referees for their constructive comments and helpful suggestions.

Conflicts of Interest: The authors declare no conflict of interest.

References

1. T'kindt, V.; Billaut, J.C. *Multicriteria Scheduling: Theory, Models and Algorithms*, 2nd ed.; Springer: Berlin/Heidelberg, Germany, 2006.
2. Agnetis, A.; Mirchandani, P.B.; Pacciareli, D.; Pacifici, A. Scheduling problems with two competing agents. *Oper. Res.* **2004**, *52*, 229–242. [CrossRef]
3. Cheng, T.C.E.; Ng, C.T.; Yuan, J.J. Multi-agent scheduling on a single machine to minimize total weighted number of tardy jobs. *Theor. Comput. Sci.* **2006**, *362*, 273–281. [CrossRef]
4. Cheng, T.C.E.; Ng, C.T.; Yuan, J.J. Multi-agent scheduling on a single machine with max-form criteria. *Eur. J. Oper. Res.* **2008**, *188*, 603–609. [CrossRef]
5. Lee, W.C.; Wang, W.J.; Shiau, Y.R.; Wu, C.C. A single-machine scheduling problem with two-agent and deteriorating jobs. *Appl. Math. Model.* **2010**, *34*, 3098–3107. [CrossRef]
6. Leung, J.Y.T.; Pinedo, M.; Wan, G.H. Competitive two-agent scheduling and its applications. *Oper. Res.* **2010**, *58*, 458–469. [CrossRef]
7. Ng, C.T.; Cheng, T.C.E.; Yuan, J.J. A note on the complexity of the problem of two-agent scheduling on a single machine. *J. Comb. Optim.* **2006**, *12*, 387–394. [CrossRef]
8. Agnetis, A.; Chen, B.; Nicosia, G.; Pacifici, A. Price of fairness in two-agent single-machine scheduling problems. *Eur. J. Oper. Res.* **2019**, *276*, 79–87. [CrossRef]
9. Zhang, Y.B.; Zhang, Z.; Liu, Z.H. The price of fairness for a two-agent scheduling game mini-mizing total completion time. *J. Comb. Optim.* **2020**. [CrossRef]
10. Mor, B.; Mosheiov, G. Minimizing maximum cost on a single machine with two competing agents and job rejection. *J. Oper. Res. Soc.* **2016**, *67*, 1524–1531. [CrossRef]

11. Gao, Y.; Yuan, J.J. Bi-criteria Pareto-scheduling on a single machine with due indices and precedence constraints. *Discret. Optim.* **2017**, *25*, 105–119. [CrossRef]
12. Agnetis, A.; Billaut, J.; Gawiejnowicz, S.; Pacciarelli, D.; Soukhal, A. *Multiagent Scheduling-Models and Algorithms*; Springer: Berlin/Heidelberg, Germany, 2014.
13. Liu, P.; Gu, M.; Li, G.G. Two-agent scheduling on a single machine with release dates. *Comput. Oper. Res.* **2019**, *111*, 35–42. [CrossRef]
14. Oron, D.; Shabtay, D.; Steiner, G. Single machine scheduling with two competing agents and equal jobs processing times. *Eur. J. Oper. Res.* **2015**, *244*, 86–99. [CrossRef]
15. Perez-Gonzalez, P.; Framinan, J. A common framework and taxonomy for multicriteria scheduling problem with interfering and competing jobs: multi-agent scheduling problems. *Eur. J. Oper. Res.* **2014**, *235*, 1–16. [CrossRef]
16. Yuan, J.J. Complexities of some problems on multi-agent scheduling on a single machine. *J. Oper. Res. Soc. China* **2016**, *4*, 379–384. [CrossRef]
17. Yuan, J.J. Complexities of four problems on two-agent scheduling. *Optim. Lett.* **2018**, *12*, 763–780. [CrossRef]
18. Blazewicz, J.; Finke, G. Minimizing mean weighted execution time loss on identical and uniform processors. *Inform. Process. Lett.* **1987**, *24*, 259–263. [CrossRef]
19. Potts, C.N.; Van Wassenhove, L.N. Single machine scheduling to minimize total late work. *Oper. Res.* **1992**, *40*, 586–595. [CrossRef]
20. Potts, C.N.; Van Wassenhove, L.N. Approximation algorithms for scheduling a single machine to minimize total late work. *Oper. Res. Lett.* **1991**, *11*, 261–266. [CrossRef]
21. Hariri, A.M.A.; Potts, C.N.; Van Wassenhove, L.N. Single machine scheduling to minimize total weighted late work. *ORSA J. Comput.* **1995**, *7*, 232. [CrossRef]
22. Sterna, M. A survey of scheduling problems with late work criteria. *Omega* **2011**, *39*, 120–129. [CrossRef]
23. Wang, D.J.; Kang, C.C.; Shiau, Y.R.; Wu, C.C. A two-agent single-machine scheduling problem with late work criteria. *Soft Comput.* **2017**, *21*, 2015–2033. [CrossRef]
24. Zhang, X.G.; Wang, Y. Two-agent scheduling problems on a single-machine to minimize the total weighted late work. *J. Comb. Optim.* **2017**, *33*, 945–955.
25. Zhang, Y.; Yuan, J.J. A note on a two-agent scheduling problem related to the total weighted late work. *J. Comb. Optim.* **2019**, *37*, 989–999. [CrossRef]
26. Lee, C.Y. Machine scheduling with an availability constraints. *J. Glob. Optim.* **1996**, *9*, 395–416. [CrossRef]

© 2020 by the authors. Licensee MDPI, Basel, Switzerland. This article is an open access article distributed under the terms and conditions of the Creative Commons Attribution (CC BY) license (http://creativecommons.org/licenses/by/4.0/).

Article

Two-Agent Pareto-Scheduling of Minimizing Total Weighted Completion Time and Total Weighted Late Work

Yuan Zhang [†], Zhichao Geng [*,†] and Jinjiang Yuan [†]

School of Mathematics and Statistics, Zhengzhou University, Zhengzhou 450001, China; zy2020@gs.zzu.edu.cn (Y.Z.); yuanjj@zzu.edu.cn (J.Y.)
* Correspondence: zcgeng@zzu.edu.cn
† These authors contributed equally to this work.

Received: 26 October 2020; Accepted: 16 November 2020; Published: 20 November 2020

Abstract: We investigate the Pareto-scheduling problem with two competing agents on a single machine to minimize the total weighted completion time of agent A's jobs and the total weighted late work of agent B's jobs, the B-jobs having a common due date. Since this problem is known to be NP-hard, we present two pseudo-polynomial-time exact algorithms to generate the Pareto frontier and an approximation algorithm to generate a $(1+\epsilon)$-approximate Pareto frontier. In addition, some numerical tests are undertaken to evaluate the effectiveness of our algorithms.

Keywords: scheduling; two agents; pareto frontier; approximation algorithms

1. Introduction

Problem description and motivation: Multi-agent scheduling has attracted an ever-increasing research interest due to its extensive applications (see the book of Agnetis et al. [1]). Among the common four problem-versions (including lexical-, positive-combination-, constrained-, and Pareto-scheduling, as shown in Li and Yuan [2]) for a given group of criteria for multiple agents, Pareto-scheduling has the most important practical value, since it reflects the effective tradeoff between the actual and (usually) conflicting requirements of different agents.

Our considered problem is formally stated as follows. Assume that two agents (A and B) compete to process their own sets of independent and non-preemptive jobs on a single machine. The set of the n_X jobs from agent $X \in \{A, B\}$ is $\mathcal{J}_X = \{J_1^X, J_2^X, \cdots, J_{n_X}^X\}$ with $\mathcal{J}_A \cap \mathcal{J}_B = \phi$. For convenience, we call a job from agent X an X-job. All jobs are available at time zero, and are scheduled consecutively without idle time due to the regularity of the objective functions as shown later. Each job J_j^X has a processing time p_j^X and a weight w_j^X. In addition, each B-job J_j^B has also a common due date d. We assume that all parameters p_j^X, w_j^X and d are known integers.

Let σ be a schedule. We use $C_j^X(\sigma)$ to denote the completion time of job J_j^X in σ. The objective function of agent A is the total weighted completion time, denoted by $\sum w_j^A C_j^A(\sigma)$, while the objective function of agent B is the total weighted late work, denoted by $\sum w_j^B Y_j^B(\sigma)$. Here, the late work $Y_j^B(\sigma)$ of job J_j^B indicates the amount processed after the due date d, specifically,

$$Y_j^B(\sigma) = \begin{cases} 0, & \text{if } C_j(\sigma) \leq d, \\ C_j^B(\sigma) - d, & \text{if } d < C_j^B(\sigma) \leq d + p_j^B, \\ p_j^B, & \text{if } C_j^B(\sigma) > d + p_j^B. \end{cases} \quad (1)$$

Following Hariri et al. (1995) [3], job J_j^B is said to be *early, partially early,* and *late* in σ, if $Y_j^B(\sigma) = 0$, $0 < Y_j^B(\sigma) < p_j^B$, and $Y_j^B(\sigma) = p_j^B$, respectively.

Falling into the category of Pareto-scheduling, the problem studied in this paper aims at generating all Pareto-optimal points (PoPs) and the corresponding Pareto-optimal schedules (PoSs) (the definitions of PoPs and PoSs will be given in Section 2) of all jobs with regard to $\sum w_j^A C_j^A$ and $\sum w_j^B Y_j^B$). Using the notations in Agnetis et al. [1], our studied scheduling problem can be denoted by $1|d_j^B = d|^{\#}(\sum w_j^A C_j^A, \sum w_j^B Y_j^B)$. For this problem, we will devise some efficient approximate algorithms.

Our considered scheduling model arises from many practical scenarios. For example, in a factory, two concurrent projects (A and B), each containing a certain amount of activities with distinct importance, have to share a limited resource. The former focuses on the mean completion time of its activities. In contrast, the latter requires its activities to be completed before the due date as much as possible, since, otherwise, the shortcomings of some key technical forces after the due date will occur and result in irretrievable loss. It is necessary to model the goal of project B as the weighted late work, that is, minimizing the parts left unprocessed before the due date. In addition, two projects naturally have to negotiate to seek a trade-off method of utilizing the common resource.

For another example, in a distribution center, two categories (A- and B-) goods are stored in a warehouse, in which the former comprises common goods and the latter comprises fresh goods with a shelf life. It is hoped that the shipping preparations for the A-goods will be completed as soon as possible. However, due to their limited shelf life, if they are transported after a certain time, the B-goods will not be fresh enough when they reach the customers. Therefore, it is reasonable to respectively model the goals of A-goods and B-goods by minimizing the total weighted completion time and the total weighted late work, and seek an efficient transportation method.

Related works and our contribution: Numerous works have addressed multi-agent scheduling problems in the literature. With the aim of this paper, we only summarize briefly some related results. Wan et al. [4] provided a strongly polynomial-time algorithm for the two-agent Pareto-scheduling problem on a single machine to minimize the number of the tardy A-jobs and the maximum cost of the B-jobs. Later, Wan et al. [5] investigated two Pareto-scheduling problems on a single machine with two competing agents and a linear-deterioration processing time: $1||^{\#}(E_{\max}^A, E_{\max}^B)$ and $1||^{\#}(\sum E_j^A, E_{\max}^B)$, where $\sum E_j^A$ is the total earliness of the A-jobs and $E_{\max}^X$ is the maximum earliness of the X-jobs. For these two problems, they respectively proposed a polynomial-time algorithm. Gao and Yuan [6] showed that the following two Pareto-scheduling problems with a positional due index and precedence constraints are both polynomially solvable: $1||^{\#}(\sum C_j^A, f_{\max}^B)$ and $1||^{\#}(f_{\max}^A, f_{\max}^B)$, where $f_{\max}^X$ indicates the maximum cost of the X-jobs. He et al. [7] extensively considered the versions of the problems in Gao and Yuan [6] with deteriorating or shortening processing times and without positional due indices and precedence constraints, and devised polynomial-time algorithms. Yuan et al. [8] showed the single-machine preemptive problem $1|r_j, pmtn|^{\#}(L_{\max}^a : L_{\max}^b)$ can be solved in a polynomial time, where $L_{\max}^X$ indicates the maximum lateness of the X-jobs. Wan [9] investigated the single-machine two-agent scheduling problem to minimize the maximum costs with position-dependent jobs, and developed a polynomial-time algorithm.

While most results on Pareto-scheduling concentrate on devising exact algorithms to obtain the Pareto frontier, there are also some methods (such as [10–14]) of developing approximate algorithms to generate the approximate Pareto frontier. Dabia et al. [10] adopted the trimming technique to derive the approximate Pareto frontier for some multi-objective scheduling problems. Yin et al. [15] considered two just-in-time (JIT) scheduling problems with two competing agents on unrelated parallel machines, in which the one agent's criterion is to maximize the weighted number of its JIT jobs, and another agent's criterion is either to maximize its maximum gains from its JIT jobs or to maximize the weighted number of its JIT jobs. They showed that the two problems are both unary NP-hard when the machine number is not fixed, and proposed either a polynomial-time algorithm or a fully polynomial-time approximation scheme (FPTAS) when the machine number is a constant. Yin et al. [16] also considered similar problems in the setting of a two-machine flow shop, and provided two pseudo-polynomial-time exact algorithms to find the Pareto frontier. Chen et al. [17] studied a multi-agent Pareto-scheduling problem in a no-wait flow shop setting, in which each agent's criterion is to maximize its own weighted number of JIT jobs. They showed that it is unary NP-hard when the number of agents is arbitrary, and presented pseudo-polynomial time algorithms and an $(1, 1-\epsilon, \ldots, 1-\epsilon)$-approximation algorithm when the number of agents is fixed.

From the perspective of methodology, as a type of optimization problem, the multi-agent scheduling problem's solution algorithms potentially allow for exploiting the optimal robot path planning by a gravitational search algorithm (Purcaru et al. [18]) and optimization based on phylogram analysis (Soares et al. [19]).

In the prophase work (Zhang and Yuan [20]), we proved that the constrained scheduling problem of minimizing the total late work of agent A's jobs with equal due dates subject to the makespan of agent B's jobs not exceeding a given upper bound, is NP-hard even if agent B has only one job. It implies the NP-hardness of our considered problem in this paper. Thus we limit the investigation to devising pseudo-polynomial-time exact algorithms and an approximation algorithm to generate the approximate Pareto frontier.

In addition, in our recent work (Zhang et al. [21]), we considered several three-agent scheduling problems under different constraints on a single machine, in which the three agents' criteria are to minimize the total weighted completion time, the weighted number of tardy jobs, and the total weighted late work. Among those problems, there are two questions related to this paper: $1|p_j^A \uparrow\downarrow w_j^A|^\#(\Sigma w_j^A C_j^A, \Sigma w_j^B Y_j^B)$, which is solved in $O(n_A n_B^2 U^A U^B)$, and $1|p_j^A \uparrow\downarrow w_j^A, d_j^B \uparrow\downarrow w_j^B|^\#(\Sigma w_j^A C_j^A, \Sigma w_j^B Y_j^B)$, which is solved in $O(n_A n_B U^A U^B)$. The notation $p_j^A \uparrow\downarrow w_j^A$ represents that the jobs of the first agent have inversely agreeable processing times and weights, i.e., the smaller the processing time for a job, the greater its weight, and the notation $d_j^B \uparrow\downarrow w_j^B$ represents that the jobs of agent B have inversely agreeable due dates and weights. U^A and U^B are the upper bounds on the criteria $\Sigma w_j^A C_j^A$ and $\Sigma w_j^B Y_j^B$, respectively. In contrast to Zhang et al. [21], in this article we remove the constraint $p_j^A \uparrow\downarrow w_j^A$ and turn to the optimization problem of B-jobs having a common due date.

The remainder of the paper is organized as follows. In Section 2, some preliminaries are provided. In Sections 3 and 4, we present two dynamic programming algorithms and an FPTAS. In Section 5, some numeral tests are undertaken to show the algorithms' efficiency. Section 6 concludes the paper and suggests the future research direction.

2. Preliminaries

For self-consistency, in this section we describe some notions and properties related to Pareto-scheduling, and we present other useful notations in the description of the algorithms in the following sections.

Definition 1. *Consider two m-vectors* $u = (u_1, u_2, \ldots, u_m)$ *and* $v = (v_1, v_2, \ldots, v_m)$.

(i) We say that **u dominates v**, denoted by $u \preceq v$, if $u_i \leq v_i$ for $i = 1, 2, \ldots, m$.
(ii) We say that **u strictly dominates v**, denoted by $u \prec v$, if $u \preceq v$ and $u \neq v$.
(iii) Given a constant $\epsilon > 0$, we say that **u ϵ-dominates v**, denoted by $u \preceq_\epsilon v$, if and only if $u_i \leq (1+\epsilon)v_i$ for $i = 1, 2, \ldots, m$.

Definition 2. *Given two agents' criteria $\gamma^A(\sigma)$ and $\gamma^B(\sigma)$, a feasible schedule σ is called **Pareto-optimal** and the corresponding objective vector $(\gamma^A(\sigma), \gamma^B(\sigma))$ is called a **Pareto-optimal point**, if no other feasible schedule π satisfies $(\gamma^A(\pi), \gamma^B(\pi)) \prec (\gamma^A(\sigma), \gamma^B(\sigma))$. All the Pareto-optimal points form the **Pareto frontier**, denoted by P.*

Let **R** be the set of the objective vectors of all feasible schedules, and **Q** be a subset of **R**.

Definition 3. *A vector $u \in Q$ is called **non-dominated** in Q, if there exists no other vector $v \in Q$ such that $v \prec u$.*

It is not difficult to see that, for the above definitions, the latter is an extension of the former, and especially when **Q** is exactly equal to **R**, all the non-dominated vectors in **Q** compose the Pareto-optimal frontier. The following lemma establishes the relationship between sets **P** and a subset $Q \subseteq R$.

Lemma 1. *For any set Q with $P \subseteq Q \subseteq R$, if O is the set including all the non-dominated vectors in Q, then $O = P$.*

Proof. By Definition 2, for each Pareto-optimal point $u \in P$, there is no other vector $v \in R$ such that $v \prec u$, and naturally, such a fact also holds for the set **Q**, since $Q \subseteq R$. Then, it follows that $P \subseteq O$ by the definition of the set **O**. Next we show that $O \subseteq P$. If not, we pick up one vector **w** from $O \setminus P$. Again by Definition 2, there is some vector $w \in P$ such that $w \prec u$. Nevertheless, this is impossible, since $w \in P \subseteq Q$ leads to no existence of such a vector **w** in **Q** by the assumption of **w** and Definition 3. Thus $O = P$. □

From Lemma 1, to generate the Pareto frontier **P**, an alternative is to first determine a set **Q** with $P \subseteq Q \subseteq R$, and then delete the dominated vectors in **Q**. Throughout the reminder of this paper, such a subset **Q** is called an ***intermediate set***. Obviously, **R** is also an intermediate set.

Definition 4. *For a given constant $\epsilon > 0$, a $(1+\epsilon)$-approximate Pareto frontier, denoted by P_ϵ, is a set of the objective vectors satisfying, for any $(\gamma^A(\sigma), \gamma^B(\sigma)) \in P$, there exists at least one objective vector $(\gamma^A(\sigma'), \gamma^B(\sigma')) \in P_\epsilon$ such that $(\gamma^A(\sigma'), \gamma^B(\sigma')) \preceq_\epsilon (\gamma^A(\sigma), \gamma^B(\sigma))$.*

Definition 5. *A family of algorithms $\{\mathcal{A}_\epsilon : \epsilon > 0\}$ is called a **fully polynomial-time approximation scheme (FPTAS)** if, for each $\epsilon > 0$, $\mathcal{A}_\epsilon$ generates a $(1+\epsilon)$-approximate Pareto frontier with a running time in the polynomial in the instance size and $1/\epsilon$.*

Besides those already mentioned in Section 1, the following notations will also be used later:

- $\mathcal{J}_j^X$: indicates the set of the first j jobs in $\mathcal{J}^X$, namely, $\mathcal{J}_j^X = \{J_1^X, J_2^X, \cdots, J_j^X\}$.
- $J_i^X \prec_\sigma J_j^{X'}$ indicates that job J_i^X immediately precedes $J_j^{X'}$ in schedule σ, where $X, X' \in \{A, B\}$.
- $s_j^X(\sigma)$ indicates the starting time of job J_j^X in σ.
- $P_{\text{sum}}^X = \sum_{j=1}^{n_X} p_j^X$ indicates the total processing time of all X-jobs.
- P_{sum} indicates the total processing time of all jobs, and $P_{\text{sum}} = P_{\text{sum}}^A + P_{\text{sum}}^B$.
- $W_{\text{sum}}^X = \sum_{j=1}^{n_X} w_j^X$ indicates the total weight of all X-jobs.

- W_{sum} indicates the total weight of all jobs, and $W_{\text{sum}} = W^A_{\text{sum}} + W^B_{\text{sum}}$.
- $p^X_{\max}$ indicates the maximum processing time of the X-jobs, namely, $p^X_{\max} = \max\{p^X_j : 1 \leq j \leq n_X\}$.
- $w^X_{\max}$ indicates the maximum weight of the X-jobs, namely, $w^X_{\max} = \max\{w^X_j : 1 \leq j \leq n_X\}$.

3. An Exact Algorithm

In this section a dynamic programming algorithm for problem $1|d^B_j = d|^{\#}(\sum w^A_j C^A_j, \sum w^B_j Y^B_j)$ is presented. For description convenience, for a given schedule σ, the job set $\mathcal{J}$ is divided into the following four subsets: $\mathcal{J}^{A_1}(\sigma) = \{J^A_j : C^A_j(\sigma) \leq d\}$, $\mathcal{J}^{A_2}(\sigma) = \{J^A_j : C^A_j(\sigma) > d\}$, $\mathcal{J}^{B_1}(\sigma) = \{J^B_j : s^B_j(\sigma) < d\}$, and $\mathcal{J}^{B_2}(\sigma) = \{J^B_j : s^B_j(\sigma) \geq d\}$. Obviously, such a partition of the job set is well defined for a given schedule.

The following lemma establishes the structural properties of the Pareto-optimal schedule.

Lemma 2. *For each Pareto-optimal point (C, Y) of problem $1|d^B_j = d|^{\#}(\sum w^A_j C^A_j, \sum w^B_j Y^B_j)$, there is a Pareto-optimal schedule σ such that*
(i) $\mathcal{J}^{A_1}(\sigma) \prec_\sigma \mathcal{J}^{B_1}(\sigma) \prec_\sigma \mathcal{J}^{A_2}(\sigma) \prec_\sigma \mathcal{J}^{B_2}(\sigma)$.
(ii) the jobs in $\mathcal{J}^{B_1}(\sigma)$ are sequenced in the non-increasing order of their weights and the jobs in $\mathcal{J}^{B_2}(\sigma)$ are sequenced arbitrarily.
(iii) the jobs in $\mathcal{J}^{A_1}(\sigma)$ and $\mathcal{J}^{A_2}(\sigma)$ are sequenced according to the weighted shortest processing time (WSPT) rule.

Proof. In Lemma 2, statement (i) can easily be observed, since the jobs in $\mathcal{J}^{B_2}(\sigma)$ are late and this will not result in any increase in their total late work when moving them to the end of the schedule, and as many A-jobs as possible can be positioned before the B-jobs in $\mathcal{J}^{B_1}(\sigma)$, provided that the last job in $\mathcal{J}^{B_1}(\sigma)$ is not late. The left two statements in Lemma 2 can easily be proved by an interchange argument and the detail is omitted here. □

Lemma 2 allows us only to consider the feasible schedules simultaneously satisfying the conditions (i)-(iii). To this end, we re-number the n_A jobs in $\mathcal{J}^A$ in the WSPT order and the n_B B-jobs in the maximum weight first (MW) order so that

$$\frac{p^A_1}{w^A_1} \leq \frac{p^A_2}{w^A_2} \leq \cdots \leq \frac{p^A_{n_A}}{w^A_{n_A}}. \tag{2}$$

$$w^B_1 \geq w^B_2 \geq \cdots \geq w^B_{n_B}. \tag{3}$$

Such a sorting takes $O(n \log n)$ time.

According to Lemma 1, the algorithm to be described adopts the strategy of first finding the intermediate set dynamically and then deleting the dominated points in it. It is necessary to mention that in the proposed algorithm we appropriately relax the conditions to find a modestly larger intermediate set. For briefly describing the dynamic programming algorithm, we introduce the following terminologies and notations.

- an *ABAB*-schedule is defined to be a schedule π for $\mathcal{I} \subseteq \mathcal{J}$ satisfying (i) $\pi = \pi_1 \prec \pi_2 \prec \pi_3 \prec \pi_4$, where among the four mutually disjointed subschedules $\pi_1, \pi_2, \pi_3,$ and π_4, the A-jobs are included in π_1 and π_3, and the B-jobs are included in π_2 and π_4; (ii) no idle time exists between the jobs in each subschedule, but this is not necessarily so between two subschedules. Moreover, the idle time between π_3 and π_4 is supposed to be long enough; (iii) the jobs in each subschedule are sequenced in the increasing order of their indices.
- an $(\overrightarrow{x}, \overleftarrow{y})$-**schedule** is defined to be an *ABAB*-schedule π for $\mathcal{J}^A_x \cup (\mathcal{J}^B \setminus \mathcal{J}^B_{y-1})$ with no idle time existing between subschedules π_2 and π_3, where $x \in \{1, 2, \ldots, n_A\}$ and $y \in \{1, 2, \ldots, n_B\}$.

- a vector (t_1, t_2, t_3, C, Y) is introduced to denote a **state of** $(\vec{x}, \overleftarrow{y})$, in which t_1, t_2, t_3, C, and Y, respectively, stand for the end point of π_1, the start point of π_2, the end point of π_3, the total weighted completion time of the A-jobs of $\mathcal{J}_x^A$, and the total weighted late work of the B-jobs of $\mathcal{J}^B \setminus \{\mathcal{J}_{y-1}^B\}$. Note that a state of $(\vec{x}, \overleftarrow{y})$ at least corresponds to some $(\vec{x}, \overleftarrow{y})$-schedule.
- $\Gamma(\vec{x}, \overleftarrow{y})$ denotes the set of all the states of $(\vec{x}, \overleftarrow{y})$.
- $\tilde{\Gamma}(\vec{x}, \overleftarrow{y})$ denotes the set obtained from $\Gamma(\vec{x}, \overleftarrow{y})$ by deleting the vectors (t_1, t_2, t_3, C, Y), for which there is another vector $(\overline{t_1}, \overline{t_2}, \overline{t_3}, \overline{C}, \overline{Y})$ with $t_1 \leq \overline{t_1}, t_2 \geq \overline{t_2}, t_3 \leq \overline{t_3}, C \leq \overline{C}$, and $Y \leq \overline{Y}$.
- Let $\mathbf{Q}_1 = \{(C, Y) : (t_1, t_2, t_3, C, Y) \in \tilde{\Gamma}(\overrightarrow{n_A}, \overleftarrow{1})\}$, and let $\tilde{\mathbf{Q}}_1$ be the set of the non-dominated vectors in $\mathbf{Q}_1$.

To solve problem $1|d_j^B = d|^\#(\sum w_j^A C_j^A, \sum w_j^B Y_j^B)$, we have to first compute $\tilde{\Gamma}(\overrightarrow{n_A}, \overleftarrow{1})$ and then obtain the Pareto-frontier $\tilde{\mathbf{Q}}_1$. This can be realized by dynamically computing the sets $\Gamma(\vec{x}, \overleftarrow{y})$ for all the possible choices of the tuple (x, y). Note that each $(\vec{x}, \overleftarrow{y})$-schedule can be obtained either by adding job J_x^A to some $(\overrightarrow{x-1}, \overleftarrow{y})$-schedule, or by adding job J_y^B to some $(\vec{x}, \overleftarrow{y+1})$-schedule. Therefore, we can informally describe our dynamic programming algorithm as follows.

Initially, set $\Gamma(\vec{0}, \overleftarrow{n_B+1}) = \{(0, t_0, t_0, 0, 0) : d - p_{\max}^B + 1 \leq t_0 \leq d + p_{\max}^B - 1\}$ and $\Gamma(\vec{x}, \overleftarrow{y}) = \emptyset$ if $(x, y) \neq (0, n_B+1)$. Then we recursively generate all the state sets $\Gamma(\vec{x}, \overleftarrow{y})$ from the previously-generated sets $\Gamma(\overrightarrow{x-1}, \overleftarrow{y})$ and $\Gamma(\vec{x}, \overleftarrow{y+1})$. Specifically,

- For each state $(t_1, t_2, t_3, C, Y) \in \Gamma(\overrightarrow{x-1}, \overleftarrow{y})$ with $\Gamma(\overrightarrow{x-1}, \overleftarrow{y}) \neq \phi$, add two states $(t_1', t_2', t_3', C', Y')$ and $(t_1'', t_2'', t_3'', C'', Y'')$ to the set $\Gamma(\vec{x}, \overleftarrow{y})$, with

$$(t_1', t_2', t_3', C', Y') = (t_1 + p_x^A, t_2, t_3, C + w_x^A(t_1 + p_x^A), Y),$$

and

$$(t_1'', t_2'', t_3'', C'', Y'') = (t_1, t_2, t_3 + p_x^A, C + w_x^A(t_3 + p_x^A), Y).$$

These two states respectively correspond to the newly obtained $(\vec{x}, \overleftarrow{y})$-schedules by scheduling job J_x^A immediately following the subschedule π_1 and immediately following the subschedule π_3, in some $(\overrightarrow{x-1}, \overleftarrow{y})$ schedule π that corresponds to the state (t_1, t_2, t_3, C, Y). Note that the first case occurs only when $t_1 + p_x^A \leq t_2$ is satisfied.

- For each state $(t_1, t_2, t_3, C, Y) \in \Gamma(\vec{x}, \overleftarrow{y+1})$, also add two two states $(t_1', t_2', t_3', C', Y')$ and $(t_1'', t_2'', t_3'', C'', Y'')$ to the set $\Gamma(\vec{x}, \overleftarrow{y})$, with

$$(t_1', t_2', t_3', C', Y') = (t_1, t_2 - p_y^B, t_3, C, Y + w_y^B \max\{t_2 - d^B, 0\}),$$

and

$$(t_1'', t_2'', t_3'', C'', Y'') = (t_1, t_2, t_3, C, Y + w_y^B p_y^B).$$

These two states respectively correspond to the newly obtained $(\vec{x}, \overleftarrow{y})$-schedules by scheduling job J_y^B immediately preceding the subschedule π_2 and immediately following the subschedule π_4, in some $(\vec{x}, \overleftarrow{y+1})$ schedule π that corresponds to the state (t_1, t_2, t_3, C, Y). Note that the first case occurs only when $t_1 \leq t_2 - p_y^B < d^B$ is satisfied.

Note that, if in the above state-generation procedures we replace sets $\Gamma(\overrightarrow{x-1}, \overleftarrow{y})$ and $\Gamma(\vec{x}, \overleftarrow{y+1})$ with sets $\tilde{\Gamma}(\overrightarrow{x-1}, \overleftarrow{y})$ and $\tilde{\Gamma}(\vec{x}, \overleftarrow{y+1})$, then the resulting set of new states, denoted by $\Gamma'(\vec{x}, \overleftarrow{y})$, may be different from $\Gamma(\vec{x}, \overleftarrow{y})$. Recall that, when deleting those dominated vectors in the sets $\Gamma(\vec{x}, \overleftarrow{y})$ and $\Gamma'(\vec{x}, \overleftarrow{y})$, the newly obtained sets are respectively denoted by $\tilde{\Gamma}(\vec{x}, \overleftarrow{y})$ and $\tilde{\Gamma}'(\vec{x}, \overleftarrow{y})$, which will be shown to be identical in the following lemma.

Lemma 3. $\tilde{\Gamma}(\overrightarrow{x},\overleftarrow{y}) = \tilde{\Gamma}'(\overrightarrow{x},\overleftarrow{y})$.

Proof. Since $\tilde{\Gamma}(\overrightarrow{x-1},\overleftarrow{y}) \subseteq \Gamma(\overrightarrow{x-1},\overleftarrow{y})$ and $\tilde{\Gamma}(\overrightarrow{x},\overleftarrow{y+1}) \subseteq \Gamma(\overrightarrow{x},\overleftarrow{y+1})$, it follows that $\Gamma'(\overrightarrow{x},\overleftarrow{y}) \subseteq \Gamma(\overrightarrow{x},\overleftarrow{y})$ by the generation procedure of the new states as described previously. If $\Gamma(\overrightarrow{x},\overleftarrow{y}) = \Gamma'(\overrightarrow{x},\overleftarrow{y})$, then naturally $\tilde{\Gamma}(\overrightarrow{x},\overleftarrow{y}) = \tilde{\Gamma}'(\overrightarrow{x},\overleftarrow{y})$. In the following, suppose that $\Gamma(\overrightarrow{x},\overleftarrow{y}) \setminus \Gamma'(\overrightarrow{x},\overleftarrow{y}) \neq \emptyset$. We next show that each state $(t'_1, t'_2, t'_3, C', Y') \in \Gamma(\overrightarrow{x},\overleftarrow{y}) \setminus \Gamma'(\overrightarrow{x},\overleftarrow{y})$ is dominated by a state $(\overline{t_1}, \overline{t_2}, \overline{t_3}, \overline{C}, \overline{Y}) \in \Gamma'(\overrightarrow{x},\overleftarrow{y})$, namely, $\overline{t_1} \leq t'_1, \overline{t_2} \geq t'_2, \overline{t_3} \leq t'_3, \overline{C} \leq C', \overline{Y} \leq Y'$.

Let π' be an $(\overrightarrow{x},\overleftarrow{y})$-schedule corresponding to $(t'_1, t'_2, t'_3, C', Y')$. According to the above discussion, there are four possibilities of deriving π' from some schedule π, which is assumed to correspond to the state (t_1, t_2, t_3, C, Y) in $\Gamma(\overrightarrow{x-1},\overleftarrow{y})$ or $\Gamma(\overrightarrow{x},\overleftarrow{y+1})$.

Case 1. π' is obtained from π by scheduling job J_x^A directly after subschedule π_1. Then $(t_1, t_2, t_3, C, Y) \in \Gamma(\overrightarrow{x-1},\overleftarrow{y})$ with $t_1 + p_x^A \leq t_2$, and there is a state $(\tilde{t}_1, \tilde{t}_2, \tilde{t}_3, \tilde{C}, \tilde{Y}) \in \tilde{\Gamma}(\overrightarrow{x-1},\overleftarrow{y})$ such that $\tilde{t}_1 \leq t_1, \tilde{t}_2 \geq t_2, \tilde{t}_3 \leq t_3, \tilde{C} \leq C$, and $\tilde{Y} \leq Y$. Let $\tilde{\pi}$ be an $(\overrightarrow{x-1},\overleftarrow{y})$-schedule corresponding to $(\tilde{t}_1, \tilde{t}_2, \tilde{t}_3, \tilde{C}, \tilde{Y})$, and let $\overline{\pi}$ be the $(\overrightarrow{x},\overleftarrow{y})$-schedule obtained from $\tilde{\pi}$ by scheduling J_x^A directly after schedule $\tilde{\pi}_1$. Let $(\overline{t_1}, \overline{t_2}, \overline{t_3}, \overline{C}, \overline{Y})$ be the state corresponding to $\overline{\pi}$. Note that the above operation to get $\overline{\pi}$ is feasible since $\tilde{t}_1 + p_x^A \leq t_1 + p_x^A \leq t_2 \leq \tilde{t}_2$. Then we have $(\overline{t_1}, \overline{t_2}, \overline{t_3}, \overline{C}, \overline{Y}) = (\tilde{t}_1 + p_x^A, \tilde{t}_2, \tilde{t}_3, \tilde{C} + w_x^A(\tilde{t}_1 + p_x^A), \tilde{Y})$. Combining with the fact that $(t'_1, t'_2, t'_3, C', Y') = (t_1 + p_x^A, t_2, t_3, C + w_x^A(t_1 + p_x^A), Y)$, we have

$$\begin{cases} \overline{t_1} = \tilde{t}_1 + p_x^A \leq t_1 + p_x^A = t'_1, \\ \overline{t_2} = \tilde{t}_2 \geq t_2 = t'_2, \\ \overline{t_3} = \tilde{t}_3 \leq t_3 = t'_3, \\ \overline{C} = \tilde{C} + w_x^A(\tilde{t}_1 + p_x^A) \leq C + w_x^A(t_1 + p_x^A) = C', \\ \overline{Y} = \tilde{Y} \leq Y = Y'. \end{cases} \quad (4)$$

Case 2. π' is obtained from π by scheduling J_x^A directly after schedule π_3. Then $(t_1, t_2, t_3, C, Y) \in \Gamma(\overrightarrow{x-1},\overleftarrow{y})$, and there is a state $(\tilde{t}_1, \tilde{t}_2, \tilde{t}_3, \tilde{C}, \tilde{Y}) \in \tilde{\Gamma}(\overrightarrow{x-1},\overleftarrow{y})$ such that $\tilde{t}_1 \leq t_1, \tilde{t}_2 \geq t_2, \tilde{t}_3 \leq t_3, \tilde{C} \leq C$, and $\tilde{Y} \leq Y$. Let $\tilde{\pi}$ be an $(\overrightarrow{x-1},\overleftarrow{y})$-schedule corresponding to $(\tilde{t}_1, \tilde{t}_2, \tilde{t}_3, \tilde{C}, \tilde{Y})$, and let $\overline{\pi}$ be the $(\overrightarrow{x},\overleftarrow{y})$-schedule obtained from $\tilde{\pi}$ by scheduling J_x^A directly after schedule $\tilde{\pi}_3$. Let $(\overline{t_1}, \overline{t_2}, \overline{t_3}, \overline{C}, \overline{Y})$ be the state corresponding to $\overline{\pi}$. Then we have $(\overline{t_1}, \overline{t_2}, \overline{t_3}, \overline{C}, \overline{Y}) = (\tilde{t}_1, \tilde{t}_2, \tilde{t}_3 + p_x^A, \tilde{C} + w_x^A(\tilde{t}_3 + p_x^A), \tilde{Y})$. Combining with the fact that $(t'_1, t'_2, t'_3, C', Y') = (t_1, t_2, t_3 + p_x^A, C + w_x^A(t_3 + p_x^A), Y)$, we have

$$\begin{cases} \overline{t_1} = \tilde{t}_1 \leq t_1 = t'_1, \\ \overline{t_2} = \tilde{t}_2 \geq t_2 = t'_2, \\ \overline{t_3} = \tilde{t}_3 + p_x^A \leq t_3 + p_x^A = t'_3, \\ \overline{C} = \tilde{C} + w_x^A(\tilde{t}_3 + p_x^A) \leq C + w_x^A(t_3 + p_x^A) = C', \\ \overline{Y} = \tilde{Y} \leq Y = Y'. \end{cases} \quad (5)$$

Case 3. π' is obtained from π by scheduling J_y^B directly before schedule π_2. Note that in this case, the condition $t_1 \leq t_2 - p_y^B < d^B$ must be satisfied. Then $(t_1, t_2, t_3, C, Y) \in \Gamma(\overrightarrow{x},\overleftarrow{y+1})$, and there is a state $(\tilde{t}_1, \tilde{t}_2, \tilde{t}_3, \tilde{C}, \tilde{Y}) \in \tilde{\Gamma}(\overrightarrow{x},\overleftarrow{y+1})$ such that $\tilde{t}_1 \leq t_1, \tilde{t}_2 \geq t_2, \tilde{t}_3 \leq t_3, \tilde{C} \leq C$, and $\tilde{Y} \leq Y$. Let $\tilde{\pi}$ be an $(\overrightarrow{x},\overleftarrow{y+1})$-schedule corresponding to $(\tilde{t}_1, \tilde{t}_2, \tilde{t}_3, \tilde{C}, \tilde{Y})$, and let $\overline{\pi}$ be the $(\overrightarrow{x},\overleftarrow{y})$-schedule obtained from $\tilde{\pi}$ by scheduling J_y^B directly before schedule π_2. Let $(\overline{t_1}, \overline{t_2}, \overline{t_3}, \overline{C}, \overline{Y})$ be the state corresponding to $\overline{\pi}$. The above

25

operation to obtain $\bar{\pi}$ is feasible. In fact, $\tilde{t}_1 \leq t_1 \leq t_2 - p_y^B$, which means there are enough spaces for J_y^B to be scheduled in. In the following we will illustrate that the condition $\tilde{t}_2 - p_y^B < d$ is satisfied.

Claim 1. *If $\tilde{t}_2 \neq t_2$, then $\tilde{t}_2 \leq d$.*

Suppose to the contrary that $\tilde{t}_2 > d$, then J_y^B is partially early or late in $\tilde{\pi}$, implying that $J_{y+1}^B, J_{y+2}^B, \ldots, J_{n_B}^B$ are all late in $\tilde{\pi}$, i.e., there is no job in $\tilde{\pi}_2$, which further suggests that $\sum_{j=1}^{x} p_j^A = \tilde{t}_1 + \tilde{t}_3 - \tilde{t}_2$. What is more, since $\tilde{Y} \leq Y$, the jobs $J_{y+1}^B, J_{y+2}^B, \ldots, J_{n_B}^B$ are also late in π, which also indicates that $\tilde{Y} = Y$ and $\sum_{j=1}^{x} p_j^A = t_1 + t_3 - t_2$. From $\tilde{t}_1 + \tilde{t}_3 - \tilde{t}_2 = t_1 + t_3 - t_2$, $\tilde{t}_1 \leq t_1$, $\tilde{t}_2 \geq t_2$, and $\tilde{t}_3 \leq t_3$ we know that $\tilde{t}_2 = t_2$ contradicts $\tilde{t}_2 \neq t_2$. Thus, $\tilde{t}_2 \leq d^B$. Claim 1 follows.

If $\tilde{t}_2 - p_y^B \geq d^B$, then $\tilde{t}_2 \neq t_2$. From Claim 1 we have $\tilde{t}_2 \leq d < d + p_y^B$, i.e., $\tilde{t}_2 - p_y^B < d$, which is a contradiction. Thus the condition $\tilde{t}_2 - p_y^B < d^B$ is satisfied and the operation to get $\bar{\pi}$ is feasible. Then we have $(\bar{t}_1, \bar{t}_2, \bar{t}_3, \bar{C}, \bar{Y}) = (\tilde{t}_1, \tilde{t}_2 - p_y^B, \tilde{t}_3, \tilde{C}, \tilde{Y} + w_y^B \max\{\tilde{t}_2 - d, 0\})$. Combining with the fact that $(t_1', t_2', t_3', C', Y') = (t_1, t_2 - p_y^B, t_3, C, Y + w_y^B \max\{t_2 - d, 0\})$, we have

$$\begin{cases} \bar{t}_1 = \tilde{t}_1 \leq t_1 = t_1', \\ \bar{t}_2 = \tilde{t}_2 - p_y^B \geq t_2 - p_y^B = t_2', \\ \bar{t}_3 = \tilde{t}_3 \leq t_3 = t_3', \\ \bar{C} = \tilde{C} \leq C = C'. \end{cases} \quad (6)$$

Next we prove that $\bar{Y} \leq Y'$. In fact, if $\tilde{t}_2 = t_2$, then $\bar{Y} = \tilde{Y} + w_y^B \max\{\tilde{t}_2 - d, 0\} \leq Y + w_y^B \max\{t_2 - d, 0\} = Y'$. If $\tilde{t}_2 \neq t_2$, then from Claim 1 we know that $\tilde{t}_2 \leq d$, and then $t_2 - d < 0$. Thus we have $\bar{Y} = \tilde{Y} \leq Y = Y'$.

Case 4. π' is obtained from π by scheduling J_y^B directly after schedule π_4. Then $(t_1, t_2, t_3, C, Y) \in \Gamma(\overrightarrow{x}, \overleftarrow{y+1})$, and there is a state $(\tilde{t}_1, \tilde{t}_2, \tilde{t}_3, \tilde{C}, \tilde{Y}) \in \tilde{\Gamma}(\overrightarrow{x}, \overleftarrow{y+1})$ such that $\tilde{t}_1 \leq t_1, \tilde{t}_2 \geq t_2, \tilde{t}_3 \leq t_3, \tilde{C} \leq C$, and $\tilde{Y} \leq Y$. Let $\tilde{\pi}$ be an $(\overrightarrow{x}, \overleftarrow{y+1})$-schedule corresponding to $(\tilde{t}_1, \tilde{t}_2, \tilde{t}_3, \tilde{C}, \tilde{Y})$, and let $\bar{\pi}$ be the $(\overrightarrow{x}, \overleftarrow{y})$-schedule obtained from $\tilde{\pi}$ by scheduling J_x^A directly after schedule $\tilde{\pi}_3$. Let $(\bar{t}_1, \bar{t}_2, \bar{t}_3, \bar{C}, \bar{Y})$ be the state corresponding to $\bar{\pi}$. Then we have $(\bar{t}_1, \bar{t}_2, \bar{t}_3, \bar{C}, \bar{Y}) = (\tilde{t}_1, \tilde{t}_2, \tilde{t}_3, \tilde{C}, \tilde{Y} + w_y^B p_y^B)$. Combining with the fact that $(t_1', t_2', t_3', C', Y') = (t_1, t_2, t_3, C, Y + w_y^B p_y^B)$, we have

$$\begin{cases} \bar{t}_1 = \tilde{t}_1 \leq t_1 = t_1', \\ \bar{t}_2 = \tilde{t}_2 \geq t_2 = t_2', \\ \bar{t}_3 = \tilde{t}_3 \leq t_3 = t_3', \\ \bar{C} = \tilde{C} \leq C = C', \\ \bar{Y} = \tilde{Y} + w_y^B p_y^B \leq Y + w_y^B p_y^B = Y'. \end{cases} \quad (7)$$

The result follows. □

Theorem 1. *Algorithm 1 solves the Pareto-frontier scheduling problem $1|d_j^B = d|^{\#}(\Sigma w_j^A C_j^A, \Sigma w_j^B Y_j^B)$ in $O(n_A n_B d P_{sum} U^A U^B)$ time.*

Algorithm 1: For problem $1|d_j^B = d|^{\#}(\sum w_j^A C_j^A, \sum w_j^B Y_j^B)$

1 Set $\Gamma(\overrightarrow{0}, \overleftarrow{n_B+1}) = \{(0, t_0, t_0, 0, 0) : d - p_{\max}^A + 1 \leq t_0 \leq d + p_{\max}^B - 1\}$ and set $\Gamma(\overrightarrow{x}, \overleftarrow{y}) = \emptyset$ if $(x, y) \neq (0, n_B + 1)$.
2 **for** $x = 0, 1, \ldots, n_A, y = n_B + 1, n_B, \ldots, 1$, **do**
3   **for** each $(t_1, t_2, t_3, C, Y) \in \Gamma(\overrightarrow{x-1}, \overleftarrow{y})$, **do**
4   **if** $0 < x \leq n_A$, **then**
5          $\Gamma(\overrightarrow{x}, \overleftarrow{y}) := \Gamma(\overrightarrow{x}, \overleftarrow{y}) \cup (t_1, t_2, t_3 + p_x^A, C + w_x^A(t_3 + p_x^A), Y)$
6   **end**
7   **if** $0 < x \leq n_A$ and $t_1 + p_x^A \leq t_2$, **then**
8          $\Gamma(\overrightarrow{x}, \overleftarrow{y}) := \Gamma(\overrightarrow{x}, \overleftarrow{y}) \cup (t_1 + p_x^A, t_2, t_3, C + w_x^A(t_1 + p_x^A), Y)$
9   **end**
10   **end**
11   **for** each $(t_1, t_2, t_3, C, Y) \in \Gamma(\overrightarrow{x}, \overleftarrow{y+1})$, **do**
12   **if** $1 \leq y < n_B + 1$, **then**
13          $\Gamma(\overrightarrow{x}, \overleftarrow{y}) := \Gamma(\overrightarrow{x}, \overleftarrow{y}) \cup (t_1, t_2, t_3, C, Y + w_y^B p_y^B)$
14   **end**
15   **if** $1 \leq y < n_B + 1$ and $t_1 \leq t_2 - p_y^B < d$, **then**
16          $\Gamma(\overrightarrow{x}, \overleftarrow{y}) := \Gamma(\overrightarrow{x}, \overleftarrow{y}) \cup (t_1, t_2 - p_y^B, t_3, C, Y + w_y^B \max\{t_2 - d, 0\})$
17   **end**
18   **end**
19   For each newly generated $\Gamma(\overrightarrow{x}, \overleftarrow{y})$, set $\Gamma(\overrightarrow{x}, \overleftarrow{y}) := \tilde{\Gamma}(\overrightarrow{x}, \overleftarrow{y})$
20 **end**
21 Generate $\tilde{Q}_1$ and, for each state $(C, Y) \in \tilde{Q}_1$, derive the corresponding optimal schedule by backtracking.

Proof. The correctness of Algorithm 1 is guaranteed by Lemma 2, Lemma 1, and Lemma 3. Here we only analyze its time complexity. The initialization step takes $O(P_{\text{sum}} + n_A n_B)$ time, which is dominated by the final time complexity of Algorithm 1. In the implementation of Algorithm 1, we guarantee that $\Gamma(\overrightarrow{x}, \overleftarrow{y}) = \tilde{\Gamma}(\overrightarrow{x}, \overleftarrow{y})$. Note that $0 \leq t_1 \leq d^B$ and $d^B - p_{\max}^A + 1 \leq t_3 \leq P_{\text{sum}}$, then each state set $\Gamma(\overrightarrow{x}, \overleftarrow{y})$ contains $O(d^B P_{\text{sum}} U^A U^B)$ states. Moreover, $\Gamma(\overrightarrow{x}, \overleftarrow{y})$ is obtained by performing at most two (constant) operations on the states in $\Gamma(\overrightarrow{x-1}, \overleftarrow{y}) \cup \Gamma(\overrightarrow{x}, \overleftarrow{y+1})$ for $x = 0, 1, \ldots, n_A$, $y = n_B + 1, n_B, \ldots, 1$. Note that the upper bounds of $\sum w_j^A C_j^A$ and $\sum w_j^B Y_j^B$ are given by $U^A = \sum_{j=1}^{n_A} w_j^A P_{\text{sum}}$ and $U^B = \sum_{j=1}^{n_B} w_j^B p_j^B$, respectively. Thus, the overall running time of Algorithm 1 is $O(n_A n_B d P_{\text{sum}} U^A U^B)$. □

4. An FPTAS

In this section, for problem $1|d_j^B = d|^{\#}(\sum w_j^A C_j^A, \sum w_j^B Y_j^B)$, we first give another dynamic programming algorithm, and then turn it into an FPTAS by the trimming technique. As for Algorithm 1, we first introduce the following terminologies and notations.

- An (x, y)-**schedule** is defined to be an $ABAB$-schedule π for $\mathcal{J}_x^A \cup \mathcal{J}_y^B$ with no idle time existing between subschedules π_1, π_2 and π_3, where $x \in \{1, 2, \ldots, n_A\}$ and $y \in \{1, 2, \ldots, n_B\}$.
- A vector $(t_1, t_2, t_3, W, k(\pi), C, Y)$ is introduced to denote a **state of** (x, y), in which t_1, t_2, t_3, W, k, C and Y, respectively, stand for the end point of π_1, the end point of π_2, the end point of π_3, the total weight of the jobs in π_3, the index of the last B-job in π_2, the total weighted completion time of the

A-jobs of $\mathcal{J}_x^A$, and the total weighted late work of the B-jobs of $\mathcal{J}_y^B$. Note that a state of (x,y) at least corresponds to some (x,y)-schedule.

- $\Gamma(x,y)$ denotes the set of all the states of (x,y).
- $\tilde{\Gamma}(x,y)$ denotes the set obtained from $\Gamma(x,y)$ by deleting the vectors $(\overline{t_1},\overline{t_2},\overline{t_3},\overline{W},k,\overline{C},\overline{Y})$, for which there is another vector (t_1,t_2,t_3,W,k,C,Y) with $t_1 \leq \overline{t_1}, t_2 \leq \overline{t_2}, t_3 \leq \overline{t_3}, W \leq \overline{W}, C \leq \overline{C}, Y \leq \overline{Y}$.
- Let $\mathbf{Q}_2 = \{(C,Y) : (t_1,t_2,t_3,W,k,C,Y) \in \tilde{\Gamma}(n_A,n_B)\}$, and let $\tilde{\mathbf{Q}}_2$ be the set of the non-dominated vectors in $\mathbf{Q}_2$.

Clearly, $\mathbf{Q}_2$ is an intermediate set. Similarly to the discussion for Algorithm 1, we can generate all the $\Gamma(x,y)$ for all the possible choices of the tuple (x,y) dynamically in the following way.

Initially, set $\Gamma(0,0) = \{(0,0,0,0,0,0,0)\}$ and $\Gamma(x,y) = \emptyset$ if $(x,y) \neq (0,0)$. Then we recursively generate all the state sets $\Gamma(x,y)$ from the previously generated sets $\Gamma(x-1,y)$ and $\Gamma(x,y-1)$. Specifically,

- For each state $(t_1,t_2,t_3,W,k,C,Y) \in \Gamma(x-1,y)$ with $\Gamma(x-1,y) \neq \emptyset$, add two states $(t_1',t_2',t_3',W',k',C',Y')$ and $(t_1'',t_2'',t_3'',W'',k'',C'',Y'')$ to the set $\Gamma(x,y)$, with

$$(t_1',t_2',t_3',W',k',C',Y') = (t_1+p_x^A, t_2+p_x^A, t_3+p_x^A, W, k, C+w_x^A(t_1+p_x^A)+Wp_x^A, Y$$
$$+w_k^B\max\{\min\{t_2+p_x^A-d^B, p_x^A\},0\}), \text{ and } (t_1'',t_2'',t_3'',W'',k'',C'',Y'')$$
$$= (t_1,t_2,t_3+p_x^A, W+w_x^A, k, C+w_x^A(t_3+p_x^A), Y).$$

These two states respectively correspond to the newly obtained (x,y)-schedules by scheduling job J_x^A immediately following the subschedule π_1 and immediately following the subschedule π_3, in some $(x-1,y)$ schedule π that corresponds to the state (t_1,t_2,t_3,W,k,C,Y). Note that the first case occurs only when $t_1+p_x^A \leq t_2$ is satisfied.

- For each state $(t_1,t_2,t_3,W,k,C,Y) \in \Gamma(x,y-1)$, also add two two states $(t_1',t_2',t_3',W',k',C',Y')$ and $(t_1'',t_2'',t_3'',W'',k'',C'',Y'')$ to the set $\Gamma(x,y)$, with

$$(t_1',t_2',t_3',W',k',C',Y') = (t_1, t_2+p_y^B, t_3+p_y^B, W, y, C+Wp_y^B, Y+w_y^B\max\{t_2+p_y^B-d^B,0\}),$$

and

$$(t_1'',t_2'',t_3'',W'',k'',C'',Y'') = (t_1,t_2,t_3,W,k,C,Y+w_y^Bp_y^B).$$

These two states respectively correspond to the newly obtained (x,y)-schedules by scheduling job J_y^B immediately after π_2 and immediately following the subschedule π_4, in some $(x,y-1)$ schedule π that corresponds to the state (t_1,t_2,t_3,W,k,C,Y). Note that the first case occurs only when $t_2 < d^B$ is satisfied.

Note that, if in the above state-generation procedures we replace sets $\Gamma(x-1,y)$ and $\Gamma(x,y-1)$ with sets $\tilde{\Gamma}(x-1,y)$ and $\tilde{\Gamma}(x,y-1)$, then the resulting set of new states, denoted by $\Gamma'(x,y)$, may be different from $\Gamma(x,y)$. Recall that, when deleting those dominated vectors in the sets $\Gamma(x,y)$ and $\Gamma'(x,y)$, the newly obtained sets are respectively denoted by $\tilde{\Gamma}(x,y)$ and $\tilde{\Gamma}'(x,y)$, which will be shown to be identical in the following lemma, and its proof is similar to that of Lemma 3.

Lemma 4. $\tilde{\Gamma}(x,y) = \tilde{\Gamma}'(x,y)$.

Theorem 2. *Algorithm 2 solves* $1|d_j^B = d|^{\#}(\sum w_j^A C_j^A, \sum w_j^B Y_j^B)$ *in* $O(n_A n_B^2 d P_{sum} W_{sum}^A U^A U^B)$ *time.*

Algorithm 2: For solving $1|d_j^B = d|^{\#}(\sum w_j^A C_j^A, \sum w_j^B Y_j^B)$

1 Set $\Gamma(0,0) = \{(0,0,0,0,0,0)\}$ and set $\Gamma(x,y) = \emptyset$ if $(x,y) \neq (0,0)$.
2 **for** $x = 0, 1, \ldots, n_A, y = 0, 1, \ldots, n_B$, **do**
3   **for** each $(t_1, t_2, t_3, W, k, C, Y) \in \Gamma(x-1, y)$, **do**
4   **if** $0 < x \leq n_A$, **then**
5           $\Gamma(x,y) := \Gamma(x,y) \cup (t_1, t_2, t_3 + p_x^A, W + w_x^A, k, C + w_x^A(t_3 + p_x^A), Y)$
6   **if** $k = 0$ or $(k \neq 0$ and $t_2 + p_x^A - d^B < p_k^B)$, **then**
7              $\Gamma(x,y) := \Gamma(x,y) \cup (t_1 + p_x^A, t_2 + p_x^A, t_3 + p_x^A, W, k, C + w_x^A(t_1 + p_x^A) + W p_x^A, Y + w_k^B \max\{\min\{t_2 + p_x^A - d, p_x^A\}, 0\})$
8   **end**
9   **end**
10   **end**
11   **for** each $(t_1, t_2, t_3, C, Y) \in \Gamma(x, y-1)$, **do**
12   **if** $0 < y \leq n_B$, **then**
13          $\Gamma(x,y) := \Gamma(x,y) \cup (t_1, t_2, t_3, W, k, C, Y + w_y^B p_y^B)$
14   **if** $t_2 < d$, **then**
15          $\Gamma(x,y) := \Gamma(x,y) \cup (t_1, t_2 + p_y^B, t_3 + p_y^B, W, y, C + W p_y^B, Y + w_y^B \max\{t_2 + p_y^B - d, 0\})$
16   **end**
17   **end**
18   **end**
19   **The elimination step:** for each newly generated $\Gamma(x,y)$, set $\Gamma(x,y) := \tilde{\Gamma}(x,y)$
20 **end**
21 Generate $\tilde{Q}_2$ and, for each state $(C, Y) \in \tilde{Q}_2$, derive the corresponding optimal schedule by backtracking.

Proof. The correctness of Algorithm 2 is guaranteed by the discussion above. Next we only analyze its time complexity. The initialization step takes $O(n_A n_B)$ time, which is dominated by the final time complexity of Algorithm 2. In the implementation of Algorithm 2, we guarantee that $\Gamma(x,y) = \tilde{\Gamma}(x,y)$. Note that $0 \leq t_1 \leq d^B$ and $d - p_{\max}^A + 1 \leq t_3 \leq P_{sum}$, $0 \leq k \leq n_B$ and $0 \leq W \leq W_{sum}^A$, then each state set $\Gamma(x,y)$ contains $O(n_B d P_{sum} W_{sum}^A U^A U^B)$ states. Moreover, $\Gamma(x,y)$ is obtained by performing at most two (constant) operations on the states in $\Gamma(x-1,y) \cup \Gamma(x, y-1)$ for $x = 0, 1, \ldots, n_A$, $y = 0, 1, \ldots, n_B$. Thus, the overall running time of Algorithm 2 is $O(n_A n_B^2 d^B P_{sum} W_{sum}^A U^A U^B)$. □

Next we turn Algorithm 2 into an FPTAS in the following way. Set $\Delta = 1 + \frac{\epsilon}{2n}$, $L^1 = \lceil \log_\Delta d \rceil$, $L^3 = \lceil \log_\Delta(P_{sum}) \rceil$, $L^W = \lceil \log_\Delta(W_{sum}^A) \rceil$, $L^A = \lceil \log_\Delta(U^A) \rceil$ and $L^B = \lceil \log_\Delta(U^B) \rceil$. Set $I_i^1 = [\Delta^{(i-1)}, \Delta^i]$ for $i = 1, 2, \ldots, L^1$, $I_i^3 = [\Delta^{(i-1)}, \Delta^i]$ for $i = 1, 2, \ldots, L^3$, $I_i^W = [\Delta^{(i-1)}, \Delta^i]$ for $i = 1, 2, \ldots, L^W$, $I_i^A = [\Delta^{(i-1)}, \Delta^i]$ for $i = 1, 2, \ldots, L^A$ and $I_i^B = [\Delta^{(i-1)}, \Delta^i]$ for $i = 1, 2, \ldots, L^B$. For $x = 0, 1, \ldots, n_A$ and $y = 0, 1, \ldots, n_B$, $\tilde{\Gamma}(x,y)$ is obtained from $\Gamma(x,y)$ by the following operation: for any two states $(t_1, t_2, t_3, W, k, C, Y)$ and $(\bar{t}_1, \bar{t}_2, \bar{t}_3, \overline{W}, k, C, Y)$ in $\Gamma(x,y)$, if (t_1, t_3, W, C, Y) and $(\bar{t}_1, \bar{t}_3, \overline{W}, \overline{C}, \overline{Y})$ fall into the same box $I_u^1 \times I_v^3 \times I_w^W \times I_p^A \times I_q^B$ for $u = 1, 2, \ldots, L^1$, $v = 1, 2, \ldots, L^3$, $w = 1, 2, \ldots, L^W$, $p = 1, 2, \ldots, L^A$ and $q = 1, 2, \ldots, L^B$ with

$t_2 \leq \overline{t_2}$, remaining the first one. Note that it takes $O(L^1 L^3 L^W L^A L^B)$ time to partition the boxes. Moreover, we define

$$\mathbf{Q}_3 = \{(C,Y) : (t_1, t_2, t_3, W, k, C, Y) \in \widehat{\Gamma}(n_A, n_B)\} \tag{8}$$

and let $\tilde{\mathbf{Q}}_3$ be the set of non-dominated vectors in $\mathbf{Q}_3$.

Theorem 3. *Algorithm 3 is an FPTAS for solving* $1|d_j^B = d|^\#(\sum w_j^A C_j^A, \sum w_j^B Y_j^B)$.

Algorithm 3: For solving $1|d_j^B = d|^\#(\sum w_j^A C_j^A, \sum w_j^B Y_j^B)$

1 Set $\Gamma(0,0) = \{(0,0,0,0,0,0,0)\}$ and set $\Gamma(x,y) = \emptyset$ if $(x,y) \neq (0,0)$.
2 **for** $x = 0, 1, \ldots, n_A, y = 0, 1, \ldots, n_B,$ **do**
3   the same operations with Algorithm 2
4   **The elimination step:** for each newly generated $\Gamma(x,y)$, set $\Gamma(x,y) := \widehat{\Gamma}(x,y)$
5 **end**
6 Generate $\tilde{\mathbf{Q}}_3$ and, for each state $(C, Y) \in \tilde{\mathbf{Q}}_3$, derive the corresponding optimal schedule by backtracking.

Proof. By induction on $z = x + y$, We prove that, for any state $(t'_1, t'_2, t'_3, W', k', C', Y') \in \Gamma(x,y)$, there is a state $(\overline{t_1}, \overline{t_2}, \overline{t_3}, \overline{W}, k', \overline{C}, \overline{Y}) \in \widehat{\Gamma}(x,y)$ such that $\overline{t_1} \leq \Delta^z t'_1, \overline{t_2} \leq t'_2, \overline{t_3} \leq \Delta^z t'_3, \overline{W} \leq \Delta^z W', \overline{C} \leq \Delta^z C'$ and $\overline{Y} \leq \Delta^z Y'$.

This is obviously true for $z = 0$. Inductively suppose that it holds up to $z - 1$. Next we show that it also holds for z. Recall that each state $(t'_1, t'_2, t'_3, W', k', C', Y') \in \Gamma(x,y)$ is derived from some state $(t_1, t_2, t_3, W, k, C, Y)$ in $\Gamma(x-1, y)$ or $\Gamma(x, y-1)$. Let π' be an (x, y)-schedule corresponding to $(t'_1, t'_2, t'_3, W', k', C', Y')$, and let π be a schedule corresponding to the state $(t_1, t_2, t_3, W, k, C, Y)$. Using the induction hypothesis, there is a state $(\widehat{t_1}, \widehat{t_2}, \widehat{t_3}, \widehat{W}, k, \widehat{C}, \widehat{Y})$ in $\widehat{\Gamma}(x-1, y)$ or $\widehat{\Gamma}(x, y-1)$ such that $\widehat{t_1} \leq \Delta^{z-1} t_1, \widehat{t_2} \leq t_2, \widehat{t_3} \leq \Delta^{z-1} t_3, \widehat{W} \leq \Delta^{z-1} W, \widehat{C} \leq \Delta^{z-1} C$, and $\widehat{Y} \leq \Delta^{z-1} Y$. Let $\widehat{\pi}$ be an $(x-1, y)$-schedule or $(x, y-1)$-schedule corresponding to $(\widehat{t_1}, \widehat{t_2}, \widehat{t_3}, \widehat{W}, k, \widehat{C}, \widehat{Y})$, and if it is feasible, let $\tilde{\pi}$ be the (x, y)-schedule obtained from $\widehat{\pi}$ by performing the same operation that we perform on π to get π'. Let $(\tilde{t_1}, \tilde{t_2}, \tilde{t_3}, \tilde{W}, \tilde{k}, \tilde{C}, \tilde{Y})$ be the state corresponding to $\tilde{\pi}$. Furthermore, there is a state $(\overline{t_1}, \overline{t_2}, \overline{t_3}, \overline{W}, \tilde{k}, \overline{C}, \overline{Y}) \in \widehat{\Gamma}(x,y)$ in the same box with $(\tilde{t_1}, \tilde{t_2}, \tilde{t_3}, \tilde{W}, \tilde{k}, \tilde{C}, \tilde{Y})$ such that $\overline{t_1} \leq \Delta \tilde{t_1}, \overline{t_2} \leq \tilde{t_2}, \overline{t_3} \leq \Delta \tilde{t_3}, \overline{W} \leq \Delta \tilde{W}, \overline{C} \leq \Delta \tilde{C}$ and $\overline{Y} \leq \Delta \tilde{Y}$. There are four possible ways to get π' from π.

Case 1. π' is obtained from π by scheduling J_x^A directly after schedule π_1. Note that in this case, the condition $t_2 + p_x^A - d^B < p_k^B$ must be satisfied. Then $(t_1, t_2, t_3, W, k, C, Y) \in \Gamma(x-1, y)$, and the operation to get $\tilde{\pi}$ is feasible since $\widehat{t_2} + p_x^A - d^B \leq t_2 + p_x^A - d^B < p_k^B$. Then we have $(\tilde{t_1}, \tilde{t_2}, \tilde{t_3}, \tilde{W}, \tilde{k}, \tilde{C}, \tilde{Y}) = (\widehat{t_1} + p_x^A, \widehat{t_2} + p_x^A, \widehat{t_3} + p_x^A, \widehat{W}, k, \widehat{C} + w_x^A(\widehat{t_1} + p_x^A) + \widehat{W} p_x^A, \widehat{Y} + w_k^B \max\{\min\{\widehat{t_2} + p_x^A - d^B, p_x^A\}, 0\})$. Combining with the fact that $(t'_1, t'_2, t'_3, W', k', C', Y') = (t_1 + p_x^A, t_2 + p_x^A, t_3 + p_x^A, W, k, C + w_x^A(t_1 + p_x^A) + W p_x^A, Y + w_k^B \max\{\min\{t_2 + p_x^A - d^B, p_x^A\}, 0\})$, we have

$$\begin{cases} \overline{t_1} \leq \Delta \tilde{t_1} = \Delta(\widehat{t_1} + p_x^A) \leq \Delta^z(t_1 + p_x^A) = \Delta^z t'_1, \\ \overline{t_2} \leq \tilde{t_2} = \widehat{t_2} + p_x^A \leq t_2 + p_x^A = t'_2, \\ \overline{t_3} \leq \Delta \tilde{t_3} = \Delta(\widehat{t_3} + p_x^A) \leq \Delta^z(t_3 + p_x^A) = \Delta^z t'_3, \\ \overline{W} \leq \Delta \tilde{W} = \Delta \widehat{W} \leq \Delta^z W = \Delta^z W', \\ \tilde{k} = k = k', \\ \overline{C} \leq \Delta \tilde{C} = \Delta(\widehat{C} + w_x^A(\widehat{t_1} + p_x^A) + \widehat{W} p_x^A) \leq \Delta^z(C + w_x^A(t_1 + p_x^A) + W p_x^A) = \Delta^z C', \\ \overline{Y} \leq \Delta \tilde{Y} = \Delta(\widehat{Y} + w_k^B \max\{\min\{\widehat{t_2} + p_x^A - d^B, p_x^A\}, 0\}) \leq \Delta^z(Y + w_k^B \max\{\min\{t_2 + p_x^A - d^B, p_x^A\}, 0\}) = \Delta^z Y. \end{cases} \tag{9}$$

Case 2. π' is obtained from π by scheduling J_x^A directly after schedule π_3. Then $(t_1, t_2, t_3, W, k, C, Y) \in \Gamma(x-1, y)$, and $\tilde{\pi}$ is clearly a feasible schedule. Then we have $(\tilde{t}_1, \tilde{t}_2, \tilde{t}_3, \tilde{W}, \tilde{k}, \tilde{C}, \tilde{Y}) = (\hat{t}_1, \hat{t}_2, \hat{t}_3 + p_x^A, \hat{W} + w_x^A, k, \hat{C} + w_x^A(\hat{t}_3 + p_x^A), \hat{Y})$. Combining with the fact that $(t'_1, t'_2, t'_3, W', k', C', Y') = (t_1, t_2, t_3 + p_x^A, W + w_x^A, k, C + w_x^A(t_3 + p_x^A), Y)$, we have

$$\begin{cases} \overline{t_1} \leq \Delta \tilde{t}_1 = \Delta \hat{t}_1 \leq \Delta^z t_1 = \Delta^z t'_1, \\ \overline{t_2} \leq \tilde{t}_2 = \hat{t}_2 \leq t_2 = t'_2, \\ \overline{t_3} \leq \Delta \tilde{t}_3 = \Delta(\hat{t}_3 + p_x^A) \leq \Delta^z(t_3 + p_x^A) = \Delta^z t'_3, \\ \overline{W} \leq \Delta \tilde{W} = \Delta(\hat{W} + w_x^A) \leq \Delta^z(W + w_x^A) = \Delta^z W', \\ \tilde{k} = k = k', \\ \overline{C} \leq \Delta \tilde{C} = \Delta(\hat{C} + w_x^A(\hat{t}_3 + p_x^A)) \leq \Delta^z(C + w_x^A(t_3 + p_x^A)) = \Delta^z C', \\ \overline{Y} \leq \Delta \tilde{Y} = \Delta \hat{Y} \leq \Delta^z Y = \Delta^z Y'. \end{cases} \quad (10)$$

Case 3. π' is obtained from π by scheduling J_y^B directly after schedule π_2. Note that in this case, the condition $t_2 < d^B$ must be satisfied. Then $(t_1, t_2, t_3, W, k, C, Y) \in \Gamma(x, y-1)$, and the operation to get $\tilde{\pi}$ is feasible since $\hat{t}_2 \leq t_2 < d^B$. Then we have $(\tilde{t}_1, \tilde{t}_2, \tilde{t}_3, \tilde{W}, \tilde{k}, \tilde{C}, \tilde{Y}) = (\hat{t}_1, \hat{t}_2 + p_y^B, \hat{t}_3 + p_y^B, \hat{W}, y, \hat{C} + \hat{W} p_y^B, \hat{Y} + w_y^B \max\{\hat{t}_2 + p_y^B - d^B, 0\})$. Combining with the fact that $(t'_1, t'_2, t'_3, W', k', C', Y') = (t_1, t_2 + p_y^B, t_3 + p_y^B, W, y, C + W p_y^B, Y + w_y^B \max\{t_2 + p_y^B - d^B, 0\})$, we have

$$\begin{cases} \overline{t_1} \leq \Delta \tilde{t}_1 = \Delta \hat{t}_1 \leq \Delta^z t_1 = \Delta^z t'_1, \\ \overline{t_2} \leq \tilde{t}_2 = \hat{t}_2 + p_y^B \leq t_2 + p_y^B = t'_2, \\ \overline{t_3} \leq \Delta \tilde{t}_3 = \Delta(\hat{t}_3 + p_y^B) \leq \Delta^z(t_3 + p_y^B) = \Delta^z t'_3, \\ \overline{W} \leq \Delta \tilde{W} = \Delta \hat{W} \leq \Delta^z W = \Delta^z W', \\ \tilde{k} = y = k', \\ \overline{C} \leq \Delta \tilde{C} = \Delta(\hat{C} + \hat{W} p_y^B) \leq \Delta^z(C + W p_y^B) = \Delta^z C', \\ \overline{Y} \leq \Delta \tilde{Y} = \Delta(\hat{Y} + w_y^B \max\{\hat{t}_2 + p_y^B - d^B, 0\}) \leq \Delta^z(Y + w_y^B \max\{t_2 + p_y^B - d^B, 0\}) = \Delta^z Y'. \end{cases} \quad (11)$$

Case 4. π' is obtained from π by scheduling J_y^B directly after schedule π_4. Then $(t_1, t_2, t_3, W, k, C, Y) \in \Gamma(x, y-1)$, and $\tilde{\pi}$ is clearly a feasible schedule. Then we have $(\tilde{t}_1, \tilde{t}_2, \tilde{t}_3, \tilde{W}, \tilde{k}, \tilde{C}, \tilde{Y}) = (\hat{t}_1, \hat{t}_2, \hat{t}_3, \hat{W}, k, \hat{C}, \hat{Y} + w_y^B p_y^B)$. Combining with the fact that $(t'_1, t'_2, t'_3, W', k', C', Y') = (t_1, t_2, t_3, W, k, C, Y + w_y^B p_y^B)$, we have

$$\begin{cases} \overline{t_1} \leq \Delta \tilde{t}_1 = \Delta \hat{t}_1 \leq \Delta^z t_1 = \Delta^z t'_1, \\ \overline{t_2} \leq \tilde{t}_2 = \hat{t}_2 \leq t_2 = t'_2, \\ \overline{t_3} \leq \Delta \tilde{t}_3 = \Delta \hat{t}_3 \leq \Delta^z t_3 = \Delta^z t'_3, \\ \overline{W} \leq \Delta \tilde{W} = \Delta \hat{W} \leq \Delta^z W = \Delta^z W', \\ \tilde{k} = k = k', \\ \overline{C} \leq \Delta \tilde{C} = \Delta \hat{C} \leq \Delta^z C = \Delta^z C', \\ \overline{Y} \leq \Delta \tilde{Y} = \Delta(\hat{Y} + w_y^B p_y^B) \leq \Delta^z(Y + w_y^B p_y^B) = \Delta^z Y'. \end{cases} \quad (12)$$

Thus, for each state (C', Y') in $\tilde{\mathbf{Q}}_2$, there is a state $(\overline{C}, \overline{Y})$ in $\mathbf{Q}_3$ such that $\overline{C} \leq \Delta^n C' \leq (1+\epsilon)C'$ and $\overline{Y} \leq \Delta^n Y' \leq (1+\epsilon)Y'$.

Next we analyze its time complexity. The initialization step takes $O(n_A n_B)$ time, which is dominated by the final time complexity of Algorithm 3. In the implementation of Algorithm 3, we guarantee that $\Gamma(x,y) = \hat{\Gamma}(x,y)$. Note that there are $O(L^1 L^3 L^W L^A L^B)$ distinct boxes and $0 \leq k \leq n_B$, then there are at most $O(n_B L^1 L^3 L^W L^A L^B)$ different states $(t_1, t_2, t_3, W, k, C, Y)$ in $\Gamma(x,y)$. Moreover, $\Gamma(x,y)$ is obtained by performing at most two (constant) operations on the states in $\Gamma(x-1,y) \cup \Gamma(x,y-1)$ for $x = 0, 1, \ldots, n_A$, $y = 0, 1, \ldots, n_B$. Thus, the overall running time of Algorithm 3 is $O(n_A n_B^2 L^1 L^3 L^W L^A L^B)$. □

5. Numerical Results

In this section some numerical results are provided to show the efficiency of our proposed algorithms. For running our optimization algorithms, we need to input the following parameters relative with the job instances: the numbers of A-jobs and B-jobs, the processing times and weights of all the jobs, and the common due date of B-jobs. By running Algorithms 1 and 2, we get the Pareto frontier. To use Algorithm 3, we need to choose the value of ϵ (>0) to get a $(1+\epsilon)$-approximate Pareto frontier. Note that for the same instance, the Pareto frontiers obtained by Algorithms 1 and 2 are the same, except that the running time of Algorithm 1 is theoretically faster than that of Algorithm 2. The closer the $(1+\epsilon)$-approximate Pareto frontier obtained by Algorithm 3 is to the curve obtained by Algorithms 1 and 2, the closer it is to the optimal solution.

We randomly generate some job instances, in which the numbers of the jobs are set to be $n = 4$ ($n_A = n_B = 2$), $n = 6$ ($n_A = n_B = 3$), and $n = 10$ ($n_A = n_B = 5$). The processing times and the weights of the jobs are randomly generated between 1 and 2. The common due date of B-jobs is set to be 5. What is more, we set $\epsilon = 1$. We ran our algorithms on these instances in a Matlab R2016b environment on an Intel(R) Core(TM) CPU, 2.50 GHz, 4 GB of RAM computer. In fact, when the number of the jobs is small, the Pareto frontier or the approximate Pareto frontier can be found relatively quickly, but when the number of the jobs increases, the running time will increase hugely. The following three Figures 1–3 present the Pareto frontier and $(1+\epsilon)$-approximate Pareto frontier generated by Algorithms 1–3. As can be seen from the three figures, the results obtained by Algorithms 1 and 2 are exactly the same. The results of Algorithm 3 are consistent with those of Algorithms 1 and 2, which may be due to the coincidence caused by the small size of the instance we chose and the few choices in the sizes of the jobs. In fact, considering that the problem we studied is NP-hard, our algorithm can only reach pseudo-polynomial-time theoretically. Therefore, our algorithm is theoretically more suitable for small-scale instances, where the sizes of the jobs are relatively uniform, which fits with the nature of such problems in real life, such as in logistics distribution centers where we use boxes of fixed sizes.

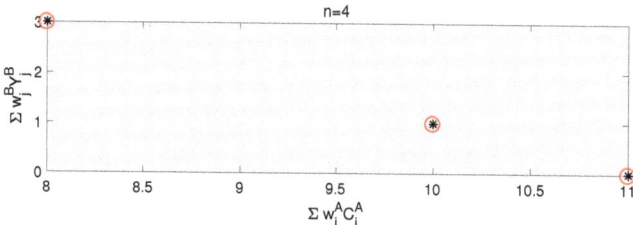

Figure 1. The black stars are the points generated by Algorithms 1 and 2, the red circles are points generated by Algorithm 3.

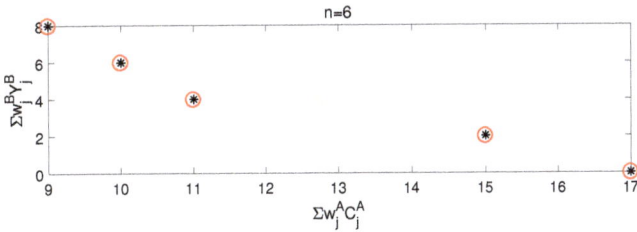

Figure 2. The black stars are the points generated by Algorithms 1 and 2, the red circles are points generated by Algorithm 3.

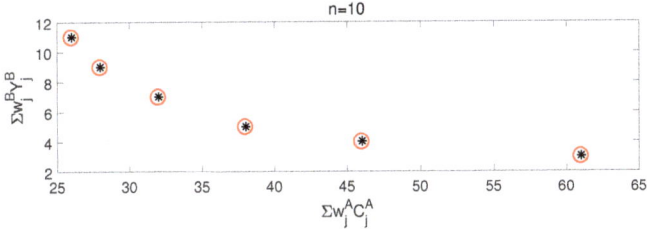

Figure 3. The black stars are the points generated by Algorithms 1 and 2, the red circles are points generated by Algorithm 3.

6. Conclusions

In this paper we investigated a Pareto-optimal problem of scheduling two agents' jobs on a single machine to minimize one agent's total weighted completion time and the other's total weighted late work. For this problem, we devised two dynamic programming algorithms to obtain the Pareto frontier, and an FPTAS to generate an approximate Pareto frontier. Some numerical results were also provided. Compared with the two problems $1|p_j^A \uparrow\downarrow w_j^A|^{\#}(\Sigma w_j^A C_j^A, \Sigma w_j^B Y_j^B)$ and $1|p_j^A \uparrow\downarrow w_j^A, d_j^B \uparrow\downarrow w_j^B|^{\#}(\Sigma w_j^A C_j^A, \Sigma w_j^B Y_j^B)$ studied in Zhang et al. [20], the constraint $p_j^A \uparrow\downarrow w_j^A$ was removed from the problem considered in this paper and we turned to the optimization problem under the condition that B-jobs had a common due date. Table 1 lists the computational complexity of the above three problems. As we can see from Table 1, the condition $p_j^A \uparrow\downarrow w_j^A$ seems to have a greater impact on the complexity result of the problem. In future research, we can try to devise more efficient approximation algorithms for our considered problem with a constant performance-ratio, and we can also study two-agent problems with other combinations of objective functions.

Table 1. Complexity of three problems.

Problems	Complexity	Reference
$1\|p_j^A \uparrow\downarrow w_j^A\|^{\#}(\Sigma w_j^A C_j^A, \Sigma w_j^B Y_j^B)$	$O(n_A n_B^2 U^A U^B)$	Zhang et al. [20]
$1\|p_j^A \uparrow\downarrow w_j^A, d_j^B \uparrow\downarrow w_j^B\|^{\#}(\Sigma w_j^A C_j^A, \Sigma w_j^B Y_j^B)$	$O(n_A n_B U^A U^B)$	Zhang et al. [20]
$1\|d_j^B = d\|^{\#}(\Sigma w_j^A C_j^A, \Sigma w_j^B Y_j^B)$	$O(n_A n_B d P_{\text{sum}} U^A U^B)$	Theorem 1

Author Contributions: Supervision, J.Y.; writing–original draft, Y.Z.; writing–review and editing, Z.G. All authors have read and agreed to the published version of the manuscript.

Funding: This research was supported in part by the NSFC under grant numbers 12071442, 11671368 and 11771406.

Conflicts of Interest: The authors declare no conflicts of interest.

Abbreviations

The following abbreviations are used in this manuscript:

MDPI   Multidisciplinary Digital Publishing Institute
DOAJ   Directory of open access journals
TLA   Three letter acronym
LD   linear dichroism

References

1. Agnetis, A.; Billaut, J.C.; Gawiejnowicz, S.; Pacciarelli, D.; Soukhal, A. *Multiagent Scheduling: Models and Algorithms*; Springer: Berlin/Heidelberg, Germany, 2014.
2. Li, S.S.; Yuan, J.J. Single-machine scheduling with multi-agents to minimize total weighted late work. *J. Sched.* **2020**, *23*, 497–512. [CrossRef]
3. Hariri, A.M.A.; Potts, C.N.; Van Wassenhove, L.N. Single machine scheduling to minimize total weighted late work. *ORSA J. Comput.* **1995**, *7*, 232–242. [CrossRef]
4. Wan, L.; Yuan, J.J.; Wei, L.J. Pareto optimization scheduling with two competing agents to minimize the number of tardy jobs and the maximum cost. *Appl. Math. Comput.* **2016**, *273*, 912–923. [CrossRef]
5. Wan, L.; Wei, L.J.; Xiong, N.X.; Yuan, J.J.; Xiong, J.C. Pareto optimization for the two-agent scheduling problems with linear non-increasing deterioration based on Internet of Things. *Future Gene. Comp. Syst.* **2017**, *76*, 293–300. [CrossRef]
6. Gao, Y.; Yuan, J.J. Bi-criteria Pareto-scheduling on a single machine with due indices and precedence constraints. *Discret. Optim.* **2017**, *25*, 105–119. [CrossRef]
7. He, C.; Leung, J. Two-agent scheduling of time-dependent jobs. *J. Comb. Optim.* **2017**, *34*, 362–377. [CrossRef]
8. Yuan, J.J.; Ng, C.T.; Cheng, T.C.E. Two-agent single-machine scheduling with release dates and preemption to minimize the maximum lateness. *J. Sched.* **2015**, *18*, 147–153. [CrossRef]
9. Wan, L. Two-Agent Scheduling to Minimize the Maximum Cost with Position-Dependent Jobs. *Discrete Dyn. Nat. Soc.* **2015**. [CrossRef]
10. Dabia, S.; Talbi, E.G.; Van Woensel, T.; De Kok, T. Approximating multi-objective scheduling problems. *Comput. Oper. Res.* **2013**, *40*, 1165–1175. [CrossRef]
11. Lee, K.; Choi, B.C.; Leung, J.Y.T.; Pinedo, M.L. Approximation algorithms for multi-agent scheduling to minimize total weighted completion time. *Inf. Process. Lett.* **2009**, *109*, 913–917. [CrossRef]
12. Legriel, J.; Guernic, C.L.; Cotton, S.; Maler, O. Approximating the pareto front of multi-criteria optimization problems. In Proceedings of the 16th International Conference on Tools and Algorithms for the Construction and Analysis of Systems held at the 13th Joint European Conferences on Theory and Practice of Software, Paphos, Cyprus, 20–28 March 2010.
13. Marinescu, R. Efficient approximation algorithms for multi-objective constraint optimization. In Proceedings of the 2nd International Conference on Algorithmic Decision Theory, Piscataway, NJ, USA, 26–28 October 2011.
14. Vassilvitskii, S.; Yannakakis, M. Efficiently computing succinct trade-off curves. *Theor. Comput. Sci.* **2005**, *348*, 334–356. [CrossRef]
15. Yin, Y.Q.; Cheng, S.R.; Cheng, T.C.E.; Wang, D.J.; Wu, C.C. Just-in-time scheduling with two competing agents on unrelated parallel machines. *Omega-Int. J. Manag. Sci.* **2016**, *63*, 41–47. [CrossRef]
16. Yin, Y.Q.; Cheng, T.; Wang, D.J.; Wu, C.C. Two-agent flowshop scheduling to maximize the weighted number of just-in-time jobs. *J. Sched.* **2017**, *20*, 313–335. [CrossRef]

17. Chen, R.X.; Li, S.S.; Li, W.J. Multi-agent scheduling in a no-wait flow shop system to maximize the weighted number of just-in-time jobs. *Eng. Optim.* **2019**, *51*, 217–230. [CrossRef]
18. Purcaru, C.; Precup, R.E.; Iercan, D.; Fedorovici, L.O.; David, R.C.; Dragan, F. Optimal robot path planning using gravitational search algorithm. *Int. J. Artif. Intell.* **2013**, *10*, 1–20.
19. Soares, A.; Râbelo, R.; Delbem, A. Optimization based on phylogram analysis. *Expert Syst. Appl.* **2017**, *78*, 32–50. [CrossRef]
20. Zhang, Y.; Yuan, J.J. A note on a two-agent scheduling problem related to the total weighted late work. *J. Comb. Optim.* **2019**, *37*, 989–999. [CrossRef]
21. Zhang, Y.; Yuan, J.J.; Ng, C.T.; Cheng, T.C.E. Pareto-optimization of three-agent scheduling to minimize the total weighted completion time, weighted number of tardy jobs, and total weighted late work. *Nav. Res. Logist.* **2020**, in press.

Publisher's Note: MDPI stays neutral with regard to jurisdictional claims in published maps and institutional affiliations.

© 2020 by the authors. Licensee MDPI, Basel, Switzerland. This article is an open access article distributed under the terms and conditions of the Creative Commons Attribution (CC BY) license (http://creativecommons.org/licenses/by/4.0/).

Article

A Hybrid Metaheuristic for the Unrelated Parallel Machine Scheduling Problem

Dung-Ying Lin * and Tzu-Yun Huang

Department of Industrial Engineering and Engineering Management, National Tsing Hua University, Hsinchu 30013, Taiwan; zxcv6564@gmail.com
* Correspondence: dylin@ie.nthu.edu.tw; Tel.: +886-(03)-574-2694

Abstract: The unrelated parallel machine scheduling problem aims to assign jobs to independent machines with sequence-dependent setup times so that the makespan is minimized. When many practical considerations are introduced, solving the resulting problem is challenging, especially when problems of realistic sizes are of interest. In this study, in addition to the conventional objective of minimizing the makespan, we further consider the burn-in (B/I) procedure that is required in practice; we need to ensure that the scheduling results satisfy the B/I ratio constrained by the equipment. To solve the resulting complicated problem, we propose a population-based simulated annealing algorithm embedded with a variable neighborhood descent technique. Empirical results show that the proposed solution strategy outperforms a commonly used commercial optimization package; it can obtain schedules that are better than the schedules used in practice, and it does so in a more efficient manner.

Keywords: unrelated parallel machine scheduling; simulated annealing; variable neighborhood descent; metaheuristic; production scheduling

Citation: Lin, D.-Y.; Huang, T.-Y. A Hybrid Metaheuristic for the Unrelated Parallel Machine Scheduling Problem. *Mathematics* **2021**, *9*, 768. https://doi.org/10.3390/math9070768

Academic Editor: Chin-Chia Wu

Received: 24 February 2021
Accepted: 29 March 2021
Published: 1 April 2021

Publisher's Note: MDPI stays neutral with regard to jurisdictional claims in published maps and institutional affiliations.

Copyright: © 2021 by the authors. Licensee MDPI, Basel, Switzerland. This article is an open access article distributed under the terms and conditions of the Creative Commons Attribution (CC BY) license (https://creativecommons.org/licenses/by/4.0/).

1. Introduction

The unrelated parallel machine scheduling problem aims to assign a set of jobs to a set of unrelated machines that can process the jobs in parallel without affecting each other. Unrelated parallel machine scheduling problems are of significant practical relevance, and they arise in many applications (i.e., the electronic assembly industry investigated in this study). However, many constraints and additional considerations need to be addressed in practice. For instance, jobs completed at machines need to go through a burn-in (B/I) process prior to being placed in service. The B/I process forces certain manufacturing failures to occur under supervised conditions so that the quality of the product can be examined and ensured. However, as B/I equipment is expensive, many companies have only a limited amount of this kind of equipment, and this becomes the bottleneck of this type of scheduling problem. In this study, we investigate the unrelated parallel machine scheduling problem that aims to minimize the makespan and maximize B/I equipment utilization. As each job has its own suitable B/I equipment, the maximization of B/I equipment utilization can be achieved by making the ratio of completed jobs match the number of B/I equipment available. Other than this additional consideration, the conventional constraints (i.e., sequence-dependent setup times) are also accommodated in the study. As the unrelated parallel machine scheduling problem possesses non-deterministic polynomial-time (NP)-hard complexity [1], solving problem instances with practical sizes is a challenging task, especially when many practical constraints need to be considered. To address this type of problem, we propose a population-based simulated annealing (PBSA) method embedded with a variable neighborhood descent (VND) heuristic to solve it. The proposed solution strategy is empirically applied to real-world problem instances. A comparison with a commercial optimization package demonstrates that the devised approach can determine the optimal schedules in small problem instances. When the problem size increases, we show that the designed approach can determine schedules that are better than the schedules used in practice, and it does so in a much more efficient manner than the commercial optimization package, which fails to obtain solutions.

The remainder of this paper is structured as follows. Section 2 critically overviews the related work regarding production schedules. Section 3 presents the mathematical formulation for the unrelated parallel machine scheduling problem considering various practical constraints. The solution strategy tailored for solving the resulting program is presented in Section 4. Empirical studies are summarized in Section 5 to demonstrate the efficiency and effectiveness of the proposed solution strategies. The final section offers conclusions and provides suggestions for future research.

2. Literature Review

In this section, we broadly categorize the production scheduling problem into single and multiple operation problems according to the production process. In a single operation problem, a job only needs to be processed/assembled once in a single and suitable machine to complete its production procedure. In a multiple operation problem, a job needs to go through various production/assembly steps. Each production/assembly step can only be performed in a specific machine. Only when a job goes through all the required steps is its production procedure completed in a multiple operation problem. Single operation problems can be further classified into single and parallel machine problems. Multiple operation problems can be further divided into flow shop, job shop and open shop problems.

2.1. Single Operation

In single operation production, the problem is considered as a single machine scheduling problem if all the jobs are completed by a single machine.

2.1.1. Single Machine Scheduling

Different research streams analyze single machine scheduling problems while considering various objective functions and constraints. Ref. [2] studied the single machine scheduling problem that aims to minimize both energy consumption and maximum tardiness, and they proposed a mixed-integer linear programming model to formulate it. A ε-constraint method that integrated local search, preprocessing, valid inequalities and solution space reduction techniques was developed to determine the Pareto optimal solution of the two obtained compromised objective functions. Ref. [3] considered the minimization of maximum costs, considering uncertain processing times. Various cost functions were analyzed, and polynomial algorithms were devised to solve the resulting problem. Ref. [4] investigated a single machine scheduling problem that takes release dates and inventory constraints into consideration. With the predetermined processing time of each job and each job's known impact on the inventory level, the work aimed to determine the optimal sequence of jobs such that the makespan was minimized. It was shown that the problem is strongly NP-hard, and a series of algorithms were proposed to tackle it.

2.1.2. Parallel Machine Scheduling

When there are multiple machines with similar functionalities and these machines can work simultaneously without affecting each other, the problem is considered a parallel machine scheduling problem. Based on the features of the employed machines, parallel machine scheduling can be further classified into identical and unrelated parallel machine scheduling problems. The machines considered in identical parallel machine scheduling problems are homogeneous, and the processing time of a job at any machine is identical. The processing times of a job at the different machines considered in the unrelated parallel machine scheduling problem, however, can be different, and the processing times at different machines are not relevant.

Identical Parallel Machines

Ref. [5] presented a mathematical model for an identical parallel machine scheduling problem in which job splitting and sequence-dependent setup times were considered. Simulated annealing and genetic algorithm metaheuristic-based approaches were proposed with various encoding and decoding methods. The encoding effectively represented the solutions compactly, while the decoding heuristically split jobs in a different manner. The numerical results showed that the proposed approaches could determine solutions of good quality. Ref. [6] investigated the parallel machine scheduling problem that takes job tooling requirements and job-dependent setup times into consideration. A biased random-key genetic algorithm integrated with a variable neighborhood descent-based local search was proposed to solve the resulting NP-hard problem. Numerical results showed that the proposed solution approach outperformed benchmark methods.

One variant of the identical parallel machine problem is the uniform parallel scheduling problem. The processing times of a job at different machines considered in the uniform parallel scheduling problem can vary. However, the processing times are proportional to each other and are at a fixed rate. Ref. [7] examined a uniform parallel-machine scheduling problem with the objective of minimizing the total resource consumption subject to a bounded makespan. They developed a metaheuristic and showed that it could outperform the particle swarm optimization heuristic and approximate the theoretical lower bound.

Unrelated Parallel Machines

For the unrelated parallel machine scheduling problem, the genetic algorithm (GA) proposed by [8] is a popular solution method that has been employed in many past studies. Compared to the situation with the conventional GA, many researchers have attempted to incorporate various enhancements to solve unrelated parallel machine scheduling problems. For instance, ref. [9] added a fast local search and local search-enhanced crossover operator for the GA. Ref. [10] proposed a hybrid GA that integrates dispatching rules (i.e., processing time-based, completion time-based and sequence-based rules) into the overall solution framework. Ref. [11] studied the unrelated parallel machine scheduling problem and derived a strategy to dynamically allocate jobs to dedicated machines so that the total earliness and tardiness times could be minimized. A modified GA with a distributed release time control mechanism was proposed and was shown to perform well. Ref. [12] introduced new decoding methods developed for the total tardiness objective within a GA solution framework, and they were able to improve the performance of the GA.

Based on simulated annealing (SA), ref. [13] introduced a sine cosine algorithm as a local search method to improve algorithmic convergence when solving unrelated parallel machine scheduling problems with sequence-dependent and machine-dependent setup times. Ref. [14] evaluated the performances of four stochastic local search methods, namely, simulated annealing, iterated local search, late acceptance hill-climbing, and step counting hill-climbing, in solving unrelated parallel machine scheduling problems with sequence-dependent setup times. These methods were compared together with the GA proposed by [9] and the heuristic devised by [15]. Empirical results showed that SA performed best in solving large problem instances. Ref. [16] targeted the unrelated parallel machine scheduling problem with a random rework and presented two mixed-integer programs. As the problem is strongly NP-hard, a genetic algorithm and a simulated annealing algorithm that utilize aggregate task estimation techniques were proposed and were shown to be effective in solving the problem. Additionally, it was reported that the simulated annealing algorithm performed better than the genetic algorithm.

In the literature, local search (LS) has also been widely used or integrated with other solution techniques in tackling production scheduling problems. The variable neighborhood descent (VND) approach is the extension of LS. The VND method explores several neighborhood structures sequentially (say, 1 to n structures) from an incumbent solution. If an improved solution is identified, the search restarts from the first structure and explores neighborhood solutions with these structures again until a prespecified stopping criterion is met [17]. Ref. [18] combined VND with mathematical programming techniques and proposed a multi-start VND method for solving unrelated parallel machine problems. The core concept was to decompose the given problem into job assignment and job sequencing subproblems. Numerical results showed that the proposed algorithm performed well. Ref. [19] integrated variable neighborhood search and SA and proposed a two-stage hybrid metaheuristic to solve the unrelated parallel machine scheduling problem. The first stage determines an initial solution with the "earlier release date first" (ERD) rules, while the second stage explores the neighborhood with various structures. It was shown in the conducted numerical experiments that the proposed solution strategy outperforms a commercial optimization package. Ref. [20] substituted local search with VND in the iterated greedy search, artificial bee colony and genetic algorithms to evaluate whether the substitution could improve the performances of these metaheuristics for solving unrelated parallel machine scheduling problems. The Taguchi robust method was used to calibrate the parameters used in the framework. Empirical results showed that replacing local search with VND indeed increases the performance of all tested metaheuristics.

Aside from the above methods, many researchers have attempted various approaches to address unrelated parallel machine scheduling problems. For instance, ref. [21] considered various practical resources in such problems and tried to solve the resulting problem with the GA and an artificial immune system (AIS). After calibrating the parameters with the Taguchi method, the AIS was shown to outperform the GA in solving large problem instances. Ref. [22] addressed machine load minimization in the unrelated parallel machine scheduling problem. A hybrid particle swarm optimization (PSO) technique and a GA were proposed, and the Taguchi method was used to calibrate the parameters. The results showed that the hybrid approach performed better than the GA, PSO algorithm or PSO algorithm with local search. Ref. [23] developed a two-stage heuristic to solve unrelated parallel machine scheduling problems with more than two machines. In the first stage, a mixed-integer linear programming model was solved to estimate the lower bound. In the second stage, a constraint programming model was employed to schedule jobs on machines.

Some past studies discussed B/I related issues (i.e., [24–28]). One of the most relevant study is [27] which considered the B/I machines as the batch processing machines. In that study, the aims were to minimize the maximum tardiness or to minimize the number of tardy jobs while considering the processing time, due date and release time constraints. Dynamic programming-based algorithms were developed to solve the resulting problems. Similarly, ref. [28] proposed a mixed-integer linear programming model that formulates the B/I requirement as a batch processing problem. Two solution heuristics, delay window-time parallel saving algorithm (DWPSA) and delay window-time generalized saving algorithm (DWGSA), were proposed to solve the proposed formulation.

2.2. Multiple Operation

There are three major types of problems in multiple production scheduling, namely, flow shop, job shop and open shop problems. In a typical flow shop scheduling problem, all jobs are required to complete an identical production process that can be processed at different machines. Conventional job shop scheduling is similar to flow shop scheduling, except that each job should complete the procedure in a specific order. In the open shop scheduling problem, jobs can be completed in random order.

2.2.1. Flow Shop scheduling Problem

Ref. [29] developed a greedy algorithm to solve the flow shop scheduling problem in two phases. The first phase, destruction, eliminates some jobs from the incumbent solution, while the second phase, construction, heuristically reinserts the eliminated jobs from the first phase into the sequence. It was shown that the greedy algorithm is easy to implement and can outperform many benchmarks. Ref. [30] studied the flow shop scheduling problem. First-in-first-out batch dispatching rules were designed to determine the initial solution, followed by mixed-integer program reoptimization techniques and local search heuristics to improve the solution quality. Ref. [31] adopted artificial immune system-based methods for solving a two-stage hybrid flow shop scheduling problem and showed that the proposed methods outperformed the existing lower bounds. Ref. [32] proposed a hybrid metaheuristic that integrated the GA and random sampling to solve the sequence-dependent flow-shop scheduling problem. Ref. [33] solved the distributed blocking flow shop scheduling problem with three hybrid iterative greedy algorithms.

2.2.2. Job Shop Scheduling Problem

Ref. [34] investigated the job shop scheduling problem to address the dynamic events that are inevitable in production environments. A GA embedded with various heuristic dispatching rules was devised to solve the problem. Ref. [35] studied the flexible job shop scheduling problem (FJSP) that allows a job to be processed at any machine from a given set. An algorithm that combines the advantages of the GA and tabu search was proposed to tackle the FJSP and was demonstrated to be effective in solving it. Ref. [36] improved the coding and decoding procedures in PSO and showed that the improvement was suitable for practical job shop scheduling. Ref. [37] utilized the GA and decentralization scheme to minimize the makespan in production scheduling. The proposed solution framework was compared with a shortest processing time rule-based approach (SPT) and was shown to outperform the SPT.

2.2.3. Open Shop Scheduling Problem

In solving the flexible open shop scheduling problem, ref. [38] proved the asymptotic optimality of the general dense scheduling (GDS) algorithm and showed that the proposed GDS-based heuristic could converge to good solutions in large problem instances. When using the minimization of the total flow time as the objective function in the open shop scheduling problem, ref. [39] adopted a GA and an ant colony optimization (ACO) method to solve the problem. For cases when the minimization of the makespan is the objective function, ref. [40] proposed using the GA to solve it.

2.3. Summary

In this work, the focus is on the unrelated parallel production scheduling problem. We summarize the studies we have discussed above in Table 1 to highlight the contribution of our work.

The obvious difference between the objective function in our work and those in past studies is the B/I procedure considered in our formulation. The differences in constraints can also be seen in the table. In terms of the solution approach, as suggested by [16], SA is an ideal choice for solving this problem. However, as the problem investigated in this work contains complicated constraints and objective functions, it is not rare to have different solutions with identical objective values (or multiple optimal solutions). As a standard SA needs the objective value to find the descent direction, having multiple solutions with identical objective values makes it difficult to identify the direction based on these values since they are all the same. To explore the neighborhood solutions in an efficient manner, we decide to adopt the population-based simulated annealing (PBSA) algorithm. Furthermore, as indicated by [20], replacing the local search process with VND can potentially increase the performances of various metaheuristics. Therefore, we further embed a VND procedure in the PBSA algorithm.

One of the most relevant studies in terms of the solution approach of our work is that of [19], which integrated SA and VNS in solving unrelated parallel machine scheduling problems. The primary difference between VNS and VND is that VNS introduces a shaking procedure into the solution framework. However, as the constraints considered in our work are rather complicated, so introducing a random shaking procedure can easily generate an infeasible solution. From our preliminary experiments, we believe that VND is more appropriate than VNS for solving our problem.

Table 1. Literature comparison.

Study	Objective	Primary Constraints	Solution Approach
[10]	Minimizing the total completion time	Setup time and production availability	GA
[21]	Minimizing the makespan	Resource constraints, sequence-dependent setup times, different release dates, machine eligibility and precedence constraints	GA and AIS
[22]	Minimizing the total machine load	Past sequence-dependent setup times, release dates, deteriorating jobs and learning effects	Integrated PSO and GA
[19]	Minimizing the makespan	Nonzero arbitrary release dates, limited additional resources, and non-anticipatory sequence-dependent setup times	Integrated SA and VNS
[16]	Minimizing the total weighted tardiness	Random rework and due dates	GA and SA
[20]	Minimizing the total weighted tardiness	Sequence- and machine-dependent setup times	VND
Current Study	Minimizing the makespan and B/I violations	Sequence-dependent setup times, different work starting times, machine eligibility, burn-in eligibility and work time limits	Integrated PBSA and VND

3. Mathematical Formulation

In this section, we formally define the problem under study, followed by the utilized notations, resulting in a mathematical and detailed explanation.

In each planning horizon, as illustrated in Figure 1, the master production schedule (MPS) outputs the jobs that need to be completed within the predefined time horizon. The jobs completed in the production lines are then sent to B/I equipment to finalize the manufacturing process. During the process, the daily production scheduling problem (DPS) aims to assign jobs to appropriate production lines so that the makespan can be minimized while meeting the B/I requirement as well as possible.

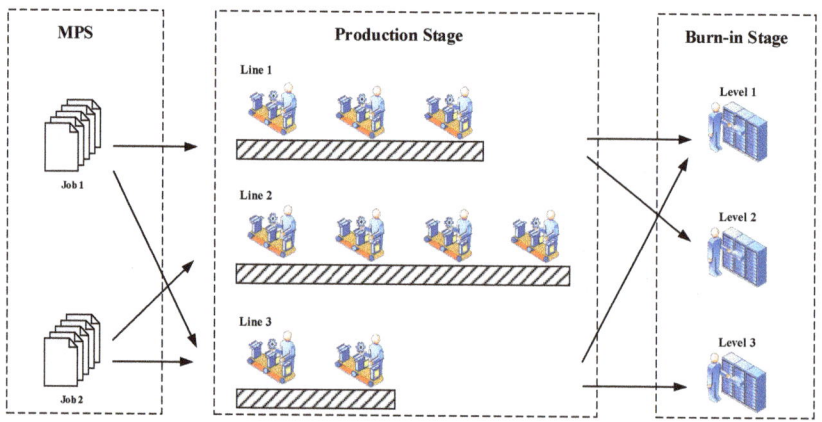

Figure 1. Illustration of the production process considered in this study.

The current study focuses on the DPS. To focus on the core issues of the problem, the following assumptions are imposed.

1. Each job contains only one kind of product. The products that need to be assembled in each job are provided by MPS.

2. If two consecutive jobs processed in one production line are of different products, setup time is required.
3. Each production line contains no job at the beginning of each day.
4. The process time of each job on each production line is given and fixed.
5. If a job begins in a production line, it will be completed without interruption.

The following practical constraints are considered in the mathematical formulation.

1. Each production line has its own maximum number of daily production hours.
2. Each product and its corresponding job have some production requirements and can only be processed in the predetermined/specified production lines.
3. Each job has its own earliest starting time and can only start after that prespecified time.
4. Each job requires a specific level of B/I equipment.
5. There is an upper bound of the total production hours for all the production lines considered together.

Notations

Based on the abovementioned statements and assumptions, the problem studied corresponds to an unrelated parallel machine scheduling problem. The mathematical formulation is revised from [9] with some modifications of the constraints to meet the requirements imposed in practice.

Sets

N	set of jobs
M	set of production lines
BI	set of burn-in levels
T	set of the planning horizon

Parameters

P_{ij}	processing time of job j on production line i
S_{ijk}	setup time between job j and job k on production line i. S_{ijk}= setup time if two consecutive jobs processed on production line i belong to different job types; $S_{ijk} = 0$ if two consecutive jobs processed on production line i belong to identical job types.
cap_{it}	maximum daily processing hours for production line i on day t.
TP_t	maximum daily processing hours for all the production lines considered together.
Q	an extremely large number
U_t	penalty parameter for burn/in ratio violations on day t
BI_1, BI_2, BI_3	target ratio of jobs assigned to B/I levels 1, 2 and 3, respectively. Suppose that a company hopes to maintain three B/I levels of 5:4:1; we can set $BI_1 = 5$, $BI_2 = 4$, and $BI_3 = 1$.
MN_j	number of products that need to be assembled in each job j. As the B/I level violations are calculated based on MN_j, this parameter is introduced.

Decision Variables

X_{ijkt}	$X_{ijkt} = 1$ if job j is processed immediately before job k on production line i on day t; $X_{ijkt} = 0$ otherwise. Note that if job j or k cannot be processed on production line i, then $X_{ijkt} = 0$.
B_{jbt}	$B_{jbt} = 1$, if job j has a B/I level of b on day t; $B_{jbt} = 0$ otherwise.
C_{ijt}	completion time of job j on production line i on day t
F_{it}	complete time for each production line i on day t
$C_{max,t}$	maximum completion time on day t

Mathematical Formulation

$$\text{Min} \sum_{t}(\alpha C_{max,t} + \beta U_t) \tag{1}$$

Subject to

$$\sum_{t \in T}\sum_{i \in M}\sum_{\substack{j \in 0 \cup N \\ j \neq k}} X_{ijkt} = 1 \qquad \forall k \in N \tag{2}$$

$$\sum_{t \in T}\sum_{i \in M}\sum_{\substack{k \in N \\ j \neq k}} X_{ijkt} \leq 1 \qquad \forall j \in N \tag{3}$$

$$\sum_{k \in N} X_{i0kt} \leq 1 \qquad \forall i \in M; \forall t \in T \tag{4}$$

$$\sum_{\substack{h \in 0 \cup N \\ h \neq k, h \neq j}} X_{ihjt} \geq X_{ijkt} \qquad \forall j,k \in N, j \neq k; \forall i \in M; \forall t \in T \tag{5}$$

$$C_{ikt} \geq C_{ijt} + S_{ijk} + P_{ik} + Q \cdot (X_{ijkt} - 1) \qquad \forall j,k \in N, j \neq k; \forall i \in M; \forall t \in T \tag{6}$$

$$F_{it} \geq C_{ijt} \qquad \forall j \in N; \forall i \in M; \forall t \in T \tag{7}$$

$$F_{it} \leq cap_{it} \qquad \forall i \in M; \forall t \in T \tag{8}$$

$$C_{max,t} \geq F_{it} \qquad \forall i \in M; \forall t \in T \tag{9}$$

$$\sum_{i \in M} F_{it} \leq TP_t \qquad \forall t \in T \tag{10}$$

$$\sum_{t \in T}\sum_{b \in BI} B_{kbt} \leq 1 \qquad \forall k \in N \tag{11}$$

$$\sum_{b \in BI} B_{kbt} = \sum_{i \in M}\sum_{j \in \{0\} \cup \{N\}} X_{ijkt} \qquad \forall k \in N, j \neq k; \forall t \in T \tag{12}$$

$$\left|\sum_{k \in N} B_{k1t} \cdot MN_k - \frac{BI_1}{BI_3}\sum_{k \in N} B_{k3t} \cdot MN_k\right|$$
$$+ \left|\sum_{k \in N} B_{k2t} \cdot MN_k - \frac{BI_2}{BI_3}\sum_{k \in N} B_{k3t} \cdot MN_k\right|$$
$$+ \left|\sum_{k \in N} B_{k1t} \cdot MN_k - \frac{BI_1}{BI_2}\sum_{k \in N} B_{k2t} \cdot MN_k\right|$$
$$= U_t \tag{13}$$

$$X_{ijkt} \in \{0,1\} \qquad \forall j,k \in N, j \neq k; \forall i \in M; \forall t \in T \tag{14}$$

$$B_{jbt} \in \{0,1\} \qquad \forall j \in N; \forall b \in BI; \forall t \in T \tag{15}$$

The objective function (1) aims to minimize the makespan ($C_{max,t}$) and the B/I level violations (U_t). Parameters α and β are the weights of the corresponding terms, and these will be calibrated in the numerical experiments.

Equation (2) ensures that each job is assigned to a production line i and that each job k on this machine has only one preceding job j. Equation (3) is the constraint that each job j has at most one succeeding job k. Each production line is assigned at most a dummy job 0 at the beginning that represents the first job of this production line. The design is described in Equation (4). Equation (5) defines the job order. If job j is assigned to production line i, there must be a preceding job h on this production line. If job j is the first job on this production line, then h must be a dummy job. However, job j may not have a succeeding job k.

Equation (6) defines the completion time C_{ikt} of the jobs. As job k should be processed after job j on production line i, the completion time of job k (C_{ikt}) must be greater than the completion time of job j (C_{ijt}) plus the corresponding setup time S_{ijk} and the processing time of job k (P_{ik}). If $X_{ijkt} = 0$, which means that job k cannot be processed immediately after job j on production line i, this constraint becomes a redundant constraint. Equation (7) ensures that the completion time of each production line (F_{it}) is greater than or equal to the completion time of any job in that production line (C_{ijt}). F_{it} will be used to calculate the objective value later.

Equation (8) sets the maximum number of processing hours cap_{it} for each production line on each day t. Equation (9) calculates the maximum completion time $C_{max,t}$ according to the statuses of all the jobs in all production lines. Equation (10) constrains the maximum number of processing hours for all the production lines considered together.

The limitation that any job can be assigned to a B/I level is enforced in Equation (11). Equation (12) establishes the relationship between a machine and the B/I level. Only if a job is assigned (i.e., any $X_{ijkt} = 1$) can a B/I level be assigned. The B/I ratio penalty U_t is calculated based on Equation (13). Let us use the first term $\left| \sum_{k \in N} B_{k1t} \cdot MN_k - \frac{BI_1}{BI_3} \sum_{k \in N} B_{k3t} \cdot MN_k \right|$ in that equation as an example to illustrate the proposed design. This term is designed as the absolute value of the number of products that need to be assembled in jobs that are assigned to B/I level 1 ($\sum_{k \in N} B_{k1t} \cdot MN_k$) minus the ratio $\frac{BI_1}{BI_3}$ of the number of products that need to be assembled in jobs that are assigned to B/I level 3 ($\sum_{k \in N} B_{k3t} \cdot MN_k$). If the result of the first term is zero, then the ratio of the jobs assigned to levels 1 and 3 exactly matches the prespecified ratio $\frac{BI_1}{BI_3}$. Otherwise, $\left| \sum_{k \in N} B_{k1t} \cdot MN_k - \frac{BI_1}{BI_3} \sum_{k \in N} B_{k3t} \cdot MN_k \right|$ can serve as the measure of how far away the assignment is from the desired value $\frac{BI_1}{BI_3}$ and can be used as a penalty in the objective function to drive the assignment to match the desired value as closely as possible. The second and third terms can be interpreted in a similar manner. Finally, Equations (14) and (15) state that the decision variable considered in the formulation is of binary value.

As shown in [1], the unrelated parallel machine scheduling problem is an NP-hard problem. The mathematical formulation presented in this section is a parallel machine scheduling problem with additional side constraints. Therefore, it is at least of NP-hard complexity. Solving such problems of practical sizes can be a challenging task. To address this issue, we propose a metaheuristic-based solution approach in the following section.

4. Solution Approach

In this study, we propose population-based simulated annealing (PBSA), which is an extension of SA [41,42]. SA is the search heuristic analogous to the process of solid physical annealing. During the annealing process, a solid is heated and cooled down slowly until it achieves the most likely crystal lattice configuration so that the resulting solid has superior structural integrity. Similar to this process, SA compares the current solution with its neighborhood solution and accepts an improved solution in each iteration.

Inferior solutions can also be accepted with a limited probability so that the search process can escape from local optima and finally approximate the global optimum. The probability of accepting inferior solutions depends on a nonincreasing temperature parameter with each iteration of the SA algorithm. In PBSA, instead of a single neighborhood solution, a population of neighborhood solutions are generated during each iteration, and only the best solution among them is used as the incumbent solution in the next iteration. With this improvement, the search for a neighborhood solution can be highly effective. To further enhance the performance of this approach, the traditional LS method employed in SA is replaced by VND in the proposed PBSA method. We next detail our critical algorithmic steps.

4.1. Initial Solution

The initialization step first sorts the jobs in descending order according to multiple attributes, namely, the earliest starting time, the number of allowable production lines and the total processing time. Then, the jobs are assigned according to the first-in-first-out (FIFO) rule to different production lines while satisfying all the assignment rules. Some of the jobs may be left unassigned after this procedure. However, the assigned and unassigned jobs together form an initial feasible solution. Note that the initial solution is designed to be the initial point for the following search procedure. The PBSA presented later can always converge to a good final solution regardless of the initial solution.

4.2. Algorithm Steps

There are four primary steps in the proposed PBSA algorithm, namely, initialization, neighborhood search, incumbent solution updating and termination.

4.2.1. Initialization

For ease of explanation, we introduce additional notations. We denote T_H, T and T_L as the highest possible, current and lowest possible temperatures, respectively. We first initialize the current temperature T as T_H and reduce T over iterations to simulate the "cooling down" process of solid physical annealing. The initial and incumbent solutions are both initialized as ∞ in the beginning. Starting from the initial solution found in the previous section, the PBSA algorithm enters the search procedure.

4.2.2. Neighborhood Search

In VND, three neighborhood search approaches/structures are employed in our solution framework. The first is single job switching (Figure 2), which switches one randomly selected job between two production lines (j_1 and j_4 in this illustrative example). The second is moving jobs from one production line to another (illustrated in Figure 3). A job (j_2 in this example) from one production line is randomly selected and inserted at the beginning of another production line. The final approach is one-to-two job switching (illustrated in Figure 4), which switches one job in a production line with two consecutive jobs in another production line. To reduce the number of setups, we sort the jobs so that jobs with the same job type can be grouped together after any of the above changes occur in any production line. Note that all the changes only take place when the scheduling rules are not violated. In other words, we explore the neighborhood solutions within the feasible region. As reported in the literature, there exist various alternative neighborhood search procedures. However, from our preliminary experiments, these three procedures yield the best performance and are incorporated in our solution framework.

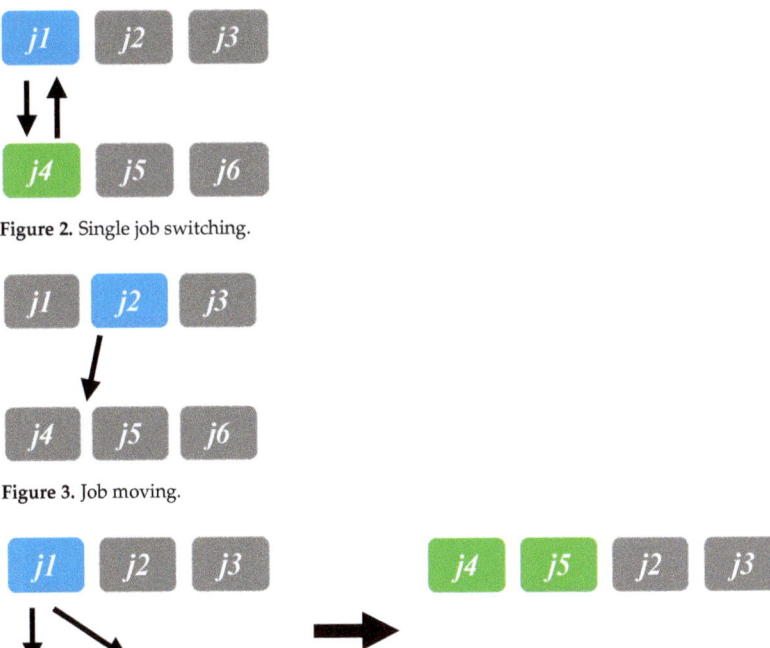

Figure 2. Single job switching.

Figure 3. Job moving.

Figure 4. One-to-two job switching.

In VND, we examine these three neighborhood structures sequentially. As we use the PBSA framework, we generate 5 neighborhood solutions by examining each neighborhood structure during each neighborhood search, and only the best solution is selected as a candidate. If a solution that is better than the incumbent solution is identified, we restart from the first structure and explore the neighborhood solutions with these structures again until a stopping criterion is met.

Specifically, we first define a maximum number of neighborhood structures that can be examined RL_n. When the neighborhood search begins, we examine each neighborhood structure sequentially, and a counter R_n is used to keep track of the number of times a neighborhood structure is examined. If a superior solution is identified, we set $R_n = 0$ and restart from the first neighborhood structure. If $R_n = RL_n$ when examining a neighborhood structure, we examine the next neighborhood structure. If all the neighborhood structures are examined, we continue to the next iteration of the PBSA algorithm.

Over the PBSA iterations, the probability of accepting inferior solutions decreases due to the lowering temperature. Therefore, we increase RL_n gradually to increase the possibility of exploring a larger solution space. On the other hand, VND increases the chance of finding a superior solution during each neighborhood search. Based on our empirical experiment, the design balances the search procedure and is effective in solving the overall problem.

4.2.3. Incumbent Solution Updating

When a neighborhood solution is superior to the incumbent solution, we update the incumbent solution. To further improve the solution quality, after this update, we search the unassigned jobs and examine whether the insertion of additional jobs into the solution is possible. If yes, we insert the jobs and use the updated solution as the incumbent solution. If insertion is not possible, the superior solution is used as the incumbent solution directly.

Other than the above updating process, there is a limited probability (denoted as P in this study) of allowing the search procedure to accept inferior solutions. This probability is calculated based on the following modified Boltzmann function [41]:

$$P = min\left\{1, e^{-\frac{\Delta}{T}}\right\}$$

In the function, $\Delta = C(x') - C(x)$ is the difference between the objective value of the current solution ($C(x')$) and that of the incumbent solution ($C(x)$). T denotes the current temperature. If a randomly generated real number γ is greater than P, the current inferior solution is accepted and becomes the incumbent solution in the next iteration. In this study, we reduce the temperature T when the number of the searches for each neighborhood structure reaches the prespecified limit. The reduction of T is controlled by:

$$T = T \times T_{scale}$$

where T_{scale} is the rate at which the temperature decreases. The value of T_{scale} is set between 0 and 1, and this causes the value of T to decrease over multiple iterations. PBSA can converge effectively with the above cooling mechanism.

4.2.4. Termination

As many companies need solutions periodically so that they can adjust their scheduling results based on the current dynamic manufacturing environment, we terminate the search procedure and report the incumbent solution when the maximum allowed computational time is reached.

4.3. Summary

Overall, the proposed PBSA can be summarized as in Figure 5. With the generated initial solution and parameter settings, the search procedure examines the neighborhood solution with VND, and controls the PBSA framework based on the obtained results until the stopping criterion is met. Note that as VND is employed in the neighborhood search procedure, the iteration counter R_n is reset to 0 only when the search procedure identifies a solution that is superior to the incumbent solution or when each neighborhood structure reaches the number of pre-specified limit. In other cases, $R_n = R_n + 1$.

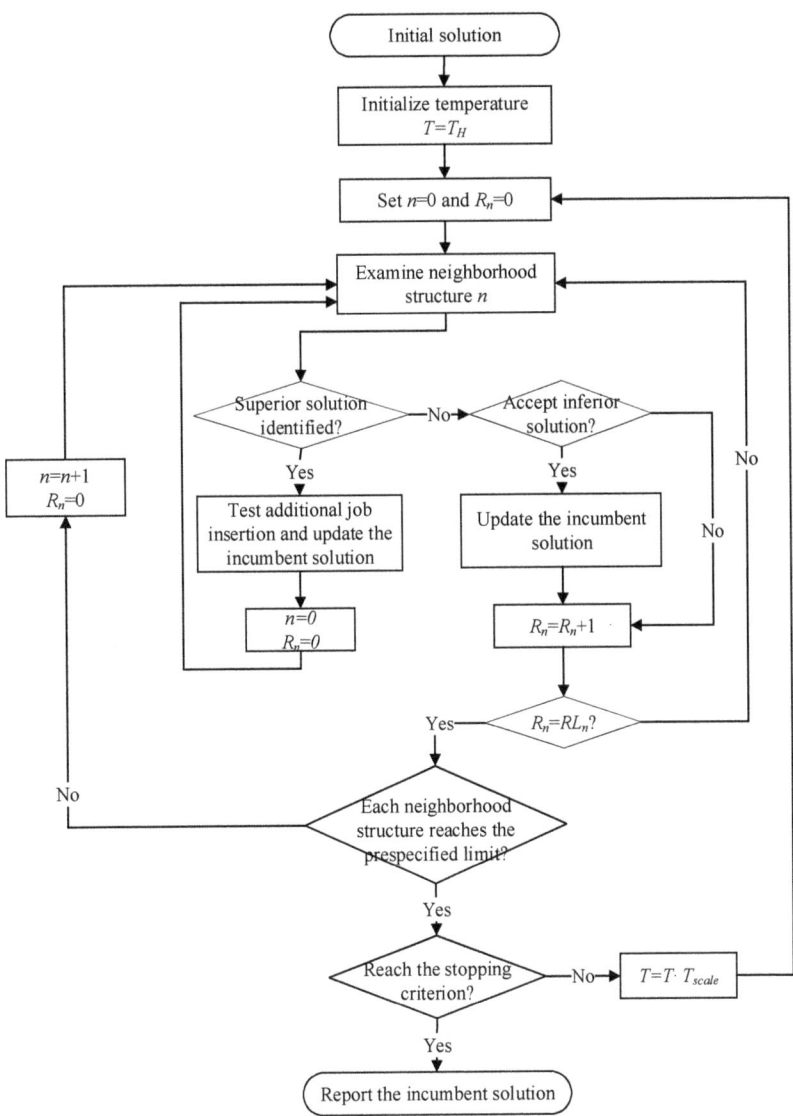

Figure 5. Algorithmic steps of the population-based simulated annealing (PBSA) approach.

5. Solution Approach

To validate the effectiveness of the proposed solution framework and evaluate its performance, PBSA is empirically applied to problem instances of different sizes. In the experiments, the solutions from a commercial optimization package Gurobi 9.1.0 are used as the benchmark. Furthermore, we conduct a sensitivity analysis on the weights imposed in the objective function to capture the impact of these parameters. The proposed PBSA heuristic is implemented in the ANSI C++ programming language. The numerical experiments for both Gurobi and our solution method are conducted on a Windows-based machine with an Intel i7-8700 CPU at 3.20 GHz and 8 GB of memory. Note that the solutions reported for the proposed PBSA algorithm are averaged over 10 runs as the search process involves randomness.

5.1. Parameter Calibration

We first perturb the parameters α and β to evaluate their impacts on the objective value. The problem instance used contains 6 machines, 5 days for the planning horizon and 150 jobs. The results are summarized in Table 2 and Figure 6.

Table 2. The impacts of α and β on the objective value.

$\alpha:\beta$	Cmax	B/I Penalty
1:0.0001	43.26	1437.65
1:0.001	43.63	1357.95
1:0.01	44.48	1308.20
1:0.1	50.67	1303.20
1:1	57.93	1298.45

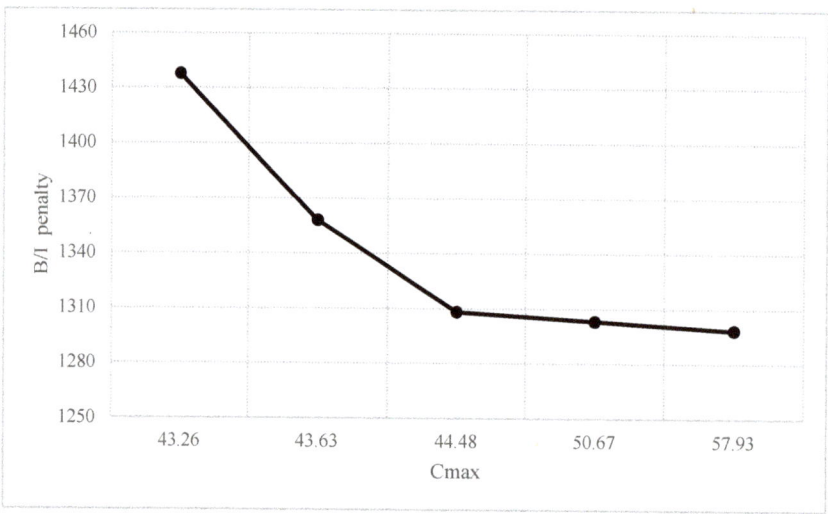

Figure 6. The tradeoff between Cmax and the burn-in (B/I) penalty.

As expected, given a fixed α, the result tends to improve the B/I penalty and worsen Cmax simultaneously when β increases. We can see the apparent tradeoff between Cmax and the B/I penalty in Figure 6. However, as $\alpha:\beta$ goes beyond 1:0.01, the improvement in the B/I penalty is only marginal and almost ignorable. Therefore, we use $\alpha = 1$ and $\beta = 0.01$ in the rest of the experiments. Furthermore, we conduct preliminary experiments with various parameter combinations of T_H, T_L, T_{scale} and RL_n using datasets obtained in practice and identify the parameter combination that performs best. The optimal values are $T_H = 25$, $T_L = 1$, $T_{scale} = 0.98$ and $RL_n = 500$. Note that the Taguchi-based method may be used to calibrate the parameters [20–22]. However, we observe that no significant improvement can be obtained by that method for our cases. Therefore, we adopt the above parameters. For practical purposes, the maximum allowed computational time is limited to 1200 s. These parameters are used throughout the remaining experiments.

5.2. Validation

For validation purposes, we compare our solutions with the Gurobi solutions, which can be considered the optimal solutions. Note that Gurobi is used to solve the formulation presented in Section 3 with the default settings. The comparison is summarized in Table 3. As shown in the table, the proposed PBSA algorithm can determine the same optimal solutions as those obtained by Gurobi, demonstrating the efficacy of the proposed solution method. When the problem size increases, Gurobi fails to obtain feasible solutions for problem instances with more than 15 jobs. However, the proposed PBSA method can still determine solutions, thereby demonstrating its scalability. Furthermore, it is noted that PBSA with the two neighborhood structures explained in Figures 2 and 3 (PBSA with 2VND) has a higher probability of finding improved solutions than PBSA with all three structures (PBSA with 3VND). It is suspected that the neighborhood structure depicted in Figure 4 makes it difficult for the search process to converge to an improved solution within the limited CPU time allowed. Let us depict the convergence of the proposed algorithm using the case with L/T/N = 6/3/100 as an example in Figure 7 to further discuss the results.

Table 3. Validation of the proposed PBSA algorithm.

L/T/N [1]	Gurobi		PBSA with 3VND [4]			PBSA with 2VND [5]		
	Objective Value	CPU [2] (s)	Objective Value	Cmax	B/I Penalty	Objective Value	Cmax	B/I Penalty
3/2/10	23.59	1.47	23.59	14.44	915.50	23.59	14.44	915.50
3/2/15	31.74	3614.93	31.74	20.60	1113.75	31.74	20.60	1113.75
4/2/10	19.12	0.14	19.12	9.97	915.50	19.12	9.97	915.50
4/2/15	27.95	10,790.65	27.95	16.81	1113.75	27.95	16.81	1113.75
4/2/20	*[3]	*	30.01	20.22	978.75	**30.00**	20.22	978.75
4/2/50	*	*	45.86	25.34	2052.25	**45.60**	25.23	2036.75
4/2/100	*	*	**58.62**	36.52	2209.50	59.69	36.86	2283.75
4/3/100	*	*	54.14	40.00	1414.00	**54.05**	39.91	1414.00
5/3/100	*	*	48.34	34.18	1416.00	**48.07**	33.88	1418.50
6/3/150	*	*	71.38	50.50	2088.00	**70.57**	50.24	2032.75
6/4/150	*	*	**79.33**	54.67	2466.00	79.34	54.68	2466.00
6/3/200	*	*	74.06	51.69	2236.50	**68.49**	49.70	1878.75
6/4/200	*	*	90.95	65.84	2511.00	**88.81**	63.70	2511.00
6/5/200	*	*	**121.60**	78.51	4308.75	122.23	78.62	4361.25

[1] L: number of production lines; T: number of scheduling days in a week; N: number of jobs. [2] CPU: computational time. [3]*: fails to determine solutions within 8 h. [4] 3VND: all three neighborhood structures explained in Figures 2–4. [5] 2VND: the neighborhood structures explained in Figures 2 and 3.

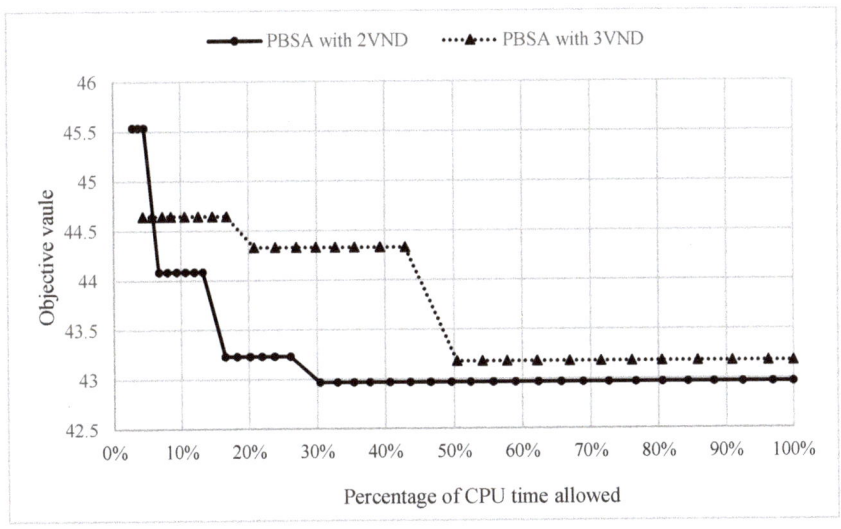

Figure 7. Convergence of the algorithm.

As the computational time of each iteration required for PBSA with 3VND increases, the number of iterations that can be performed within the limited CPU time decreases, resulting in poor convergence when compared with that of PBSA with 2VND. Although PBSA can potentially explore large solution spaces, we still recommend PBSA with 2VND for practical purposes. As the actual production environment is rather dynamic, faster convergence to an ideal solution seems to be the most attractive option for most of the companies we encounter.

We next evaluate the impact of the numbers of production lines and scheduling days on the performance of the proposed algorithm. As Gurobi failed to determine feasible solutions in most of the cases, we only summarize our solutions in Figure 8.

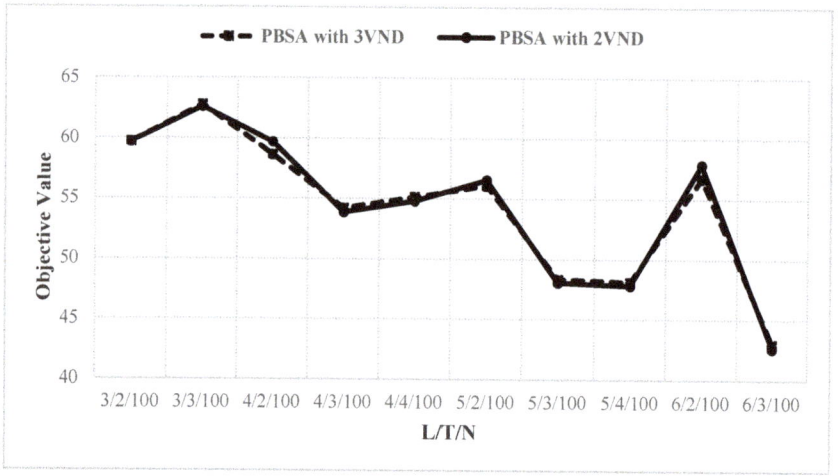

Figure 8. Impacts of the numbers of production lines and scheduling days.

Nevertheless, PBSA with 2VND outperforms PBSA with 3VND in most cases. Note that, contrary to the mathematical formulation that assumes all jobs should be assigned, the proposed PBSA algorithm has the flexibility to allow for unscheduled jobs if some jobs cannot be inserted into the production schedule. For this reason, the objective values do not increase/decrease monotonically with increases in the numbers of production lines and scheduling days.

5.3. Practical Application Scenarios

Using the proposed PBSA algorithm with 2VND, we experiment with three application scenarios, namely, scenarios with various B/I ratios (summarized in Table 4), additional constraints (summarized in Table 5) and superhot runs.

From Table 4, we can see that the B/I ratio of 5:5:0 dominates 5:4:1 and 6:3:1 in terms of the objective values and number of scheduled jobs (NSJ) yielded. In the case where L/T/N = 6/3/100, the B/I penalty can reach zero, indicating that the ratio perfectly matches 5:5:0. In other words, it is possible to fully utilize the expensive B/I equipment, should we adjust the ratio properly.

In real-world applications, many supervisors may ask for additional constraints based on their past experiences, as they believe that the additional constraints can improve the scheduling performance of the algorithm. In Table 5, we introduce the additional constraint to fully utilize long production lines as much as possible, as many supervisors may think continuous processing without interruption can reduce the setup times and can improve the overall performance. However, the effect of this additional constraint drastically reduces the number of jobs that can be scheduled with the same setup and contradicts their intuition. Therefore, it is suggested to consider only the necessary constraints for practical purposes.

Table 4. The impacts of various B/I ratios.

	5:4:1				6:3:1				5:5:0			
L/T/N	Obj. Value	Cmax	B/I Penalty	NSJ [1]	Obj. Value	Cmax	B/I Penalty	NSJ	Obj. Value	Cmax	B/I Penalty	NSJ
3/2/100	59.70	39.45	2025.00	75	66.89	39.44	2745.00	75	39.54	39.51	3.00	76
3/3/100	62.62	42.37	2025.00	76	69.88	42.43	2745.00	76	42.63	42.52	11.00	77
4/2/100	59.69	36.86	2283.75	63	62.28	37.23	2505.00	65	38.51	37.96	55.00	68
4/3/100	54.05	39.91	1414.00	99	62.11	39.19	2292.00	99	39.86	39.78	8.00	100
4/4/100	54.78	40.64	1414.00	100	62.47	39.55	2292.00	99	39.82	39.80	2.00	100
5/2/100	56.55	36.82	1973.25	64	65.94	38.12	2782.00	64	38.31	37.73	58.00	66
5/3/100	48.07	33.88	1418.50	100	55.28	32.36	2292.00	99	33.34	33.06	28.00	100
5/4/100	47.82	33.68	1414.00	100	55.35	32.43	2292.00	99	33.60	33.32	28.00	100
6/2/100	57.84	35.75	2209.25	62	63.05	35.87	2718.00	65	37.00	36.64	36.00	66
6/3/100	42.61	28.46	1415.50	100	51.47	28.55	2292.00	100	29.06	29.06	0.00	100

[1] NSJ: number of scheduled jobs.

Table 5. Impacts of additional constraints.

	Original Problem				Problem with an Additional Constraint			
L/T/N	Obj. Value	Cmax	B/I Penalty	NSJ	Obj. Value	Cmax	B/I Penalty	NSJ
3/2/100	59.70	39.45	2025.00	75	55.79	38.47	1732.50	57
3/3/100	62.62	42.37	2025.00	76	55.79	38.47	1732.50	57
4/2/100	59.69	36.86	2283.75	63	55.23	37.12	1811.25	58
4/3/100	54.05	39.91	1414.00	99	51.55	38.95	1260.00	88
4/4/100	54.78	40.64	1414.00	100	50.34	37.74	1260.00	88
5/2/100	56.55	36.82	1973.25	64	58.38	34.03	2434.25	59
5/3/100	48.07	33.88	1418.50	100	46.44	32.81	1363.00	97
5/4/100	47.82	33.68	1414.00	100	47.21	33.58	1363.50	97
6/2/100	57.84	35.75	2209.25	62	51.85	30.23	2161.50	60
6/3/100	42.61	28.46	1415.50	100	44.28	30.20	1408.50	99

To manufacture products with a sudden surge in demand, customers may place urgent orders. For products such as this, the manufacturer may initiate a superhot run and charge a higher price for manufacturing the product. For such urgent orders, we believe that the proposed solution framework offers a possible pricing mechanism by charging based on the impact of the order on the original schedule. We illustrate this concept by inserting three urgent orders in the cases summarized in Table 6.

Table 6. The impact of superhot runs.

	Original Order				Original Order with Superhot Runs			
L/T/N	Obj. Value	Cmax	B/I Penalty	NSJ [1]	Obj. Value	Cmax	B/I Penalty	NSJ
6/4/200	88.81	63.70	2511.00	164	89.16	62.16	2699.50	160
6/5/200	122.23	78.62	4361.25	184	137.09	83.96	5312.50	179

[1] NSJ: number of scheduled jobs.

As seen from the cases, there are 4 and 5 jobs that cannot be completed within this planning horizon due to the superhot runs. The impacts of the superhot runs on the rest of the jobs are 2.5% and 2.7%, respectively, so this provides an ideal guide for pricing based on superhot runs.

6. Concluding Remarks

In this study, we developed a PBSA algorithm for the unrelated parallel machine scheduling problem considering B/I constraints. The proposed PBSA algorithm integrates the advantages of SA and VND and was implemented for practical applications. Numerical results have shown that the proposed PBSA approach can solve the abovementioned problem optimally for small problem instances and is scalable to solve problem instances of realistic sizes. In practice, there are many industries that encounter the unrelated parallel machine scheduling problem (i.e., the electronic assembly industry investigated in the current study). For the unrelated parallel machine scheduling problem, we found that the PBSA and VND can solve problems of various practical sizes efficiently.

Although encouraging results are obtained, this research can be extended in several directions. First, it is apparent that this research can be extended to investigate the multi-objective optimization problem since there are two objective functions considered. Some practical techniques can be employed to approximate the Pareto-optimal solution set (i.e., [43]). Second, as each job can be split into smaller jobs, we can find the optimal splitting strategy for the jobs such that the overall scheduling performance can be improved. Third, as the production environment is highly uncertain and stochastic, processing time uncertainty can be incorporated in future research. Fourth, instead of minimizing the deviation from the desired burn-in levels, future research can consider the optimal assignment of jobs to the exact grids in burn-in machines, which may potentially improve the scheduling results further. Finally, as the production environment changes rapidly, developing a system that can rapidly respond to the dynamic and changing environment with the basis of this study can be another useful and interesting extension.

Author Contributions: Data curation, T.-Y.H.; Investigation, D.-Y.L.; Methodology, D.-Y.L. and T.-Y.H.; Software, T.-Y.H.; Validation, D.-Y.L.; Writing—original draft, D.-Y.L. All authors have read and agreed to the published version of the manuscript.

Funding: This research was partially funded by the Ministry of Science and Technology, Taiwan, ROC grant number MOST 108-2410-H-007-097 -MY4.

Institutional Review Board Statement: Not applicable.

Informed Consent Statement: Not applicable.

Data Availability Statement: Data sharing not applicable.

Conflicts of Interest: The authors declare no conflict of interest

Abbreviations

B/I	burn-in
PBSA	population-based simulated annealing
VND	variable neighborhood descent
GA	genetic algorithm
SA	simulated annealing
LS	local search
ERD	earlier release date first
AIS	artificial immune system
PSO	particle swarm optimization
DWPSA	delay window-time parallel saving algorithm
DWGSA	delay window-time generalized saving algorithm
FJSP	flexible job shop scheduling problem
GDS	general dense scheduling
ACO	colony optimization
SPT	shortest processing time
MPS	master production schedule
DPS	daily production scheduling problem
NSJ	number of scheduled jobs

References

1. Karp, R.M. Reducibility among combinatorial problems. In *Complexity of Computer Computations*; Springer: Berlin/Heidelberg, Germany, 1972; pp. 85–103.
2. Che, A.; Wu, X.Q.; Peng, J.; Yan, P.Y. Energy-efficient bi-objective single-machine scheduling with power-down mechanism. *Comput. Oper. Res.* **2017**, *85*, 172–183. [CrossRef]
3. Fridman, I.; Pesch, E.; Shafransky, Y. Minimizing maximum cost for a single machine under uncertainty of processing times. *Eur. J. Oper. Res.* **2020**, *286*, 444–457. [CrossRef]
4. Davari, M.; Ranjbar, M.; De Causmaecker, P.; Leus, R. Minimizing makespan on a single machine with release dates and inventory constraints. *Eur. J. Oper. Res.* **2020**, *286*, 115–128. [CrossRef]
5. Kim, J.G.; Song, S.; Jeong, B. Minimising total tardiness for the identical parallel machine scheduling problem with splitting jobs and sequence-dependent setup times. *Int. J. Prod. Res.* **2020**, *58*, 1628–1643. [CrossRef]
6. Soares, L.C.R.; Carvalho, M.A.M. Biased random-key genetic algorithm for scheduling identical parallel machines with tooling constraints. *Eur. J. Oper. Res.* **2020**, *285*, 955–964. [CrossRef]
7. Lin, S.W.; Ying, K.C. Uniform Parallel-Machine Scheduling for Minimizing Total Resource Consumption With a Bounded Makespan. *IEEE Access* **2017**, *5*, 15791–15799. [CrossRef]
8. Holland, J.H. *Adaptation in Natural and Artificial Systems. An Introductory Analysis with Applications to Biology, Control and Artificial Intelligence*; ANAS: Roma, Italy, 1975.
9. Vallada, E.; Ruiz, R. A genetic algorithm for the unrelated parallel machine scheduling problem with sequence dependent setup times. *Eur. J. Oper. Res.* **2011**, *211*, 612–622. [CrossRef]
10. Joo, C.M.; Kim, B.S. Hybrid genetic algorithms with dispatching rules for unrelated parallel machine scheduling with setup time and production availability. *Comput. Ind. Eng.* **2015**, *85*, 102–109. [CrossRef]
11. Cheng, C.Y.; Huang, L.W. Minimizing total earliness and tardiness through unrelated parallel machine scheduling using distributed release time control. *J. Manuf. Syst.* **2017**, *42*, 1–10. [CrossRef]
12. Yu, C.L.; Semeraro, Q.; Matta, A. A genetic algorithm for the hybrid flow shop scheduling with unrelated machines and machine eligibility. *Comput. Oper. Res.* **2018**, *100*, 211–229. [CrossRef]
13. Jouhari, H.; Lei, D.M.; Al-qaness, M.A.A.; Abd Elaziz, M.; Ewees, A.A.; Farouk, O. Sine-Cosine Algorithm to Enhance Simulated Annealing for Unrelated Parallel Machine Scheduling with Setup Times. *Mathematics* **2019**, *7*, 1120. [CrossRef]
14. Santos, H.G.; Toffolo, T.A.M.; Silva, C.; Vanden Berghe, G. Analysis of stochastic local search methods for the unrelated parallel machine scheduling problem. *Int. Trans. Oper. Res.* **2019**, *26*, 707–724. [CrossRef]
15. Cota, L.P.; Haddad, M.N.; Souza, M.J.F.; Coelho, V.N. AIRP: A heuristic algorithm for solving the unrelated parallel machine scheduling problem. In Proceedings of the 2014 IEEE Congress on Evolutionary Computation (Cec), Beijing, China, 6–11 July; pp. 1855–1862.
16. Wang, X.M.; Li, Z.T.; Chen, Q.X.; Mao, N. Meta-heuristics for unrelated parallel machines scheduling with random rework to minimize expected total weighted tardiness. *Comput. Ind. Eng.* **2020**, *145*, 106505. [CrossRef]
17. Hansen, P.; Mladenović, N. An introduction variable neighborhood search. In *Meta-Heuristics*; Springer: Berlin/Heidelberg, Germany, 1999; pp. 433–458.
18. Fleszar, K.; Charalambous, C.; Hindi, K.S. A variable neighborhood descent heuristic for the problem of makespan minimisation on unrelated parallel machines with setup times. *J. Intell. Manuf.* **2011**, *23*, 1949–1958. [CrossRef]
19. Al-Harkan, I.M.; Qamhan, A.A. Optimize Unrelated Parallel Machines Scheduling Problems With Multiple Limited Additional Resources, Sequence-Dependent Setup Times and Release Date Constraints. *IEEE Access* **2019**, *7*, 171533–171547. [CrossRef]
20. Marinho Diana, R.O.; de Souza, S.R. Analysis of variable neighborhood descent as a local search operator for total weighted tardiness problem on unrelated parallel machines. *Comput. Oper. Res.* **2020**, *117*. [CrossRef]
21. Afzalirad, M.; Rezaeian, J. Resource-constrained unrelated parallel machine scheduling problem with sequence dependent setup times, precedence constraints and machine eligibility restrictions. *Comput. Ind. Eng.* **2016**, *98*, 40–52. [CrossRef]
22. Mir, M.S.S.; Rezaeian, J. A robust hybrid approach based on particle swarm optimization and genetic algorithm to minimize the total machine load on unrelated parallel machines. *Appl. Soft. Comput.* **2016**, *41*, 488–504. [CrossRef]
23. Fleszar, K.; Hindi, K.S. Algorithms for the unrelated parallel machine scheduling problem with a resource constraint. *Eur. J. Oper. Res.* **2018**, *271*, 839–848. [CrossRef]
24. He, Y.H.; Wang, L.B.; Wei, Y.; He, Z.Z. Optimisation of burn-in time considering the hidden loss of quality deviations in the manufacturing process. *Int. J. Prod. Res.* **2017**, *55*, 2961–2977. [CrossRef]
25. Aghaee, N.; Peng, Z.B.; Eles, P. Temperature-Gradient-Based Burn-In and Test Scheduling for 3-D Stacked ICs. *IEEE Trans. Very Large Scale Integr. (VlSl) Syst.* **2015**, *23*, 2992–3005. [CrossRef]
26. Kim, Y.D.; Kang, J.H.; Lee, G.E.; Lim, S.K. Scheduling Algorithms for Minimizing Tardiness of Orders at the Burn-in Workstation in a Semiconductor Manufacturing System. *IEEE Trans. Semicond. Manuf.* **2011**, *24*, 14–26. [CrossRef]
27. Lee, C.Y.; Uzsoy, R.; Martinvega, L.A. Efficient Algorithms for Scheduling Semiconductor Burn-in Operations. *Oper. Res.* **1992**, *40*, 764–775. [CrossRef]
28. Pearn, W.L.; Hong, J.S.; Tai, Y.T. The burn-in test scheduling problem with batch dependent processing time and sequence dependent setup time. *Int. J. Prod. Res.* **2013**, *51*, 1694–1706. [CrossRef]

29. Ruiz, R.; Stutzle, T. A simple and effective iterated greedy algorithm for the permutation flowshop scheduling problem. *Eur. J. Oper. Res.* **2007**, *177*, 2033–2049. [CrossRef]
30. Wang, I.L.; Yang, T.H.; Chang, Y.B. Scheduling two-stage hybrid flow shops with parallel batch, release time, and machine eligibility constraints. *J. Intell. Manuf.* **2012**, *23*, 2271–2280. [CrossRef]
31. Komaki, G.M.; Teymourian, E.; Kayvanfar, V. Minimising makespan in the two-stage assembly hybrid flow shop scheduling problem using artificial immune systems. *Int. J. Prod. Res.* **2016**, *54*, 963–983. [CrossRef]
32. Costa, A.; Cappadonna, F.A.; Fichera, S. A hybrid genetic algorithm for minimizing makespan in a flow-shop sequence-dependent group scheduling problem. *J. Intell. Manuf.* **2017**, *28*, 1269–1283. [CrossRef]
33. Ying, K.C.; Lin, S.W. Minimizing Makespan in Distributed Blocking Flowshops Using Hybrid Iterated Greedy Algorithms. *IEEE Access* **2017**, *5*, 15694–15705. [CrossRef]
34. Kundakci, N.; Kulak, O. Hybrid genetic algorithms for minimizing makespan in dynamic job shop scheduling problem. *Comput. Ind. Eng.* **2016**, *96*, 31–51. [CrossRef]
35. Li, X.Y.; Gao, L. An effective hybrid genetic algorithm and tabu search for flexible job shop scheduling problem. *Int. J. Prod. Econ.* **2016**, *174*, 93–110. [CrossRef]
36. Ding, H.J.; Gu, X.S. Improved particle swarm optimization algorithm based novel encoding and decoding schemes for flexible job shop scheduling problem. *Comput. Oper. Res.* **2020**, *121*, 15. [CrossRef]
37. Malega, P.; Rudy, V.; Kanasz, R.; Gazda, V. Decentralized optimization of the flexible production lines. *Adv. Prod. Eng. Manag.* **2020**, *15*, 267–276. [CrossRef]
38. Bai, D.Y.; Zhang, Z.H.; Zhang, Q. Flexible open shop scheduling problem to minimize makespan. *Comput. Oper. Res.* **2016**, *67*, 207–215. [CrossRef]
39. Ciro, G.C.; Dugardin, F.; Yalaoui, F.; Kelly, R. Open shop scheduling problem with a multi-skills resource constraint: A genetic algorithm and an ant colony optimisation approach. *Int. J. Prod. Res.* **2016**, *54*, 4854–4881. [CrossRef]
40. Hosseinabadi, A.A.R.; Vahidi, J.; Saemi, B.; Sangaiah, A.K.; Elhoseny, M. Extended Genetic Algorithm for solving open-shop scheduling problem. *Soft Comput.* **2019**, *23*, 5099–5116. [CrossRef]
41. Kirkpatrick, S.; Gelatt, C.D.; Vecchi, M.P. Optimization by Simulated Annealing. *Science* **1983**, *220*, 671–680. [CrossRef] [PubMed]
42. Metropolis, N.; Rosenbluth, A.W.; Rosenbluth, M.N.; Teller, A.H.; Teller, E. Equation of State Calculations by Fast Computing Machines. *J. Chem. Phys.* **1953**, *21*, 1087–1092. [CrossRef]
43. Lin, D.Y.; Xie, C. The Pareto-optimal Solution Set of the Equilibrium Network Design Problem with Multiple Commensurate Objectives. *Netw Spat. Econ.* **2011**, *11*, 727–751. [CrossRef]

Article

No-Idle Flowshop Scheduling for Energy-Efficient Production: An Improved Optimization Framework

Chen-Yang Cheng [1,†], Shih-Wei Lin [2,3,4,†], Pourya Pourhejazy [1,†], Kuo-Ching Ying [1,*,†] and Yu-Zhe Lin [1,5]

[1] Department of Industrial Engineering and Management, National Taipei University of Technology, Taipei 106, Taiwan; cycheng@ntut.edu.tw (C.-Y.C.); pourya@ntut.edu.tw (P.P.); s86445710993@gmail.com (Y.-Z.L.)
[2] Department of Information Management, Chang Gung University, Taoyuan 333, Taiwan; swlin@mail.cgu.edu.tw
[3] Department of Neurology, Linkou Chang Gung Memorial Hospital, Taoyuan 333, Taiwan
[4] Department of Industrial Engineering and Management, Ming Chi University of Technology, New Taipei 243, Taiwan
[5] Taiwan Semiconductor Manufacturing Company Limited, Hsinchu Science Park, Hsinchu 30078, Taiwan
* Correspondence: kcying@ntut.edu.tw
† These authors contributed equally to this work; Shih-Wei Lin is the co-first author.

Abstract: Production environment in modern industries, like integrated circuits manufacturing, fiberglass processing, steelmaking, and ceramic frit, is characterized by zero idle-time between inbound and outbound jobs on every machine; this technical requirement improves energy efficiency, hence, has implications for cleaner production in other production situations. An exhaustive review of literature is first conducted to shed light on the development of no-idle flowshops. Considering the intractable nature of the problem, this research also develops an extended solution method for optimizing the Bi-objective No-Idle Permutation Flowshop Scheduling Problem (BNIPFSP). Extensive numerical tests and statistical analysis are conducted to evaluate the developed method, comparing it with the best-performing algorithm developed to solve the BNIPFSP. Overall, the proposed extension outperforms in terms of solution quality at the expense of a longer computational time. This research is concluded by providing suggestions for the future development of this understudied scheduling extension.

Keywords: production management; energy-efficiency; scheduling; no-idle flowshop; metaheuristics

Citation: Cheng, C.-Y.; Lin, S.-W.; Pourhejazy, P.; Ying, K.-C.; Lin, Y.-Z. No-Idle Flowshop Scheduling for Energy-Efficient Production: An Improved Optimization Framework. *Mathematics* **2021**, *9*, 1335. https://doi.org/10.3390/math9121335

Academic Editors: Chin-Chia Wu and Win-Chin Lin

Received: 4 May 2021
Accepted: 7 June 2021
Published: 9 June 2021

Publisher's Note: MDPI stays neutral with regard to jurisdictional claims in published maps and institutional affiliations.

Copyright: © 2021 by the authors. Licensee MDPI, Basel, Switzerland. This article is an open access article distributed under the terms and conditions of the Creative Commons Attribution (CC BY) license (https://creativecommons.org/licenses/by/4.0/).

1. Introduction

Ecological restoration and reduced carbon emission have become major global priorities [1]. Local governments have put forward regulatory measures and policies to enforce energy-saving initiatives. These measures are predominantly formed around emission taxation and trading of emission credits, which help bring the overall emissions below the target baseline [2]. The Australian carbon reduction policy, the so-called safeguard mechanism, and the EU Emission Trading System are prime examples of reducing the negative impacts of business activities from electricity generation and mining to transportation, construction, and manufacturing.

The manufacturing sector is one of the primary energy consumers and the largest polluter with its share being more than 31 percent of the overall energy consumption and 36 percent of carbon dioxide emissions [3]. To address this issue, supply chain sustainability, in particular, the green process design practices, has been mainly focused on reducing energy consumption in logistics [4], production, and consumption phases as well as the use of renewable energies [5]. Providing on-site energy production, like solar panels and biogas fuel cells, reducing facilities' carbon footprint by replacing lighting and energy control systems, applying energy efficiency standards in the construction of new buildings, and the installation of modern supplements for the use of sustainable resources is the primary green practices reported in the literature [6].

Minimizing the costs associated with the machines' energy consumption and the resulting pollutants have been at the center of the green manufacturing studies. Considering non-processing and processing energy consumption in the production facilities [7], energy-efficiency has been explored from the operational management perspective, i.e., how to cluster jobs to minimize non-value-adding operations [8] and when to turn on/off to reduce machines idle-time, which speed level to operate, and how to plan peak and off-peak production process to save energy [9]; Production scheduling as an operational strategic tool is complex and requires additional measures to account for less tangible operational aspects.

The operations-related performance measures, more particularly those pertinent to processing energy consumption, have been the subject of many scheduling studies to account for sustainability in the production management context. Piroozfard et al. [10] introduced a multi-objective flexible job-shop scheduling problem, minimizing carbon footprint and the total late work criterion. Minimizing the makespan and total carbon emission in production environments with unrelated parallel machines was examined by Zheng and Wang [11]. Safarzadeh and Niaki [12] addressed the total green cost and the makespan finding the Pareto optimal solutions in uniform parallel machine environments. The trade-off between makespan and energy consumption in two-machine flowshop [13], hybrid flowshop [14], and unrelated parallel machine [15], and job-shop scheduling environments [16] are among the other notable contributions at the intersection of energy-efficiency and production schedule. These studies aimed to improve energy efficiency through a soft optimization approach focusing on minimizing processing costs and energy consumption. That is, a trade-off enables the decision-makers to choose between cost-effectiveness or responsiveness and energy efficiency. Although such a flexible approach is suitable in the current regulatory situation, plausibly more restricted regulations in the future urge optimization approaches that minimize non-processing energy consumption considering operational strategic measures and energy cost strategies [17]. Li et al. [18] suggested defining a limitation on the energy consumption of each machine while minimizing the makespan and the total completion time. Scheduling problems with the no-idle time between the in-coming and out-going jobs on the machines is an alternative solution to effectively reduce energy wastage in the production sector. On the other hand, the technical characteristics of modern industries, like steelmaking [19], integrated circuits manufacturing, fiberglass processing, and ceramic frit [20] require a no-idle situation. Given flowshop production as the most common process model in the manufacturing sector [21] and the significance of energy costs in the flowshops, no-idle flowshop scheduling has received recent recognition among production management scholars.

The successful implementation of policy-driven mechanisms for mandating carbon emissions depends on the effective consideration of the corporate priorities, like cost-effectiveness and responsiveness, to ensure the firms' competitiveness [22]. This situation is of high significance to address conflicting operational objectives within the no-idle production scheduling agenda that enforces maximal energy efficiency. To the best of the author's knowledge, no published journal papers have addressed the bi-objective optimization of no-idle flowshops. This study extends the energy-efficient production scheduling literature by a two-fold contribution. First, an exhaustive review of the no-idle flowshop scheduling literature is conducted to explore the developments and gaps in modern industry scheduling. Second, a Hybrid Iterated Greedy (HIG) algorithm is developed to effectively solve the bi-objective variant of no-idle flowshops while ensuring the robustness of the outcomes. The three-field $\alpha|\beta|\gamma$ notation of Graham et al. [23] is used for referring to the Bi-objective No-Idle Permutation Flowshop Scheduling Problem (BNIPFSP) as $F_m|prmu, no-idle|\alpha \cdot C_{max} + \beta \cdot \sum F_j$ in the remainder of this article. In this notation system, F_m shows the flowshop production environment with the set of given jobs being processed by a set of available machines in the same order. In the second part of the notation, *prmu* determines the permutation setting to show that the sequence of jobs is the same on all machines, and specifies that there is no idle time between inbound and outbound jobs on every machine. Finally, $\alpha \cdot C_{max} + \beta \cdot \sum F_j$ determines the weighted sum of makespan and total flowtime criteria.

The rest of this manuscript is organized into four sections. A comprehensive review of the literature is provided in Section 2. The methodology, including the extended mathematical formulation and the solution algorithm, is elaborated in Section 3. The numerical analysis comes next, in Section 4, to analyze the effectiveness of the developed solution approach. Finally, concluding remarks and directions for future research on no-idle scheduling close this research work in Section 5.

2. Literature Review

Considering the recent surge in the number of articles, a comprehensive review on no-idle flowshop scheduling and its solution methods is timely. This section reviews the published works indexed in Google Scholar. For this purpose, searching the keywords "no-idle" and "Flowshop" resulted in a total of 33 articles among which, 25 were perceived as relevant; of the relevant items, five conference papers [24–28] and two theses [29,30] were found. The journal articles are then analyzed considering the number of machines, the studied performance indicator, and the proposed solution approach suggested by Ribas et al. [31] and Neufeld et al. [32].

Computers and Operations Research and Expert Systems with Applications contributed the most to this extension of scheduling problems with two published works. With five contributions, Tasgetiren is the most prominent author, followed by Rossi with three published works. Notably, half of the contributions in no-idle flowshop scheduling are published in or after 2019, all of which are explored in the production context. A summary of the published works is provided in Appendix A with the detailed review elaborated below.

No-idle scheduling was the first time introduced by Cepek et al. [33,34] to minimize total completion time in a two-machine flowshop production environment. This seminal scheduling problem inspired more than 20 research contributions thus far, contributing to solution algorithms and/or new mathematical extensions to No-Idle Permutation Flowshop Scheduling Problem (NIPFSP). Narain and Bagga [35] developed a Branch and Bound solution method to minimize the average flowtime in a two-machine flowshop environment. Wang et al. [36] incorporated no-wait job-related constraints into the no-idle flexible flowshops. Later studies were focused on flowshop settings with m machines. Tasgetiren et al. [37] and [38] developed Differential Evolution and Discrete Artificial Bee Colony algorithms, respectively, to minimize total tardiness in NIPFSP. Tabu Search algorithm was later adopted by Ren et al. [39] to minimize the maximum completion time (makespan). Tasgetiren et al. [40] proposed a hybrid Differential Evolution and variable local search, which improved the makespan values obtained by the earlier studies.

More recent studies are rather focused on proposing novel methods and variants in the scheduling procedure. Lu [41] explored the time-dependent learning effect and deteriorating jobs in NIPFSP, minimizing the makespan criterion. Pagnozzi and Stützle [42] developed an automatic algorithm configuration approach for solving single-objective permutation flowshops. The mixed-no-idle flowshop variant was introduced by Pan and Ruiz [43] to minimize makespan using a basic Iterated Greedy (IG) algorithm. Rossi and Nagano [44,45] explored the mixed-no-idle and sequence-dependent setup time settings and minimized total flowtime using Beam Search algorithms. The same authors developed a constructive heuristic for mixed-NIPFSP with sequence-dependent setup times [46]. In a similar contribution, Nagano et al. [47] developed a constructive heuristic to solve the basic NIPFSP considering total flowtime. Zhao et al. [48] and Riahi et al. [49] developed Discrete Water Wave Optimization (DWWO) and IG, respectively, for minimizing total tardiness in NIPFSP. Benders decomposition was also tested to solve mixed-no-idle flowshops considering the makespan criterion [20]. Most recently, Zhao et al. [50] proposed a new variant to the DWWO algorithm to solve distributed assembly no-idle flowshop scheduling problems considering maximum assembly completion time. Despite its usefulness, no published journal papers are found that addresses the bi-objective variant of NIPFSPs. Motivated by this gap, we propose a new formulation and solution algorithm to contribute to energy-efficient production scheduling using bi-objective no-idle flowshops.

3. Methods

3.1. Mathematical Formulation

This study extends the Mixed-Integer Programming (MIP) formulation developed by Ruiz and Stützle [51] to account for two conflicting optimization objectives, i.e., maximum completion time (makespan) and total flowtime. The former is a measure to enhance resource utilization, while the latter measure minimizes work-in-process inventory. The indices, parameters, and decision variables listed in Table 1 are used to model the $F_m|prmu, no-idle|\alpha \cdot C_{\max} + \beta \cdot \sum F_j$ scheduling problem.

Table 1. Mathematical notations.

Symbol	Definition
n	Number of jobs at hand
m	Number of available machines
j, k	Job tag and its position index in the sequence vector, i.e., $\pi[j]; j, k \in \{0, 1, 2, \ldots, n\}$
i	Machine tag; $i \in \{1, 2, \ldots, m\}$
$P_{j,i}$	Processing time of job j on machine i
$X_{j,k}$	Binary decision variable, $= 1$ if job j is positioned at index k of the vector; $= 0$, otherwise
$C_{k,i}$	Integer decision variable, the completion time of the job assigned to position k on machine i
F_j	The total flowtime of job j

We now elaborate on the MIP formulation of the $F_m|prmu, no-idle|\alpha \cdot C_{\max} + \beta \cdot \sum F_j$ problem. The objective function in

$$\text{Minimize } z = \alpha \cdot C_{\max} + \beta \cdot \sum_{j=0}^{n} F_j \quad (1)$$

minimizes the weighted sum of the makespan and total flowtime values, which are commensurable. The former part of the objective function, $C_{\max}$, will be calculated using the no-idle calculation mechanism presented in the following sub-section, and the latter part, $\sum_{j=0}^{n} F_j$, is determined through the constraint calculations. The objective function is subject to the constraints below. Binary decision variables are used in:

$$\sum_{k=1}^{n} X_{j,k} = 1, \forall j \in \{1, 2, \ldots, n\} \quad (2)$$

where index k represents $\pi[j]$ for the sake of readability. This constraint is defined to restrict the jobs from being assigned to more than one machine. Besides, each job should occupy one and only one position in the job sequence, as demonstrated in:

$$\sum_{j=1}^{n} X_{j,k} = 1, \forall k \in \{1, 2, \ldots, n\} \quad (3)$$

The completion time of the job in position k on the machine i must be greater than or equal to the completion time of the job on the previous machine, i.e., $i-1$, plus the processing time of the same job on the machine i. These are modeled in:

$$C_{k,1} \geq \sum_{j=1}^{n} X_{j,k} \cdot P_{j,1}, \forall k \in \{1, 2, \ldots, n\} \quad (4)$$

$$C_{k,i} \geq C_{k,i-1} + \sum_{j=1}^{n} X_{j,k} \cdot P_{j,i}, \forall k \in \{1, 2, \ldots, n\}, i \in \{2, \ldots, m\} \tag{5}$$

where the former equation refers to the first machine, and the latter equation is defined for the rest of the machinery. Similarly, the completion time of a job should correspond to that of the earlier job on the same machine in:

$$C_{k,i} \geq C_{l,i} + \sum_{j=1}^{n} X_{j,k} \cdot P_{j,i}; i \in \{1, 2, \ldots, m\}, \forall k \in \{2, \ldots, n\}, \forall l \in \{1, \ldots, k-1\} \tag{6}$$

where the time of the job placed at the position k of the job sequence vector on machine i corresponds to that of its immediate earlier job at the position $k-1$ on machine i. On this basis, the completion time of the job processed on the last machine considering its flowtime is defined in:

$$\sum_{k=1}^{n} C_{k,i} \cdot X_{j,k} = F_j, \forall i = m, \forall j \in \{1, 2, \ldots, n\} \tag{7}$$

where the flowtime value in the objective function is defined. Finally, the variable types are demonstrated in:

$$\begin{aligned} X_{j,k} \in \{0,1\}, \forall j,k \in \{1, 2, \ldots, n\} \\ C_{k,i} \geq 0, \forall k \in \{1, 2, \ldots, n\}, i \in \{1, 2, \ldots, m\} \\ F_j \geq 0, \forall j \in \{1, 2, \ldots, n\} \end{aligned} \tag{8}$$

where the completion and total flowtime variables cannot accept negative values, and the job position variable only accepts binary values.

3.2. No-Idle Calculation Mechanism

To ensure that there is zero idle time throughout the production process, one should regulate each machine's first job's commencement. For this purpose, each machine's start time, S_i, is defined in:

$$S_i = S_{i-1} + \max_{h=1:n} \left(\sum_{j=1}^{h} P_{\pi[j],i-1} - \sum_{k=1}^{h-1} P_{\pi[j],i} \right), i \in \{1, \ldots, m\}, S_1 = 0 \tag{9}$$

where $P_{\pi[j],i}$ represents the processing time of the job assigned to the position j of the sequence vector π on machine i. S_{i-1} determines the start time on the previous machine. Once the start time of every machine is known, the completion time of the first job on the machine i can be calculated using:

$$C_{\pi[1],i} = S_i + P_{\pi[1],i}, i \in \{1, \ldots, m\} \tag{10}$$

where it equals the summation of the corresponding start time of the machine i and the processing time of the first job in the job sequence vector, $P_{\pi[1],i}$. Next, the completion time of the job assigned to the position j of the sequence vector π, which is processed on the machine i, is defined in:

$$C_{\pi[j],i} = C_{\pi[j-1],i} + P_{\pi[j],i}, j \in \{2, \ldots, n\}, i \in \{1, \ldots, m\} \tag{11}$$

where $C_{\pi[j],i}$ is equal to the completion time of job position in $j-1$ of job sequence vector π on machine i, $C_{\pi[j-1],i}$, plus the processing time of the job positioned in j, $P_{\pi[j],i}$. Finally, the makespan value is calculated using:

$$C_{\max} = C_{\pi[n],m} \tag{12}$$

where $C_{\pi[n],m}$ represents the completion time of the last job in sequence vector π, which is processed on the last machine. Therefore, $C_{\max} = S_m + \sum_{k=1}^{n} P_{k,m}$. An illustrative example is provided in Figure 1 to clarify the computational steps of calculating the completion time in the no-idle flowshop.

machine (i)	job (j)			
	1	2	3	4
1	3	6	6	5
2	4	5	6	5
3	4	5	4	6

Step 1	$S_2 = S_1 + \max\left((3-0), ([3+6]-4), ([3+6+5]-[4+5])\right)$ $= 0 + \max(3,5,5) = 5$ $S_3 = S_2 + \max\left((4-0), ([4+5]-4), ([4+5+5]-[4+5])\right)$ $= 5 + \max(4,5,5) = 10$
Step 2	$C_{1,3} = S_3 + p_{1,3}$ $= 10 + 4 = 14$
Step 3	$C_{2,3} = S_3 + \sum_{k=1}^{2} p_{k,3}$ $= 10 + 4 + 5 = 19$ $C_{4,3} = S_3 + \sum_{k=1}^{3} p_{k,3}$ $= 10 + 4 + 5 + 6 = 25$

Figure 1. Illustrative example on the calculation of the completion time in no-idle flowshops.

3.3. Solution Algorithm

The IG algorithm was introduced by Ruiz and Stützle [52] to solve permutation flowshops. The computational procedure of IG is inspired by human behavior when wanting a lot more of something in a greedy manner. The successful track record of the IG algorithms in solving flowshop problems inspired us to extend it for solving the $F_m|prmu, no-idle|\alpha \cdot C_{max} + \beta \cdot \sum F_j$ problem. The pseudocode of the HIG algorithm is provided in Figure 2, followed by the details on the major computational elements. It is worthwhile mentioning that the proposed modifications are adjustable and can be effectively adapted for other application areas.

3.3.1. Solution Initialization and Decoding

Solutions are decoded as a permutation of n numbers, each of which represent a job, with the processing sequence being similar on m machines. Taking the job sequence 3 − 6 − 2 − 4 − 5 − 1 as an example, the solution is symbolized by a vector, (3 6 2 4 5 1), where six jobs should be processed following the specified order on every machine. To generate the initial solution, the well-known constructive heuristic algorithm introduced by Nawaz, Enscore, Ham (NEH; [53]), which is known as one of the best constructive heuristics for solution initialization of the flowshop problems, is preferred to random solution generation to ensure a better initial approximation. The NEH considers average processing time as a priority rule for arranging the jobs. The destruction and construction module presented in the next sub-section uses the outcomes of NEH to improve the solution quality.

HIG (d)
1) *begin*
2) π // Initialization procedure using NEH
3) $\pi_{best} = \pi$;
4) $\pi' = \pi$;   // Destruction, construction procedure
5) $\pi'' = LS(\pi')$;   // Local Search procedure
6)   if $fitness(\pi'') < fitness(\pi')$, then
7) $\pi = \pi''$;
8)   else
9) $\Delta = fitness(\Pi^{new}) - fitness(\Pi^{best})$, $T \leftarrow \delta \times T$;
10)   if $random \leq \exp(-\Delta E / T)$, then
11) $\pi = \pi''$;
12)   *endif*
13)   *endif*
14) *return* π

Figure 2. Pseudocode of the Hybrid Iterated Greedy.

3.3.2. Destruction and Construction Methods

This study applies a random destruction method with no limits to facilitate a greater level of disturbance in the search procedure. The randomly extracted jobs, which equals the destruction count (*d*), will then be saved in a separate array to be considered in the construction procedure. A customized construction method for sorting and inserting the removed jobs is developed to improve the effectiveness of the search procedure while ensuring the feasibility of the resulting new solution. This approach is explained below with an illustrative example of this procedure provided in Figure 3.

Step 1. Remove the last job from Π and name it *a*.
Step 2. Insert *a* into Π before the last job. Name the jobs before and after *a* as $a - k$ and $a + k$, respectively.
Step 3. Remove job $a - k$ and rename it to *b*.
Step 4. Insert *b* next to the first job in *a* and name the jobs before and after *b* as $b - k$ and $b + k$, respectively.
Step 5. Insert $b - k$ right before *a*.
Step 6. Select $b + k$ and move it to the position before $a - k$.
Step 7. Select $a + k$ and move it to the position after *b*.

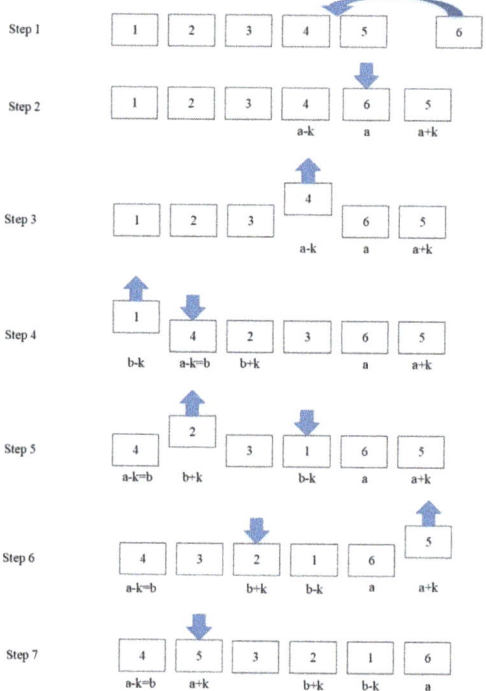

Figure 3. The customized construction method for no-idle permutation flowshops considering $k = 1$.

3.3.3. Local Search Method

After a new solution has resulted from the iterative and greedy construction procedure, a local search mechanism should be applied to search for further improvements. For this purpose, a pre-determined number of non-repetitive random job extraction and insertion, named as the local search count (γ), is used to find fitter solutions. If there is an improvement as a result of applying the local search procedure, the procedure will be continued; otherwise, it will be terminated. The pseudocode of the local search procedure is provided in Figure 4.

3.3.4. Acceptance and Stopping Conditions

Once the current best solution (Π^{best}) and the new solution (Π^{new}) are known, the search algorithm should determine if there is an improvement in the fitness value. If the new solution is of better quality than the current best solution, i.e., a smaller weighted sum of the total flowtime and makespan values $fitness(\Pi^{new}) < fitness(\Pi^{best})$ has resulted, the new solution becomes the current-best solution, $\Pi^{best} = \Pi^{new}$. Otherwise, a mechanism is required to decide whether or not to accept a new solution that is worse or similar to the current best solution.

Inspired by the Simulated Annealing algorithm [54], the cooling mechanism is used to regulate the acceptance condition. In the approach suggested by Ruiz and Stützle [52], the fitness values associated with the current and best solutions are considered to calculate the relative change in the solution quality, i.e., $\Delta = fitness(\Pi^{new}) - fitness(\Pi^{best})$. Given Δ and the initial temperature, T_0, as the algorithm parameter, the current temperature T decreases proportionately to the cooling coefficient, i.e., $T \leftarrow \delta \times T$, where $0 < \delta < 1$ is the cooling rate. Finally, the acceptance probability, calculated using $P = \exp(-\Delta E/T)$, should be compared with a random number to determine whether to accept a poor-performing solution. This mechanism is particularly useful to avoid premature convergence and getting trapped in the local optima. Unchanged fitness value, i.e., $\Delta = 0$, for a certain number of iterations, signals the termination of the algorithm. The algorithm terminates when the current best solution remains do not improve for a certain number of iterations.

LS (π)

1 Procedure_Local_Search (Π)
2   improve = True;
3   while(improve) do
4   improve = False;
5   for i=1?to n do
6   remove a random job from π without repitition;
7   Π' = best permutation obtained by inserting the job to all possible positions;
8   if $Fitness(P') < Fitness(P)$ then
9          $\Pi = \Pi'$;
10   improve = True;
11   endif
12   endfor
13   endwhile
14   return Π
15 end

Figure 4. Pseudocode of the local search procedure.

4. Results

This section begins with an elaboration on the configuration of the test bank and the algorithm calibration experiment. Numerical results and statistical analysis are then provided to compare the HIG performance with Hybrid Tabu Search (HTS; [25]). It consists of short- and long-term phases, with the short-term phase focusing on a local search and the long-term phase improving concentration and diversification and help escape the local best solutions. HTS applies the NEH [53] for solution initialization, and the 'swap' and 'insert' moves as the disturbance mechanism. Besides, three other variants of IG, denoted by the IG_1, IG_2, and IG_3 algorithms, are included to enrich the numerical experiments and provide insights into the impact of various computational elements in solving the problem. It helps explore what element of the proposed extension contributes most to the possible breakthrough. IG_1 and IG_2 apply the basic construction method, while IG_3 uses the customized construction method developed in our study. On the other hand, IG_1 and IG_3 do not have a local search mechanism, while IG_2 applies a perturbation mechanism similar to HIG. All the algorithms are coded and compiled using a personal computer with the following specs; Intel (R) Core (TM) i7 CPU 3.4 GHz, 8 GB RAM, and Windows 7 operating system.

The widely-used scheduling dataset developed by Tillard [55] is used to benchmark HIG against the best-performing algorithms in the literature developed to solve the $F_m|prmu, no - idle|\alpha.C_{max} + \beta.\sum F_j$ problem. This dataset consists of 12 job/machine combinations considering three configuration groups: (1) $n \in \{20, 50, 100\}$ jobs and $m \in \{5, 10, 20\}$ machines; (2) $n = 200$ jobs and $m \in \{10, 20\}$ machines; (3) $n = 500$ jobs and $m = 20$ machines. Ten distinct instances for each combination make a total of 120 instances for the final experiments.

The calibration experiment is conducted in two phases using random test instances. First, the best configuration is determined considering a limited set of alternatives. Next, the set of parameters adjacent to the selected configuration in the first phase will be explored to check if a better combination of parameters can be found. For this purpose, the Relative Percentage Deviation (RPD) shown in Equation (13) is considered to compare the resulting fitness values where smaller values are preferred, and RPD = 0 demonstrates the best solution. In this equation, $Fitness^*$ refers to the fitness value obtained by each of the solution algorithms and $Fitness_{best}$ is the best result.

$$\text{RPD} = \frac{Fitness_{best} - Fitness^*}{Fitness^*} \times 100 \qquad (13)$$

Random instances are used to determine the parameters of the IG_1, IG_2, IG_3, and HIG algorithms. The calibration test results are summarized in Table 2 for these algorithms. On this basis, the algorithm parameters are set to $d = 2$, $\gamma = 20$, and $\delta = 0.9$ to conduct the final experiments. To ensure a fair comparison, a termination condition similar to that of the HTS algorithm, which is applied by Ren et al. [25], is considered. That is, the algorithm terminates when the current best solution remains unchanged for 100 consecutive iterations.

Table 2. Calibration results analysis (best in **bold**).

Algorithm	Parameter	Phase I			Phase II		
		I	II	III	I	IV	V
IG1	d	2	4	6	2	1	3
	γ	20	40	60	20	10	30
	δ	0.9	0.8	0.7	0.9	0.95	0.85
	ARPD	**0.56**	1.35	1.61	**0.78**	2.78	1.26
IG2		I	II	III	I	IV	V
	d	2	4	6	2	1	3
	γ	20	40	60	20	10	30
	δ	0.9	0.8	0.7	0.9	0.95	0.85
	ARPD	**0.48**	0.65	0.78	**0.68**	0.90	1.15
IG3		I	II	III	I	IV	V
	d	2	4	6	2	1	3
	γ	20	40	60	20	10	30
	δ	0.9	0.8	0.7	0.9	0.95	0.85
	ARPD	**0.32**	0.99	1.02	**0.38**	3.92	0.88
HIG		I	II	III	I	IV	V
	d	2	4	6	2	1	3
	γ	20	40	60	20	10	30
	δ	0.9	0.8	0.7	0.9	0.95	0.85
	ARPD	**0.33**	0.48	0.87	**0.44**	4.99	0.49

Considering the calibrated parameter values and an equal importance weight ($\alpha = \beta = 0.5$) within a priori performance articulation scheme, the best-found solutions of the instances solved using HTS [25] are compared to those of the HTS, IG_1, IG_2, and IG_1 algorithms. The results are summarized in Tables 3 and 4. Except for the instance with 20 machines and 10 jobs (20 × 10), where HTS performs slightly better than HIG, the rest of the best solutions are yielded by HIG.

We first analyze the results considering various workloads and operating scales. Considering different numbers of jobs in the first set of instances, Table 3 shows that HIG performs better than HTS. The difference in performance becomes more significant with an increase in the workload. The IG_2 algorithm is also superior to the HTS, considering the first set of test instances, showing that integrating the local search mechanism contributes significantly to the success of the developed algorithm. The solutions obtained by the IG_1, IG_2, IG_3, and HIG algorithms across all test instances are then compared separately in Table 4, considering all test instances where HIG yields the best results in all cases.

Table 3. Best-found solutions considering the first set of test instances (best in **bold**).

Instance ($m \times n$)	HTS	IG_1	IG_2	IG_3	HIG
20 × 5	9437.0	9549.5	9401.0	9566.5	**9324.5**
20 × 10	**17,456.0**	17,716.5	17,616.5	17,539.0	17,472.5
20 × 20	29,319.0	29,412.5	29,149.5	29,724.5	**29,148.5**
50 × 5	47,719.0	46,597.0	46,717.5	47,199.5	**45,895.5**
50 × 10	58,610.0	61,247.0	58,375.5	58,445.0	**57,582.5**
50 × 20	107,007.0	110,308.0	105,657.0	108,775.5	**105,555.0**
100 × 5	160,401.5	162,367.0	157,927.0	158,705.5	**155,420.0**
100 × 10	214,438.5	206,815.5	202,818.0	207,382.5	**201,117.5**
100 × 20	337,888.5	335,717.0	330,472.0	330,243.5	**324,920.5**
200 × 10	685,369.0	690,565.5	657,372.5	671,582.5	**656,480.0**
200 × 20	1,003,945.5	1,006,795.0	991,009.0	1,011,902.5	**989,277.5**
500 × 20	4,456,166.0	4,540,516.5	4,449,380.0	4,494,197.5	**4,448,496.5**

Table 4. Best-found solutions across all test instances (updates in **bold**).

Instance ($m \times n$)	IG_1	IG_2	IG_3	HIG
20 × 5	9239.05	9088.05	9111.90	**9031.65**
20 × 10	15,238.65	15,054.10	15,065.25	**14,999.45**
20 × 20	29,966.20	29,591.90	29,628.70	**29,422.00**
50 × 5	43,076.70	42,166.05	42,448.70	**41,957.35**
50 × 10	62,506.85	60,375.95	60,962.05	**59,921.50**
50 × 20	111,451.80	108,472.15	109,087.45	**107,926.95**
100 × 5	144,552.95	141,478.15	142,699.50	**140,742.30**
100 × 10	209,197.70	204,574.05	207,545.50	**203,718.20**
100 × 20	327,279.80	320,069.55	320,393.40	**316,219.80**
200 × 10	697,101.30	674,801.50	684,276.40	**674,297.30**
200 × 20	990,265.30	965,298.95	974,714.40	**962,675.70**
500 × 20	4,764,325.25	4,687,959.00	4,704,317.20	**4,680,702.80**

In an overall analysis, Tables 5 and 6 provide the Average Relative Percentage Deviation (ARPD) values for different workloads and machines, respectively. The RPD analysis shows the overall impact of the number of machinery and workload on the performance of the algorithm. It is evident that HIG obtains meaningfully better solutions than the HTS, IG_1, IG_2, and IG_3 algorithms when solving the $F_m|prmu, no-idle|\alpha \cdot C_{max} + \beta \cdot \sum F_j$ problem across different operational situations. Given the RPD analysis, it is expected that HIG's superiority to the current-best-performing algorithm, HTS, will be even more significant for industry-scale applications.

A statistical test is conducted to check whether the resulting improvement in the best-found solutions is significant. The null hypothesis is that the HIG algorithm does not outperform the HTS algorithm when solving the $F_m|prmu, no-idle|\alpha \cdot C_{max} + \beta \cdot \sum F_j$ problem. The *t*-test results are summarized in Table 7. Considering 120 test instances, the *p*-value is supportive of rejecting the null hypothesis. That is, with 95 percent of confidence, we can claim that HIG is superior to the current-best-performing algorithm in the literature of BNIPFSP, i.e., the HTS algorithm. It is also observed that the proposed extension shows a significant improvement in the performance of the algorithm when compared to all three variants of the IGs.

Table 5. The Relative Performance Deviation considering various workloads (best in **bold**).

Workload (n)	Machinery (m)	HTS	IG_1	IG_2	IG_3	HIG
20	5	1.21	2.41	0.82	2.60	0.00
	10	0.00	1.49	0.92	0.48	0.09
	20	0.58	0.91	0.00	1.98	0.00
	Overall	0.60	1.60	0.58	1.68	**0.03**
50	5	3.97	1.53	1.79	2.84	0.00
	10	1.78	6.36	1.38	1.50	0.00
	20	1.38	4.50	0.10	3.05	0.00
	Overall	2.38	4.13	1.09	2.46	**0.00**
100	5	3.21	4.47	1.61	2.11	0.00
	10	6.62	2.83	0.85	3.12	0.00
	20	3.99	3.32	1.71	1.64	0.00
	Overall	4.61	3.54	1.39	2.29	**0.00**
200	10	4.40	5.19	0.14	2.30	0.00
	20	1.48	1.77	0.18	2.29	0.00
	Overall	2.94	3.48	0.16	2.29	**0.00**
500	20	0.17	2.07	0.02	1.03	**0.00**

Table 6. Average Relative Performance Deviation considering operating scale (best in **bold**).

Machinery (m)	Workload (n)	HTS	IG_1	IG_2	IG_3	HIG
5	20	1.21	2.41	0.82	2.60	0.00
	50	3.97	1.51	1.79	2.84	0.00
	100	3.21	4.47	1.61	2.11	0.00
	Overall	2.79	2.80	1.40	2.51	**0.00**
10	20	0.00	1.49	0.92	0.48	0.09
	50	1.78	6.36	1.38	1.50	0.00
	100	6.62	2.83	0.85	3.12	0.00
	200	4.40	5.19	0.14	2.30	0.00
	Overall	3.20	3.97	0.81	1.84	**0.02**
100	20	0.58	0.91	0.00	1.98	0.00
	50	1.38	4.50	0.10	3.05	0.00
	100	3.99	3.32	1.71	1.64	0.00
	200	1.48	1.77	0.18	2.29	0.00
	500	0.17	2.07	0.02	1.03	0.00
	Overall	1.26	2.09	0.33	1.66	**0.00**

Table 7. Paired *t*-test analysis of the performance differences under 0.95 confidence interval.

Instance	Average	StD	DoF	T Stat	One-Tail		Two-Tail	
					t-Critical	*p*-Value	*t*-Critical	*p*-Value
HIG Vs. HTS	7255.58	8435.13	11	2.86	1.79	0.0078	2.20	0.0157

StD: Standard Deviation, S.E.: Standard Error of the Mean, DoF: Degree of Freedom.

As a final step to the numerical analysis, the best-found solutions to all 120 test instances are recorded in Appendix B. The updated values are highlighted in bold font. Notably, 119 out of 120 best-found solutions are yielded by the HIG algorithm. The resulting values can be used in future studies to benchmark the prospect solution algorithms for solving the $F_m|prmu, no-idle|\alpha \cdot C_{max} + \beta \cdot \sum F_j$ problem.

5. Conclusions

Energy efficiency in the production sector requires well-informed operations management decisions in addition to the use of modern equipment, smart lighting and control systems, and the standard construction of facilities. Production scheduling is a prime example of planning tools that facilitate the successful implementation of green initiatives for reducing the carbon footprint. This study contributes to the energy-efficient production scheduling literature developing a mathematical model and a solution algorithm to address the gap identified in the comprehensive literature review. Extensive numerical analysis using a well-known dataset showed that almost all of the best-found solutions are yielded by the HIG algorithm. The statistical test of significance confirmed that HIG performs significantly better than the benchmark algorithm when solving the $F_m | prmu, no-idle | \alpha \cdot C_{max} + \beta \cdot \sum F_j$ problem.

Despite its effectiveness in solving the BNIPFSPs, the proposed solution algorithm is limited in that it applies a priori preference articulation approach for reconciliation of the makespan and total flowtime. To address this limitation, the following directions can be pursued. First, one can extend the Iterated Greedy algorithm to work with the Pareto Front approach to provide a comprehensive set of optimum solutions and trade-offs. Second, other multi-objective optimization algorithms can be adapted to solve this intractable scheduling extension. The third suggestion for future research includes adopting the Concept of Stratification and Incremental Enlargement to solve the problem's dynamic variant. In doing so, one can also account for operational parameter uncertainties and the possibility of rejecting a job or partially accepting a batch of jobs. Finally, the no-idle setting needs more attention in other production settings to contribute to energy efficiency literature.

Author Contributions: C.-Y.C.: Conceptualization, Methodology, Software. S.-W.L.: Conceptualization, Methodology, Software, Funding acquisition. P.P.: Investigation, Writing—Original draft, Writing—Revision. K.-C.Y.: Supervision, Conceptualization, Methodology. Y.-Z.L.: Formal analysis. All authors have read and agreed to the published version of the manuscript.

Funding: This research was partially supported by the Ministry of Science and Technology, Taiwan, under Grant MOST 109-2221-E-027-073/Most-109-2410-H-182-009-MY3, and in part by the Linkou Chang Gung Memorial Hospital under Grant BMRPA19.

Institutional Review Board Statement: Not applicable.

Informed Consent Statement: Not applicable.

Data Availability Statement: Not applicable.

Conflicts of Interest: The authors declare no conflict of interest.

Appendix A

Table A1. Summary of the No-Idle Flowshop Scheduling Literature.

Title	Authors (Year)	Publication	Scheduling Extension	Objective Function
Note: On the Two-Machine No-Idle Flowshop Problem	Cepek et al. (2000)	Naval Research Logistics	No-idle permutation flowshop	Total completion time
Flowshop/no-idle scheduling to minimize the mean flowtime	Narain and Bagga (2005)	Australia and New Zealand Industrial and Applied Mathematics (ANZIAM)	No-idle permutation flowshop	Average flowtime
No-wait flexible flowshop scheduling with no-idle machines	Wang et al. (2005)	Operations Research Letters	No-wait flexible flowshop with no-idle machines	Makespan

Table A1. Cont.

Title	Authors (Year)	Publication	Scheduling Extension	Objective Function
A differential evolution algorithm for the no-idle flowshop scheduling problem with total tardiness criterion	Tasgetiren et al. (2011)	International Journal of Production Research	No-idle permutation flowshop	Total tardiness
Tabu search algorithm for no-idle flowshop scheduling problems	Ren et al. (2010)	Computer Engineering and Design	No-idle permutation flowshop	Makespan
A DE Based Variable Iterated Greedy Algorithm for the No-Idle Permutation Flowshop Scheduling Problem with Total Flowtime Criterion	Tasgetiren et al. (2011)	Conference	No-idle permutation flowshop	Total flowtime
Hybrid Tabu Search Algorithm for bi-criteria no-idle permutation flow shop scheduling problem	Ren et al. (2011)	Conference	Bi-objective no-idle permutation flowshop	Makespan and total flowtime
A new heuristic method for minimizing the makespan in a no-idle permutation flowshop	Nagano & Branco (2012)	Conference	No-idle permutation flowshop	Makespan
A discrete artificial bee colony algorithm for the no-idle permutation flowshop scheduling problem with the total tardiness criterion	Tasgetiren et al. (2013b)	Applied Mathematical Modelling	No-idle permutation flowshop	Total tardiness
A variable iterated greedy algorithm with differential evolution for the no-idle permutation flowshop scheduling problem	Tasgetiren et al. (2013a)	Computers & Operations Research	No-idle permutation flowshop	Makespan
Metaheuristics for the no-idle permutation flowshop scheduling problem	Büyükdağlı (2013)	Thesis	No-idle permutation flowshop	-
An effective iterated greedy algorithm for the mixed no-idle permutation flowshop scheduling problem	Pan and Ruiz (2014)	OMEGA	Mixed no-idle permutation flowshop	Makespan
Research on no-idle permutation flowshop scheduling with time-dependent learning effect and deteriorating jobs	Lu (2016)	Applied Mathematical Modelling	No-idle permutation flowshop scheduling with time-dependent learning effect and deteriorating jobs	Makespan
Heuristics for the mixed no-idle flowshop with sequence-dependent setup times and total flowtime criterion	Rossi and Nagano (2019a)	Expert Systems with Applications	Mixed no-idle permutation flowshop with SDST	Total flowtime

Table A1. *Cont.*

Title	Authors (Year)	Publication	Scheduling Extension	Objective Function
Heuristics for the mixed no-idle flowshop with sequence-dependent setup times	Rossi and Nagano (2019b)	Journal of the Operational Research Society	Mixed no-idle permutation flowshop with SDST	Makespan
High-performing heuristics to minimize flowtime in no-idle permutation flowshop	Nagano et al. (2019)	Engineering Optimization	No-idle permutation flowshop	Total flowtime
A Variable Iterated Local Search Algorithm for Energy-Efficient No-idle Flowshop Scheduling Problem	Tasgetiren et al. (2019)	Conference	Bi-objective no-idle permutation flowshop	Makespan and total energy consumption
A contribution for the mixed no-idle flowshop scheduling problem with sequence-dependent setup times: analysis and solutions procedures	Rossi (2020)	Thesis	Mixed no-idle flowshop with sequence-dependent setup times	-
A hybrid discrete water wave optimization algorithm for the no-idle flowshop scheduling problem with total tardiness criterion	Zhao et al. (2020)	Expert Systems with Applications	No-idle permutation flowshop	Total tardiness
A new iterated greedy algorithm for no-idle permutation flowshop scheduling with the total tardiness criterion	Riahi et al. (2020)	Computers & Operations Research	No-idle permutation flowshop	Total tardiness
Benders decomposition for the mixed no-idle permutation flowshop scheduling problem	Bektaş et al. (2020)	Journal of Scheduling	Mixed no-idle permutation flowshop	Makespan
Heuristics and metaheuristics for the mixed no-idle flowshop with sequence-dependent setup times and total tardiness minimization	Rossi and Nagano (2020)	Swarm and Evolutionary Computation	Mixed no-idle permutation flowshop with sequence-dependent setup times	Total tardiness
A Novel General Variable Neighborhood Search through Q-Learning for No-Idle Flowshop Scheduling	Oztop et al. (2020)	Conference	No-idle permutation flowshop	Makespan
Automatic design of hybrid stochastic local search algorithms for permutation flowshop problems with additional constraints	Pagnozzi and Stützle (2021)	Operations Research Perspectives	No-idle permutation flowshop	Makespan
A cooperative water wave optimization algorithm with reinforcement learning for the distributed assembly no-idle flowshop scheduling problem	Zhao et al. (2021)	Computers & Industrial Engineering	Distributed assembly no-idle flow-shop scheduling problem	Maximum assembly completion time

Appendix B

Table A2. Best-Found Solutions (BFS) across All Test Instances (Updates in **Bold**).

	Inst. ($m \times n$)	BFS		Inst. ($m \times n$)	BFS		Inst. ($m \times n$)	BFS		Inst. ($m \times n$)	BFS		Inst. ($m \times n$)	BFS
1	20 × 5	9324.5	2	20 × 5	8768.5	3	20 × 5	9004.0	4	20 × 5	9201.0	5	20 × 5	9840.0
	20 × 10	17,456.0		20 × 10	15,359.5		20 × 10	15,311.0		20 × 10	15,060.5		20 × 10	12,932.0
	20 × 20	29,148.5		20 × 20	27,458.5		20 × 20	29,330.0		20 × 20	27,500.5		20 × 20	30,123.5
	50 × 5	45,895.5		50 × 5	40,849.5		50 × 5	39,540.0		50 × 5	40,900.0		50 × 5	46,931.0
	50 × 10	57,582.5		50 × 10	56,394.5		50 × 10	55,894.5		50 × 10	60,125.5		50 × 10	53,752.0
	50 × 20	105,555.0		50 × 20	119,208.5		50 × 20	95984.5		50 × 20	113,445.5		50 × 20	97,119.0
	100 × 5	155,420.0		100 × 5	134,879.0		100 × 5	133,638.0		100 × 5	133,875.5		100 × 5	147,159.0
	100 × 10	201,117.5		100 × 10	178,403.0		100 × 10	212,533.0		100 × 10	191,444.5		100 × 10	195,680.5
	100 × 20	324,920.5		100 × 20	297,506.0		100 × 20	310,170.5		100 × 20	345,987.0		100 × 20	355,364.0
	200 × 10	656,480.0		200 × 10	752,145.0		200 × 10	729,595.0		200 × 10	618,158.0		200 × 10	667,624.5
	200 × 20	989,277.5		200 × 20	962,653.5		200 × 20	944,299.5		200 × 20	903,733.0		200 × 20	987,579.5
	500 × 20	4,448,496.5		500 × 20	4,550,718		500 × 20	4,927,311		500 × 20	4,443,889		500 × 20	485,7293
6	20 × 5	9974.0	7	20 × 5	7746.5	8	20 × 5	8809.5	9	20 × 5	9435.5	10	20 × 5	8213.0
	20 × 10	15,218.5		20 × 10	13,795.0		20 × 10	15,206.0		20 × 10	15,190.0		20 × 10	14,449.5
	20 × 20	30,508.0		20 × 20	28,473.5		20 × 20	29,676.5		20 × 20	29,177.0		20 × 20	32,824.0
	50 × 5	40,562.5		50 × 5	43,979.0		50 × 5	41,450.0		50 × 5	39,433.0		50 × 5	40,033.0
	50 × 10	67,395.5		50 × 10	61,381.0		50 × 10	65,124.5		50 × 10	59,999.0		50 × 10	61,566.0
	50 × 20	116,283.0		50 × 20	113,469.0		50 × 20	102,066.0		50 × 20	104,491.5		50 × 20	111,647.5
	100 × 5	135,117.0		100 × 5	158,846.0		100 × 5	130,140.5		100 × 5	138,971.0		100 × 5	139,377.0
	100 × 10	198,459.5		100 × 10	216,278.5		100 × 10	214,725.5		100 × 10	21,8945.0		100 × 10	201,118.0
	100 × 20	294,212.0		100 × 20	310,486.0		100 × 20	300,453.0		100 × 20	321,981.0		100 × 20	301,118.0
	200 × 10	698,950.5		200 × 10	6,123,41.5		200 × 10	705,257.0		200 × 10	601,637.0		200 × 10	700,784.5
	200 × 20	976,708.0		200 × 20	1,025,032		200 × 20	1,003,762		200 × 20	919,347.5		200 × 20	914,365.5
	500 × 20	446,0524		500 × 20	4,829,931		500 × 20	4,979,367		500 × 20	4,414,894		500 × 20	489,4606

References

1. Zhu, Z.-S.; Liao, H.; Cao, H.-S.; Wang, L.; Wei, Y.-M.; Yan, J. The differences of carbon intensity reduction rate across 89 countries in recent three decades. *Appl. Energy* **2014**, *113*, 808–815. [CrossRef]
2. Sutherland, B.R. Tax Carbon Emissions and Credit Removal. *Joule* **2019**, *3*, 2071–2073. [CrossRef]
3. Agency, I.E. *Tracking Industrial Energy Efficiency and CO_2 Emissions*; OECD: Paris, France, 2007; ISBN 9789264030169.
4. Pourhejazy, P.; Kwon, O.K.; Lim, H. Integrating Sustainability into the Optimization of Fuel Logistics Networks. *KSCE J. Civ. Eng.* **2019**, *23*, 1369–1383. [CrossRef]
5. Zhang, H.C.; Kuo, T.C.; Lu, H.; Huang, S.H. Environmentally conscious design and manufacturing: A state-of-the-art survey. *J. Manuf. Syst.* **1997**, *16*, 352–371. [CrossRef]
6. Pourhejazy, P.; Kwon, O.K. A Practical Review of Green Supply Chain Management: Disciplines and Best Practices. *J. Int. Logist. Trade* **2016**, *14*, 156–164. [CrossRef]
7. Peng, C.; Peng, T.; Zhang, Y.; Tang, R.; Hu, L. Minimising non-processing energy consumption and tardiness fines in a mixed-flow shop. *Energies* **2018**, *11*, 3382. [CrossRef]
8. Cheng, C.-Y.; Pourhejazy, P.; Ying, K.-C.; Lin, C.-F. Unsupervised Learning-based Artificial Bee Colony for minimizing non-value-adding operations. *Appl. Soft Comput.* **2021**, *105*, 107280. [CrossRef]
9. Wu, X.; Sun, Y. A green scheduling algorithm for flexible job shop with energy-saving measures. *J. Clean. Prod.* **2018**, *172*, 3249–3264. [CrossRef]
10. Piroozfard, H.; Wong, K.Y.; Wong, W.P. Minimizing total carbon footprint and total late work criterion in flexible job shop scheduling by using an improved multi-objective genetic algorithm. *Resour. Conserv. Recycl.* **2018**, *128*, 267–283. [CrossRef]
11. Zheng, X.-L.; Wang, L. A collaborative multiobjective fruit fly optimization algorithm for the resource constrained unrelated parallel machine green scheduling problem. *IEEE Trans. Syst. Man Cybern. Syst.* **2016**, *48*, 790–800. [CrossRef]
12. Safarzadeh, H.; Niaki, S.T.A. Bi-objective green scheduling in uniform parallel machine environments. *J. Clean. Prod.* **2019**, *217*, 559–572. [CrossRef]
13. Mansouri, S.A.; Aktas, E.; Besikci, U. Green scheduling of a two-machine flowshop: Trade-off between makespan and energy consumption. *Eur. J. Oper. Res.* **2016**, *248*, 772–788. [CrossRef]
14. Zhang, B.; Pan, Q.; Gao, L.; Li, X.; Meng, L.; Peng, K. A multiobjective evolutionary algorithm based on decomposition for hybrid flowshop green scheduling problem. *Comput. Ind. Eng.* **2019**, *136*, 325–344. [CrossRef]
15. Cota, L.P.; Coelho, V.N.; Guimarães, F.G.; Souza, M.J.F. Bi-criteria formulation for green scheduling with unrelated parallel machines with sequence-dependent setup times. *Int. Trans. Oper. Res.* **2021**, *28*, 996–1017. [CrossRef]

16. Jiang, T.; Zhang, C.; Sun, Q.-M. Green job shop scheduling problem with discrete whale optimization algorithm. *IEEE Access* **2019**, *7*, 43153–43166. [CrossRef]
17. Aghelinejad, M.; Ouazene, Y.; Yalaoui, A. Complexity analysis of energy-efficient single machine scheduling problems. *Oper. Res. Perspect.* **2019**, *6*, 100105. [CrossRef]
18. Li, K.; Zhang, X.; Leung, J.Y.-T.; Yang, S.-L. Parallel machine scheduling problems in green manufacturing industry. *J. Manuf. Syst.* **2016**, *38*, 98–106. [CrossRef]
19. Niu, S.; Song, S.; Chiong, R. A Distributionally Robust Scheduling Approach for Uncertain Steelmaking and Continuous Casting Processes. *IEEE Trans. Syst. Man Cybern. Syst.* **2021**. [CrossRef]
20. Bektaş, T.; Hamzadayı, A.; Ruiz, R. Benders decomposition for the mixed no-idle permutation flowshop scheduling problem. *J. Sched.* **2020**, *23*, 513–523. [CrossRef]
21. Ding, J.-Y.; Song, S.; Gupta, J.N.D.; Wang, C.; Zhang, R.; Wu, C. New block properties for flowshop scheduling with blocking and their application in an iterated greedy algorithm. *Int. J. Prod. Res.* **2016**, *54*, 4759–4772. [CrossRef]
22. Foumani, M.; Smith-Miles, K. The impact of various carbon reduction policies on green flowshop scheduling. *Appl. Energy* **2019**, *249*, 300–315. [CrossRef]
23. Graham, R.L.; Lawler, E.L.; Lenstra, J.K.; Kan, A.H.G.R. Optimization and Approximation in Deterministic Sequencing and Scheduling: A Survey. In *Annals of Discrete Mathematics*; Elsevier: Amsterdam, The Netherlands, 1979; Volume 5, pp. 287–326.
24. Tasgetiren, M.F.; Pan, Q.-K.; Wang, L.; Chen, A.H.-L. A DE based variable iterated greedy algorithm for the no-idle permutation flowshop scheduling problem with total flowtime criterion. In Proceedings of the International Conference on Intelligent Computing, Zhengzhou, China, 11–14 August 2011; pp. 83–90.
25. Ren, W.-J.; Duan, J.-H.; Zhang, F.; Han, H.; Zhang, M. Hybrid Tabu Search Algorithm for bi-criteria No-idle permutation flow shop scheduling problem. In Proceedings of the 2011 Chinese Control and Decision Conference (CCDC), Mianyang, China, 23–25 May 2011; pp. 1699–1702.
26. Nagano, M.S.; Branco, F.J.C. A new heuristic method for minimizing the makespan in a no-idle permutation flowshop. In Proceedings of the Simposio Brasileiro de Pesquisa Operacional, Rio de Janeiro, Brazil, 24–28 September 2012.
27. Fatih Tasgetiren, M.; Öztop, H.; Gao, L.; Pan, Q.K.; Li, X. A Variable Iterated Local Search Algorithm for Energy-Efficient No-idle Flowshop Scheduling Problem. *Procedia Manuf.* **2019**, *39*, 1185–1193. [CrossRef]
28. Oztop, H.; Tasgetiren, M.F.; Kandiller, L.; Pan, Q.K. A Novel General Variable Neighborhood Search through Q-Learning for No-Idle Flowshop Scheduling. In Proceedings of the 2020 IEEE Congress on Evolutionary Computation (CEC), Glasgow, UK, 19–24 July 2020. [CrossRef]
29. Rossi, F.L. A Contribution for the Mixed No-Idle Flowshop Scheduling Problem with Sequence-Dependent Setup Times: Analysis and Solutions Procedures. Ph.D. Thesis, Universidade de São Paulo, São Paulo, Brazil, 2019.
30. Buyukdagli, O. Metaheuristics for the No-Idle Permutation Flowshop Scheduling Problem. Master Thesis, Yasar University, Bornova, Turkey, 2013.
31. Ribas, I.; Leisten, R.; Framiñan, J.M. Review and classification of hybrid flow shop scheduling problems from a production system and a solutions procedure perspective. *Comput. Oper. Res.* **2010**, *37*, 1439–1454. [CrossRef]
32. Neufeld, J.S.; Gupta, J.N.D.; Buscher, U. A comprehensive review of flowshop group scheduling literature. *Comput. Oper. Res.* **2016**, *70*, 56–74. [CrossRef]
33. Cepek, O.; Okada, M.; Vlach, M. Minimizing total completion time in a two-machine no-idle flowshop. *Res. Rep.* **1998**, *98*, 1–23.
34. Čepek, O.; Okada, M.; Vlach, M. Note: On the Two-Machine No-Idle Flowshop Problem. *Nav. Res. Logist.* **2000**, *47*, 353–358. [CrossRef]
35. Narain, L.; Bagga, P.C. Flowshop/no-idle scheduling to minimise the mean flowtime. *ANZIAM J.* **2005**, *47*, 265–275. [CrossRef]
36. Wang, Z.; Xing, W.; Bai, F. No-wait flexible flowshop scheduling with no-idle machines. *Oper. Res. Lett.* **2005**, *33*, 609–614. [CrossRef]
37. Tasgetiren, M.F.; Pan, Q.K.; Suganthan, P.N.; Jin Chua, T. A differential evolution algorithm for the no-idle flowshop scheduling problem with total tardiness criterion. *Int. J. Prod. Res.* **2011**, *49*, 5033–5050. [CrossRef]
38. Fatih Tasgetiren, M.; Pan, Q.K.; Suganthan, P.N.; Oner, A. A discrete artificial bee colony algorithm for the no-idle permutation flowshop scheduling problem with the total tardiness criterion. *Appl. Math. Model.* **2013**, *37*, 6758–6779. [CrossRef]
39. Ren, W.-J.; Pan, Q.-K.; Han, H.-Y. Tabu search algorithm for no-idle flowshop scheduling problems. *Comput. Eng. Des.* **2010**, *31*, 5071–5074.
40. Fatih Tasgetiren, M.; Pan, Q.K.; Suganthan, P.N.; Buyukdagli, O. A variable iterated greedy algorithm with differential evolution for the no-idle permutation flowshop scheduling problem. *Comput. Oper. Res.* **2013**, *40*, 1729–1743. [CrossRef]
41. Lu, Y.Y. Research on no-idle permutation flowshop scheduling with time-dependent learning effect and deteriorating jobs. *Appl. Math. Model.* **2016**, *40*, 3447–3450. [CrossRef]
42. Pagnozzi, F.; Stützle, T. Automatic design of hybrid stochastic local search algorithms for permutation flowshop problems with additional constraints. *Oper. Res. Perspect.* **2021**, *8*, 100180.
43. Pan, Q.-K.; Ruiz, R. An effective iterated greedy algorithm for the mixed no-idle permutation flowshop scheduling problem. *Omega* **2014**, *44*, 41–50. [CrossRef]
44. Rossi, F.L.; Nagano, M.S. Heuristics for the mixed no-idle flowshop with sequence-dependent setup times and total flowtime criterion. *Expert Syst. Appl.* **2019**, *125*, 40–54. [CrossRef]

45. Rossi, F.L.; Nagano, M.S. Heuristics and metaheuristics for the mixed no-idle flowshop with sequence-dependent setup times and total tardiness minimisation. *Swarm Evol. Comput.* **2020**, *55*, 100689. [CrossRef]
46. Rossi, F.L.; Nagano, M.S. Heuristics for the mixed no-idle flowshop with sequence-dependent setup times. *J. Oper. Res. Soc.* **2019**, 1–27. [CrossRef]
47. Nagano, M.S.; Rossi, F.L.; Martarelli, N.J. High-performing heuristics to minimize flowtime in no-idle permutation flowshop. *Eng. Optim.* **2019**, *51*, 185–198. [CrossRef]
48. Zhao, F.; Zhang, L.; Zhang, Y.; Ma, W.; Zhang, C.; Song, H. A hybrid discrete water wave optimization algorithm for the no-idle flowshop scheduling problem with total tardiness criterion. *Expert Syst. Appl.* **2020**, *146*, 113166. [CrossRef]
49. Riahi, V.; Chiong, R.; Zhang, Y. A new iterated greedy algorithm for no-idle permutation flowshop scheduling with the total tardiness criterion. *Comput. Oper. Res.* **2020**, *117*, 104839. [CrossRef]
50. Zhao, F.; Zhang, L.; Cao, J.; Tang, J. A cooperative water wave optimization algorithm with reinforcement learning for the distributed assembly no-idle flowshop scheduling problem. *Comput. Ind. Eng.* **2021**, *153*, 107082. [CrossRef]
51. Ruiz, R.; Vallada, E.; Fernandez-Martinez, C. Scheduling in flowshops with no-idle machines. In *Computational Intelligence in Flow Shop and Job Shop Scheduling*; Springer: Berlin/Heidelberg, Germany, 2009; pp. 21–51.
52. Ruiz, R.; Stützle, T. A simple and effective iterated greedy algorithm for the permutation flowshop scheduling problem. *Eur. J. Oper. Res.* **2007**, *177*, 2033–2049. [CrossRef]
53. Nawaz, M.; Enscore, E.E., Jr.; Ham, I. A heuristic algorithm for the m-machine, n-job flow-shop sequencing problem. *Omega* **1983**, *11*, 91–95. [CrossRef]
54. Osman, I.; Potts, C. Simulated annealing for permutation flow-shop scheduling. *Omega* **1989**, *17*, 551–557. [CrossRef]
55. Taillard, E. Benchmarks for basic scheduling problems. *Eur. J. Oper. Res.* **1993**, *64*, 278–285. [CrossRef]

Article

Improving the Return Loading Rate Problem in Northwest China Based on the Theory of Constraints

Wen-Tso Huang [1], Cheng-Chang Lu [2] and Jr-Fong Dang [3,*]

[1] Department of Business Administration, Chung Yuan Christian University, Chung Li District, Taoyuan City 320314, Taiwan; wthuang@cycu.edu.tw
[2] Financial Management Programme, Beijing Institute of Technology, Zhuhai, College of Global Talents, No.6, Jinfeng Rd., Tangjiawan, Zhuhai 519088, China; cheng-chang.lu@cgt.bitzh.edu.cn
[3] Department of Industrial Engineering and Systems Management, Feng Chia University, 100 Wenhwa Road, Taichung 407802, Taiwan
* Correspondence: jfdang@fcu.edu.tw

Abstract: This paper introduces how to improve the return loading rate problem by integrating the Sub-Tour reversal approach with the method of the Theory of Constraints (TOC). The proposed model generates the initial solution derived by the Sub-Tour reversal approach in phase 1 and then applies TOC to obtain the optimal solution, meeting the goal of improving the return loading rate to more than 50% and then lowering the total transportation distance in phase 2. To see our model capability, this study establishes an original distribution layout to compare the performance of the Sub-Tour reversal approach with our model, based on the simulation data generated by the Monte Carlo simulation. We also conduct the pair t-test to verify our model performance. The results show that our proposed model outperforms the Sub-Tour reversal approach in a significant manner. By utilizing the available data, our model can be easily implemented in the real world and efficiently seeks the optimal solutions.

Keywords: TOC; return loading rate; logistics; total transport distance; northwest China

1. Introduction

Northwest China is fruitful, and its agriculture is productive. Northwest China includes the Shaanxi, Gansu, Qinghai, Ningxia, and Xinjiang provincial administrative regions. According to the vast distances of these five provinces and the actual situation regarding the demand for logistics and transport infrastructure, most companies face a return loading rate problem. The problem results from the logistics constructions and cold chain logistics lagging behind needs, restricting the development of agriculture in the region. Most companies schedule the forward loading rate of transport vehicles approximating to 99% from supplier to customer to gain operational efficiency and corporate profits. However, in reality, those companies cannot well schedule the backward loading rate of the transport vehicles, due to the lack of needs from the customer returning to supplier (Subulan et al. [1]; Soysal et al. [2]; Kim and Lee [3]; Konstantakopoulos et al. [4]). This encouraged us to study the transport routes assisting enterprises in reducing logistics costs, improving operational efficiency and ultimately maximizing corporate profits, especially logistics costs, accounting for a large proportion of a company's total expenses. The establishment of improving the backward loading rate of transport vehicles can benefit industries and the northwest region. Thus, in this paper, we focus primarily on two factors: the total transport distance and the return loading rate, both of which are determinants in the design of transport routes. The total transport distance problem plays an important role in transport optimization since the minimization of the total transportation distance contributes to transport efficiency. In addition, the definition of return loading rate refers to the proportion of unused load capacity to the total load rate. In reality, there are many logistics operators whose side-pursuit is to shorten the delivery time; thus, the time for transport vehicles to design return routes is limited, which causes vehicles to return with no other goods, and leads to poor operational efficiency.

To solve the problem, we propose a two-phase solution procedure to derive the optimal solution. In phase 1, we apply the Sub-Tour reversal approach to obtain the initial solution based on Hillier and Lieberman [5]. By the available results, we further utilize the method of the Theory of Constraints (TOC) in phase 2 to quickly and accurately find the crux of the impact of the total transport distance and the return loading rate. To improve the return loading rate, we relax the total transportation distance adjustment to achieve optimization. The resulting outcome shows that the return loading rate is more than 50%, which results from the actual needs of northern China. Furthermore, this paper utilizes the simulation data to validate our model and adopts the Monte Carlo simulation method to test the deriving solutions. The reason why we apply Monte Carlo simulation is because Moroko and Caflisch [6] stated that numerically simulated stochastic processes can be done well by discretizing the process into small time steps and applying pseudo-random sequences to simulate the randomness. Huang et al. [7] indicated its efficiency and wide scope of applicability. This encouraged us to apply the Monte Carlo simulation to derive all simulation data throughout this paper. Based on the resulting outcomes, we know that the TOC effectively obtains the best routes design, and the standard return loading rate optimization objectives provide evidence of the superiority of the TOC method.

The remainder of this paper is organized as follows. In Section 2, a review of the literature related to the Sub-Tour reversal method, the TOC method and their operational performance measures is presented. Section 3 describes the Sub-Tour reversal model and the TOC method applied in this study. Section 4 presents the return loading rate problem in Northwest China and the simulation results. Finally, Section 5 concludes and points out the directions for future research.

2. Literature Review

With respect to the subject of logistics, numerous related issues and areas have been studied, such as route optimization, distribution center network layout and vehicle return issues. Efficient logistics management can be achieved if there is an understanding of the pros and cons of the concept. We systematically review the previous studies so as to capture the academic perspectives. Among them, distance is a primary concern of logistics. Daganzo [8] developed a simple formula to predict the distance traveled by fleets of vehicles with respect to physical distribution problems involving a depot and its area of influence. However, taking other factors into consideration, reverse logistics efficiency perhaps provides a better solution. Subulan et al. [1] claimed that reverse logistics and product recovery options, such as recycling, remanufacturing and reusing, are important issues due to the environmental and economic issues as well as the legal regulations. Kim and Lee [3] considered network design, capacity planning and vehicle routing for collection systems in reverse logistics. Dobos [9] stated that the aim of a reverse logistics system is to find optimal inventory policies with special structures, as they assumed that demand is a known continuous function in a given planning horizon and that the return rate of used items is a given function. Accordingly, they found that there is a constant delay between use and return processes. Ljungberg and Gebresenbet [10] mapped out city-center goods distribution in Uppsala and Sweden to see the possibility of reducing cost, congestion, and environmental impact by coordinating good distribution. Qualitative and quantitative data were collected via questionnaires, interviews and measurements at loading and unloading zones of retail shops. Soysal et al. [2] developed a multi-objective linear programming model for a generic beef logistics network problem. The objectives of the model are to minimize the total logistics costs and the total amount of greenhouse gas emissions due to transportation operations. Guo et al. [11] applied the Genetic Algorithm (GA) to solve the route design problem of China. As we can see, there are lots of studies invested into the logistics problems. One can refer to Konstantakopoulos et al. [4] for a detailed literature review in this field. In addition, Wang et al. [12] stated that the importance of logistics and supply chain has been amplified due to COVID-19. They proposed a hybrid multi-criteria model to evaluate third-party logistics (3PL). Duan et al. [13] claimed that agriculture decision support systems (DSSs) play an important role in improving agribusiness productivity. Thus, they presented a multicriteria analysis approach for evaluating and selecting the most appropriate agriculture DSS for sustainable agribusiness. Jiang and Zhou [14] established a supply chain utility model and discussed three different situations of supply chain members since the reasonable distribution can be a vital part in the supply chain. Paksoy et al. [15] developed a closed-loop supply chain model, describing the trade-offs between various costs considering emissions and transportations. They constructed the model in the form of linear programming formulations. Fahimnia et al. [16] studied the cost implications and carbon reduction potentials of the carbon-pricing scheme in Australia. A non-linear optimization model was constructed to depict the trade-off between transportation costs and the costs of carbon emission and fuel consumption. Özceylan et al. [17] integrated both strategic and tactical decisions among the closed-loop supply chain. The strategic level decisions consider the amounts of goods flowing in the supply chain, and tactical decisions concern balancing disassembly lines in the reverse supply chain. Özceylan et al. [18] mentioned that the increasing worldwide environmental and social concerns motivate manufacturers and consumers to implement recycling strategies. They proposed a linear programming to solve for the reverse material flows and further integrated results to forward the supply chain. Çil et al. [19] developed a mixed-model assembly line balancing (MMALB) problem with the collaboration between human workers and robots. They formulated the problem as a mixed-integer linear programming (MILP) model and further implemented the bee algorithm (BA) and artificial bee colony (ABC) algorithm to derive the solutions to a large-scale problem. Miraç and Özceylan [20] stated that the United Nations Humanitarian Response Depot UNHRD enables humanitarian actors to pre-position and stockpile relief items and support equipment for swift delivery in emergency situations. There are two different mathematical models to solve the minimization distance and maximization of the users covered. We find that the pro of these papers can be the optimal solution derived in an efficient manner, due to the single objective. The con of these papers can be the lack of taking other factors into consideration simultaneously. Therefore, the examination of previous studies indicates that there is a need to develop a model investigating not only the costs, but also the loading rate to solve the problem as mentioned earlier.

To define the result of the optimization, we introduce the loading rate of the vehicle factor. Based on the duality of the distribution costs, profits of logistics and reverse logistics to improve results, Ryu and Hyun [21] put forward an optimal modeling system that uses the push system and grouping method of effective logistics cost. We note that the TOC method can be utilized to solve the problem as mentioned, due to its efficiency. Lee et al. [22] presented an alternative method that enhances the system performance by the method of TOC. With the enhancement, they expected that the TOC methodology can be adopted by more companies, especially those that have the same characteristics. Chang and Huang [23] proposed an enhanced TOC for application in a re-entrant flow shop in which job processing times are generated from a discrete uniform distribution in which machine breakdowns are subject to an exponential distribution. As we can see, the TOC is a proven, useful approach for problems related to logistics. In this study, we consider two essential logistics factors, namely, the total transportation distance and the loading rates of transport vehicles. This paper utilizes these factors to show that the TOC further optimizes the results of the Sub-Tour reversal method and to determine the degree of improvement resulted from TOC.

3. Methodology

This paper integrates the Sub-Tour reversal and the TOC methods to optimize the logistics in improving the return loading rate problem. In phase 1, according to Hillier and Lieberman's [5] research, the Sub-Tour reversal algorithm is a useful algorithm for finding the shortest distance. The Sub-Tour reversal algorithm selects the order of some distribution centers of nodes, then reverses the visit orders and adjusts the visit orders when visiting the cities. Then, it selects the maximum reduced distance and the smallest value as the optimal solution for the ranking among the data. This Sub-Tour reversal algorithm may consist of as few as two cities. In phase 2, the TOC is utilized to optimize the return loading rate problem. The transportation performance assessment is then composed of two major factors as the total transportation distance and the loading rates of transport vehicles when returning to the initial distribution center. The TOC procedure applies only the lowest return loading rate to select the capacity-constrained resource city with the lowest return loading rate then replaces the non-capacity-constrained resource city with the highest return loading rate as the optimal solution. The Sub-Tour reversal algorithm is a common method for finding the shortest path because it can accurately and scientifically find the shortest path of the total transportation distance between established demand points. As such, it is an appropriate model for north China to apply to solve general logistics network design issues in northwest China. The TOC, in phase 2, is a research method based on bottleneck orientation. In the following, this study introduces the concept of a capacity-constrained resource (CCR) oriented in the TOC and then further optimizes the Sub-Tour reversal model results. Unlike the Sub-Tour reversal model, to find the shortest path of the total transportation distance, the TOC considers the key effect of the CCR orientation. It should be noted that it not only reduces the total transportation distance, but also improves the return transport vehicle loading rate to more than 50% for the actual goal of northern China.

3.1. Establishment of a Mathematical Model

The notations used in this paper are given as follows:

C : The standard transport capacity for refrigerated vehicles assuming for 20 tons per vehicle;
N : The number of distribution centers and demand cities;
a_n : The demands to be transported to the destination distribution center;
b_n : The demands to be transported return to the starting distribution center;
L_n : The loading rate of a single demand city n;
$M_{1,n}$: Total forward path demand of overall cities located between starting distribution center and transit distribution center;
$M_{2,n}$: Total return path demand of overall cities located between transit distribution center and destination distribution;
NV_m : Numbers of vehicles on route m;
τ_j : Single demand city j of the lowest loading rate L_n;
RLR_m : Return loading rates of refrigerated trucks driving route m returning to the starting distribution center;

We first established a mathematical model to describe the return loading rate problem as the following equation.

$$M_{1,n} = \sum_{1}^{n} a_n, \tag{1}$$

where $M_{1,n}$ denotes the total forward path demands of demand points from a_1 to an, located between the starting distribution center and transit distribution center.

$$M_{2,n} = \sum_{1}^{n} b_n, \tag{2}$$

where $M_{2,n}$ denotes the total return path demands of demand points from b_1 to b_n located between the transit distribution center and destination.

$$NV_m = \frac{M_{1,n}}{C} \tag{3}$$

The total numbers of vehicles on route m can be obtained from the total forward path demands divided by the standard transport capacity per vehicle.

$$NV_m = \lceil NV_m \rceil \tag{4}$$

If the number of vehicles on route n is not an integer number, we use the ceiling function maps NV_m to the least integer greater than or equal to NV_m.

$$RLR_m = \left(1 - \frac{(NV_m \cdot C - M_{2,n})}{(NV_m \cdot C)}\right) \cdot 100\% \tag{5}$$

The RLR_m of refrigerated trucks are the numbers of vehicles multiplied by the standard transport capacity per vehicle minus the total return path demands and then divided by the total capacity of vehicles.

3.2. Sub-Tour Reversal Model

The idea of the Sub-Tour reversal algorithm is to select the sub-sequence of some visiting cities, then simply reverse the visit sub-sequence of the cities and adjust the total visit sequence when lowering the total transportation distance of the visiting cities [5]. In this paper, there are two phases involved in optimizing logistics routes. Phase 1 utilizes the Sub-Tour reversal algorithm to find the shortest distance and then to calculate the corresponding loading rate. Suppose that if the return loading rate of more than 50% is not achieved, phase 2 applies the TOC to reach the optimal level.

After the initial mathematical model is obtained, we perform the following Sub-Tour reversal algorithm to find the shortest path of the total transportation distance. The concepts of the Sub-Tour reversal algorithm are as follows:

Step 1. Initialization: Select any feasible route as an initial solution. This initial solution does not need to pass through all of the cities but must pass through at least $N/2$ demand cities.

Step 2. Repeated: For the present solution, consider all possible sub-path reverse journeys that can be performed (except reversing the entire path) and then select the maximum reduced distance as a new solution (in case of a tie, make an arbitrary decision). Each execution is performed no more than three times, including initialization, repeated 1 and repeated 2.

Step 3. Stop rule: When there is no path to reverse to improve the current solution, stop and accept the best answer. Stop after three executions and select the smallest value as the optimal solution. Repeat up to three times. Choose the shortest route from the three repeated executions as the current optimal solution.

3.3. TOC Model

Suppose that the resulting sub-sequencing outcomes derived in phase 1 cannot meet the return loading rate by more than 50%. We perform the following TOC model to optimize the return loading rate problem. The principle of the TOC is to find out the CCR city embedded in the optimal transport route, and then the delivery vehicles substitute the CCR city to pass through non-CCRs cities to optimize the return transport vehicle loading rate. The CCR city in this study refers to the lowest return loading rate city among all of the transport points, whereas the non-CCRs cities are those with higher return loading rates. Accordingly, the non-CCRs cities can effectively replace a CCR city to optimize the results. The ideal of the TOC is to remove the CCR city and to replace it with the higher loading rate of the non-CCR cities. By doing so, we can quickly locate the CCR (i.e., lowest loading rate) city and accurately find the limitations of the route. After some iteration via replacing, the average loading rates of all cities can be balanced, achieving a return loading rate of more than 50%. The specific steps of the TOC are as follows:

Step 1. Find the CCR city. In Equation (6), the L_n represents the single loading rate of each demand city. The L_n of a single demand city can be obtained by 1 minus the standard transport capacity per vehicle minus its demand for return path, b_n, divided by the vehicle's capacity. The CCR city is the demand city with the lowest single return loading rate and calculate by equation (7). The TOC intends to select the lowest return loading rate of the demand cities from 1 to n and is assigned a CCR city.

$$L_n = \left[1 - \frac{(C - b_n)}{C}\right] \cdot 100\% \tag{6}$$

$$\tau_j = \text{Min}\{L_1, L_2, L_3, \ldots, L_n\}, \forall j \tag{7}$$

Step 2. Avoid the CCR city by deleting the demand city with the lowest single return loading rate in the current optimal solution.

Step 3. Successively replace CCRs cities having the lowest single return loading rate with non-CCRs cities and assess to determine a feasible solution that leads to the highest return loading rate and shortest route possible; substitute the CCR city with non-CCR cities if necessary.

Step 4. Stop if the return loading rate is more than 50%. Otherwise, go to step 1.

4. Results and Discussions

4.1. Problem Description

As mentioned earlier, northwest China is fruitful, and its agriculture is productive. Northwest China includes the Shaanxi, Gansu, Qinghai, Ningxia, Xinjiang provincial administrative regions. In Table 1, because of the cold chain logistics costs, we assume that one can operate the routes as shown in Figure 1, where distribution of the distance of five provinces in northwest China is on a scale of 1:80 km. In addition, the actual distance data for cold chain logistics between 15 cities are measured from the Google map. We determine three cities, 1, 7, and 13, as our distribution centers, marked as red circles in Figure 1. Distribution centers 1 and 7 are located in the western region, and distribution center 13 is closer to the eastern region. It is necessary to consider the proximity of three distribution centers to other demands, and we mark the distribution centers with red circles in Figure 1. Connections between any two of the distribution centers indicate that it is feasible for these two distribution centers to connect to each other. Contrastingly, if there is no connection, it means that it is not feasible for these two distribution centers to connect. Digitals on the lines represent the distances between any two adjacent cities. To simplify the calculation, the real distances are divided by 80, and the unit of measurement is kilometers (km). In this study, we apply the Monte Carlo simulation to generate the demands of forward and return for cold chain logistics, following the uniform distribution as shown in Table 2. We further investigate the return loading rate problem in two manners, Long-Route and Short-Route. Table 3 lists the Long-Route parameter setup derived by Table 2, and we denote A_n as 10 demanding cities: 1, 4, 5, 6, 7, 8, 9, 10, 11, 13. The a_n and b_n represent the demands of the forward and return cities such that a_n represents the demand to unload at this demand city; b_n represents the demands to be returned to the starting distribution center (node 1) and may also pass the interim distribution center (node 7). L_n represents the single point loading rate of each demand city calculated by Equation (7).

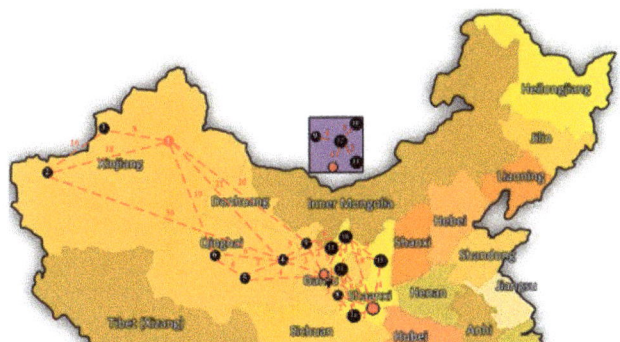

Figure 1. The northwest regional distribution centers and cities' distance (Km = 1:80).

4.2. Mathematical Modeling

In this section, we utilize two scenarios to illustrate our proposed model. The first one is the Long-Route scenario, starting from city Wu-lu-mu-qi (city 1) to city Xi-an (end city 13) and returning to city Wu-lu-mu-qi. The second one is the Short-Route scenario, starting from city Lan-zhou (city 7) to city Xi-an (end city 13) and returning to city Lan-zhou.

4.2.1. The Illustrative Example of the Long-Route Scenario

Based on Equations (1)–(5), an initial mathematical model of return loading rate problem can be obtained by using entries of columns (1) to (5) in Table 4. According to the initialization in Figure 1, the initial solution is 1-6-4-7-8-13-10-11-9-1 in the Long-Route scenario. In Table 4, due to the initial, random Long-Route pass through 10 cities, we measure $M_{1,10}$ = city 6 + city 4 + city 8 = 11 + 20 + 15 = 46, $M_{2,10}$ = city 10 + city 11 + city 9 = 6 + 10 + 5 = 21, NV_{10} = 46/20 = 2.3 and maps NV_{10} to the least integer = 3. Finally, $RLR_1 = [1 - (3 \cdot 20 - 21)/(3 \cdot 20)] \cdot 100\% = 35\%$. The initial distance from city 1 = 19 + 6 + 3 + 3 + 4 + 8 + 4 + 6 + 20 = 73 km and iteration 1 is completed.

Finally, we execute the step of the repeat and stop rule from the Sub-Tour reversal algorithm until iteration 6 is completed to find the optimal solution. When there is no path to reverse to improve the current solution in iteration 6 as 1-6-4-7-8-13-11-10-9-1, stopping and accepting the best answer of the distance is 70 km, and RLR_1 is 35% as the current optimal solution. We detail all the iterations in Table 4. The current optimal Long-Route in phase 1 is obtained; however, the return loading rate is not in excess of 50%. Then, we perform the following TOC model to optimize the return loading rate problem. The TOC finds the CCR city through the current, optimal Long-Route, where the demand city with the lowest single return loading rate is. According to Equation (6), L_n can be calculated in column 4 of Table 3. We get the following result:

$$\tau_{10} = \text{Min } \{-, 30\%, 45\%, -, 40\%, -, 50\%, 30\%, 25\%, -\} = 25\% = \text{city 9}.$$

In order to substitute CCR city, we use non-CCRs cities 4 and 5 successively to replace CCR city 9 in iteration 6. Then, we obtain iteration 7 as 1-6-5-7-8-13-11-10-4-1. The solution of the total transport distance is 72 km and the current best RLR_7 is 60%, more than 50%. Thus, we obtain the optimize solution as 1-6-5-7-8-13-11-10-4-1 from the initial solution as shown in Figure 2.

Table 1. Distances distribution of five provinces in northwest China (km = 1:80).

Province	City Node	(1)	(2)	(3)	(4)	(5)	(6)	(7)	(8)	(9)	(10)	(11)	(12)	(13)	(14)	(15)
Xinjiang	(1)	0	18	8	21	23	19	24	27	20	25	26	24	31	30	30
	(2)	18	0	16	36	34	30	38	42	35	41	41	39	46	44	45
	(3)	8	16	0	30	29	25	32	36	29	33	35	32	40	38	38
Qinghai	(4)	21	36	30	0	2	6	3	6	5	8	7	8	10	8	11
	(5)	23	34	29	2	0	4	4	8	4	9	8	7	12	10	13
	(6)	19	30	25	6	4	0	9	12	9	14	13	12	16	14	17
Gansu	(7)	24	38	32	3	4	9	0	3	4	5	4	4	8	6	8
	(8)	27	42	36	6	8	12	3	0	7	7	3	6	4	2	7
	(9)	20	35	29	5	4	9	4	7	0	6	6	4	11	9	9
Ningxia	(10)	25	41	33	8	9	14	5	7	6	0	4	3	8	8	5
	(11)	26	41	35	7	8	13	4	3	6	4	0	3	5	4	6
	(12)	24	39	32	8	7	12	4	6	4	3	3	0	8	6	6
Shanxi	(13)	31	46	40	10	12	16	8	4	11	8	5	8	0	2	4
	(14)	30	44	38	8	10	14	6	2	9	8	4	6	2	0	3
	(15)	30	45	38	11	13	17	8	7	9	5	6	4	3	3	0

Note (1): The 15 cities are (1) Wu-lu-mu-qi, (2) Ka-shi, (3) I-li, (4) Xi-ning, (5) Hai-nan, (6) Hai-xi,(7) Lan-zhou, (8) Tian-shui, (9) Wu-wei, (10) Yin-chuan, (11) Gu-yuan, (12) Zhong-wei, (13) Xi-an, (14) Bao-ji, (15) Yan-an. Note (2): Nodes 1, 7, and 13 as our distribution centers.

Table 2. The demands of the forward and return cities of the uniform distribution in the Monte Carlo simulation.

Demand Point (A_n)	Forward Demands (a_n)	Return Demands (b_n)
[8,10]	[10,20]	[1,10]

Table 3. The demands of the forward and return cities of the Long-Route example in the Monte Carlo simulation.

Demand Point (A_n)	Forward Demands (a_n)	Return Demands (b_n)	Load Rate (L_n) %
1	-	-	-
4	20	10	50%
5	12	7	35%
6	11	6	30%
7	-	-	-
8	15	8	40%
9	14	5	25%
10	10	6	30%
11	18	10	50%
13	-	-	-

Note: Exclusive of nodes 1, 7, and 13 as our distribution centers.

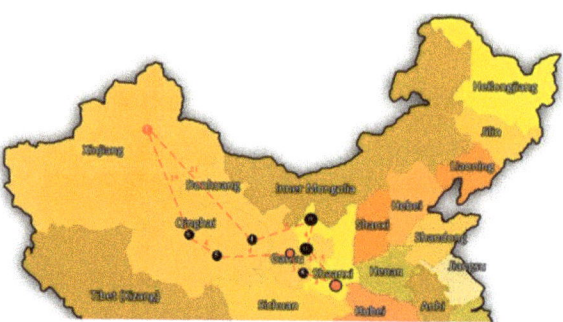

Figure 2. The optimizing solution of Long-Route scenario in TOC (Km = 1:80).

Table 4. The illustrative example of the Long-Route in TOC model.

		Sub-Tour Reversal Model					TOC Model			
	(1)	(2)	(3)	(4)	(5)	(6)	(7)	(8)	(9)	(10)
Iterations	Routes	M_1	M_2	NV_m	Distance (km)	RLR_m (%)	Routes	CCR	Distance (km)	RLR_m (%)
1	1-**6-4**-7-8-13-10-11-9-1	46	21	3	73	35%	-	-	-	-
2	1-**4-6**-7-8-13-10-11-9-1	46	21	3	75	35%	-	-	-	-
3	1-4-6-7-8-13-**11-10**-9-1	46	21	3	72	35%	-	-	-	-
4	1-4-6-7-8-13-**10-11**-9-1	46	21	3	75	35%	-	-	-	-
5	1-4-6-7-8-13-**11-10**-9-1	46	21	3	72	35%	-	-	-	-
6	1-**6-4**-7-8-13-11-10-9-1	46	21	3	70 *	35% *	1-6-4-7-8-13-11-10-9-1	City 9 (25%)	-	-
7	-	-	-	-	70	35%	1-6-5-7-8-13-11-10-**4**-1	City 4 (50%)	72	60% *

4.2.2. The Illustrative Example of the Short-Route Scenario

To introduce our proposed solution procedure, we demonstrate the Short-Route in Table 5, where we denote A_n as 8 demanding cities: 4, 7, 9, 10, 11, 12, 13, 15. Now, suppose that the initial solution is 7-9-10-15-12-11-13-7 in the Short-Route scenario. In Table 6, due to the initial, random Short-Route pass through 8 cities, we measured $M_{1,8}$ = city 9 + city 10 + city 15 + city 12 + city 11 = 80, $M_{2,10}$ = city 7 = 0, NV_8 = 80/20 = 4, and maps NV_8 to the least integer 4. Finally, $RLR_1 = [1 - (4 \cdot 20 - 0)/(4 \cdot 20)] \cdot 100\% = 0\%$. The initial distance from city 7 is 37 km, and iteration 1 is complete. Finally, we perform the Sub-Tour reversal algorithm until iteration 5 is done to find the optimal solution. When there is no path to reverse to improve the current solution in iteration 5 as 7-9-10-12-15-13-11-7, stopping and accepting the best answer of the distance is 32 km, and RLR_1 is 23.75% as the current optimal solution. We present all the iterations in Table 6. The current optimal Short-Route in phase 1 is obtained; however, the return loading rate of more than 50% is not met. We perform the TOC model to optimize the return loading rate problem. The TOC finds the CCR city through the current optimal Short-Route, where the demand city is with the lowest single return loading rate in Table 5. Then, we get the following:

$$\tau_8 = \text{Min } \{-, 15\%, 25\%, 30\%, 35\%, -, 35\%, -\} = 15\% = \text{city } 9.$$

In order to avoid CCR cities, we use non-CCR city 4 to replace CCR city 9 in iteration 6. We obtain iteration 6 as 7-4-10-13-15-12-11-7. The solution of the total transport distance is 36 km and the current best RLR_6 is 47.5%, somehow less than 50%. Thus, we obtain the optimal solution as 7-4-10-13-15-12-11-7 from the initial solution as shown in Figure 3.

Table 5. The demands of the forward and return cities of the Short-Route example in the Monte Carlo simulation.

Demand Point (A_n)	Forward Demands (a_n)	Return Demands (b_n)	Load Rate (L_n) %
4	18	9	45%
7	-	-	-
9	13	3	15%
10	15	5	25%
11	19	7	35%
12	14	6	30%
13	-	-	-
15	19	7	35%

Table 6. The illustrative example of the Short-Route in the TOC model.

	Sub-Tour Reversal Model						TOC Model			
	(1)	(2)	(3)	(4)	(5)	(6)	(7)	(8)	(9)	(10)
Iterations	Routes	M_1	M_2	NV_m	Distance (km)	RLR_m (%)	Routes	CCR	Distance (km)	RLR_m (%)
1	7-9-10-**15-12**-11-13-7	80	0	4	37	0%	-	-	-	-
2	7-9-10-**12-15**-11-13-7	80	0	4	37	0%	-	-	-	-
3	7-9-10-12-15-**13-11**-7	61	19	4	32	23.75%	-	-	-	-
4	7-9-10-**15-12**-13-11-7	61	19	4	32	23.75%	-	-	-	-
5	7-9-10-**12-15**-13-11-7	61	19	4	32	23.75% *	7-**9**-10-12-15-13-11-7	City 9 (15%)	-	-
6	-	-	-	-	32	23.75%	7-**4**-10-13-15-12-11-7	City 4 (45%)	36	47.5% *

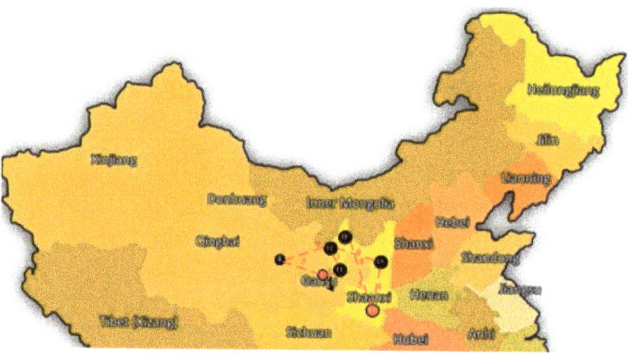

Figure 3. The optimizing solution of Short-Route scenario in TOC (km = 1:80).

4.2.3. Discussion

To solve the return loading rate problem, this study further investigates the Long-Route and Short-Route scenarios. In the Long-Route scenario, Table 3 shows the demands of the forward and return cities, adopting the Monte Carlo simulation. To substitute CCR city 9, we use non-CCR city 4 with a loading rate of 50%, successively, to replace CCR city 9 with a loading rate of 25% and then determine a feasible solution that leads to the highest return loading rate of 60%. In Table 4, after iteration 6 of the Long-Route, the current best distance is 70 km, which is less than 73 km in the initial solution. This can be achieved by using the Sub-Tour reversal model and indicates a 4.1% reduction from the initial solution. However, the current best return loading rate (RLR_6) remains the same as in the initial solution, i.e., 35%. The expected goal of a return loading rate of more than 50% is not achieved by using the Sub-Tour reversal model. Therefore, we introduce the TOC model to optimize the return loading rate. By using the TOC model, we remove CCR city 9 and replace it with the higher return loading rate, i.e., non-CCR city 4. The final results indicate that as we consider the CCR city, the distance and its return loading rate are higher than the initial solution and outperform the Sub-Tour reversal model. In Figure 2, we obtain the optimal Long-Route, starting from Wu-lu-mu-qi (city 1) to Hai-xi (city 6), Hai-nan (city 5), Lan-zhou (city 7), Tian-shui (city 8), Xi-an (city 13), Gu-yuan (city 11), Yin-chuan (city 10), Xi-ning (city 4) and returning to Wu-lu-mu-qi (city 1). The optimal Long-Route derives the total transport distance of 72 km with the best return loading rate of 60%. In the Short-Route scenario, Table 5 shows the demands of the forward and return cities by applying the Monte Carlo simulation. To substitute CCR city 9, we use non-CCR city 4 with a loading rate of 45%, successively, to replace the CCR city 9 with a loading rate of

15% and then determine a feasible solution that leads to the highest return loading rate of 47.5%.

In Table 6, after iteration 1 of the Short-Route, the current best distance decreases from the initial solution of 37 km to 32 km. The current best return loading rate increases from the initial solution of 0% to 23.75%, for a 23.75% improvement via the Sub-Tour reversal model. With the goal of improving the return loading rate to more than 50%, this study achieves the goal by using the TOC model because of the final loading rate being 47.5%. The former route can be derived by the Sub-Tour reversal model but fails to meet expectations according to the results of a distance of 32 km and a return loading rate of 23.75%. Accordingly, this study achieves its goal by using the TOC model such that the final results can be a distance equal to 36 km and a return loading rate equal to 47.5%. In Figure 3, we obtain the optimal Short-Route starting from Lan-zhou (city 7) to Xi-ning (city 4), Yin-chuan (city 10), Xi-an (city 13), Yan-an (city 15), Zhong-wei (city 12), Gu-yuan (city 11) and returning to Lan-zhou (city 7). The optimal Short-Route gets the total transport distance of 36 km with the best return loading rate of 47.5%.

To see our model capability, this paper utilizes the simulation data to validate the TOC model and adopts the Monte Carlo simulation method to test the deriving solutions. Based on the resulting outcomes in Tables 7–9, the TOC model effectively obtains the best routes design. Later, we discuss the simulation and its statistics in detail.

4.2.4. The Simulation and Its Statistics

Furthermore, this paper utilizes the simulation data to validate our model and adopts the Monte Carlo simulation method, following the uniform distribution in Table 2 to test the deriving solutions. It simulates 30 replications of datasets for certain demand cities (A_n) 10 and 8, forward (a_n) with a range between 10 and 20 tons and a return (b_n) with a range between 1 and 10 tons in Table 2. By our proposed model, we summarize our results as shown in Tables 7 and 8.

We know that the average return loading rate derived by the Sub-Tour reversal model is 24.07% and by TOC model, it is 67.33%, based on northwest China's Long-Route scenario shown in Table 7. This study further employs the pair t-test to see whether the mean difference is significant or not. Based on the data, the t statistic is 18.3190 and the p value is less than 0.05. It means that there exists a significant difference among the two models, and our model leads to an improvement of 43.26% compared to that obtained by the Sub-Tour reversal model. Following the same procedure, the TOC model results in an improvement of 44.02%, comparing to that obtained by the Sub-Tour reversal model from the Short-Route scenario in Table 8. The t statistic is 11.0204 and the p value is less than 0.05. Obviously, our proposed solution procedure can be applied to solve the return loading rate problem in an efficient manner. The resulting outcomes are arranged in Table 9. By the resulting outcomes, our proposed solution procedure successfully integrates the Sub-Tour reversal model and TOC. The results show the superiority of the TOC model in solving issues in logistics and answering the question mentioned by previous studies (see Schragenheim and Dettmer [24]; Lee et al. [13]; Chang and Huang [14]; Benavides and Landeghem [25]; Chakravorty and Hales [26]). The TOC model achieves the logistics goal for increasing the return loading rate by increasing by more than 50%. Logistics companies would benefit substantially from the application of the TOC model.

Table 7. The 30 replications simulation results of the Long-Route scenario in the Monte Carlo simulation.

	Sub-Tour Reversal Model						TOC Model			
	(1)	(2)	(3)	(4)	(5)	(6)	(7)	(8)	(9)	(10)
Runs	Routes	M_1	M_2	NV_m	Distance (km)	RLR_m (%)	Routes	CCR	Distance (km)	RLR_m (%)
1	1-4-12-7-8-13-11-14-5-1	42	22	3	82	36.67%	1-10-6-7-5-13-11-4-12-1	City 10	106	72.5%
2	1-2-4-10-14-13-12-9-8-1	49	10	3	118	16.67%	1-2-6-10-7-13-12-11-4-1	City 2	114	72.5%
3	1-11-6-7-3-13-14-8-4-1	46	17	3	151	28.33%	1-11-6-7-2-13-12-4-10-1	City 2	148	57.5%
4	1-7-8-12-6-13-2-15-14-1	41	10	3	185	16.67%	1-7-15-6-8-13-4-11-12-1	City 8	109	72.5%
5	1-8-2-14-4-13-6-3-12-1	49	19	3	228	31.67%	1-8-2-7-6-13-4-11-12-1	City 8	176	72.5%
6	1-2-11-3-12-13-5-8-4-1	65	17	4	181	21.25%	1-2-5-7-6-13-11-12-4-1	City 2	118	72.5%
7	1-6-7-12-5-13-10-14-2-1	45	12	3	129	20%	1-10-7-6-5-13-12-11-4-1	City 10	94	72.5%
8	1-9-4-15-10-13-12-8-14-1	49	16	3	95	26.67%	1-6-7-15-10-13-12-11-4-1	City 10	88	72.5%
9	1-15-9-2-4-13-11-14-12-1	49	25	3	159	41.67%	1-7-9-2-6-13-11-4-12-1	City 9	153	72.5%
10	1-12-14-2-5-13-10-4-8-1	61	15	4	169	18.75%	1-10-7-2-5-13-12-4-11-1	City 2	163	72.5%
11	1-11-12-7-4-13-5-3-2-1	49	11	3	121	18.33%	1-5-2-7-6-13-11-12-4-1	City 2	157	72.5%
12	1-3-9-5-2-13-15-10-12-1	54	15	3	157	25%	1-7-9-5-2-13-4-11-12-1	City 9	156	72.5%
13	1-7-8-11-3-13-4-5-6-1	45	23	3	140	38.33%	1-7-5-6-2-13-4-11-12-1	City 2	152	72.5%
14	1-10-11-9-7-13-6-5-14-1	39	19	2	107	47.5%	1-10-11-9-7-13-12-5-4-1	City 9	85	62.5%
15	1-7-11-9-15-13-12-14-6-1	44	22	3	94	36.67%	1-7-2-9-15-13-12-14-4-1	City 9	153	62.5%
16	1-12-3-11-5-13-10-7-8-1	69	5	4	154	6.25%	1-7-2-10-5-13-11-12-4-1	City 2	161	72.5%
17	1-12-7-3-2-13-5-11-15-1	48	18	3	178	30%	1-5-7-8-2-13-12-11-4-1	City 8	157	72.5%
18	1-10-15-11-7-13-9-4-8-1	42	11	3	97	18.33%	1-10-15-8-7-13-11-4-12-1	City 8	92	72.5%
19	1-6-3-9-4-13-8-14-15-1	53	9	3	127	15%	1-6-7-9-8-13-4-11-12-1	City 9	87	72.5%
20	1-15-7-10-8-13-9-14-2-1	35	8	2	136	20%	1-15-7-10-8-13-12-11-2-1	City 8	124	52.50%
21	1-5-15-3-4-13-8-9-11-1	59	11	3	157	18.33%	1-5-6-2-7-13-12-4-11-1	City 2	152	72.5%
22	1-3-11-14-5-13-6-9-4-1	66	17	4	120	21.25%	1-3-2-7-5-13-12-11-4-1	City 2	117	72.5%
23	1-15-10-3-5-13-2-14-4-1	57	18	3	228	30%	1-15-10-2-7-13-11-14-4-1	City 15	160	65%
24	1-7-10-15-6-13-8-4-5-1	36	17	2	102	42.5%	1-7-10-15-6-13-12-4-5-1	City 15	99	62.5%
25	1-3-4-10-6-13-11-5-14-1	51	22	3	129	36.67%	1-7-4-10-6-13-11-5-14-1	City 10	118	55%
26	1-11-9-2-15-13-14-3-7-1	54	9	3	180	15%	1-7-9-2-15-13-11-3-4-1	City 9	203	57.5%
27	1-10-8-4-6-13-7-2-9-1	43	2	3	161	3.33%	1-10-8-7-6-13-12-2-4-1	City 8	164	52.5%
28	1-5-12-6-14-13-3-2-7-1	62	5	4	174	6.25%	1-5-7-6-2-13-3-4-12-1	City 2	214	55%
29	1-14-3-11-10-13-12-6-7-1	62	16	4	168	20%	1-7-3-2-10-13-12-6-11-1	City 2	180	65%
30	1-11-8-3-15-13-7-6-2-1	60	9	3	139	15%	1-7-8-2-15-13-11-6-4-1	City 8	163	67.5%
					Avg. 145.53km	24.07%		Avg.	138.77km	67.33%

Table 8. The 30 replications simulation results of the Short-Route scenario in the Monte Carlo simulation.

	Sub-Tour Reversal Model						TOC Model			
	(1)	(2)	(3)	(4)	(5)	(6)	(7)	(8)	(9)	(10)
Runs	Routes	M_1	M_2	NV_m	Distance (km)	RLR_m (%)	Routes	CCR	Distance (km)	RLR_m (%)
1	7-15-8-12-13-14-10-7	45	10	3	44	16.67%	7-15-8-10-13-14-11-7	City 8	40	47.5%
2	7-12-14-10-13-15-11-7	47	12	3	40	20%	7-8-13-10-14-12-11-7	City 8	36	95%
3	7-14-15-12-13-10-11-7	52	14	3	39	23.33%	7-10-13-8-15-12-11-7	City 8	37	95%
4	7-11-8-9-13-14-10-7	39	10	2	40	16.67%	7-10-8-13-9-14-11-7	City 9	44	80%
5	7-8-9-12-13-11-15-7	42	12	3	41	20%	7-8-10-13-9-11-15-7	City 9	49	60%
6	7-10-12-9-13-8-11-7	42	11	3	34	18.33%	7-10-13-8-9-12-11-7	City 9	35	95%
7	7-8-12-14-13-11-9-7	47	10	3	32	16.67%	7-8-14-9-13-12-11-7	City 9	40	47.5%
8	7-9-11-8-13-10-12-7	39	13	2	32	32.5%	7-9-10-8-13-11-12-7	City 9	33	47.5%

Table 8. Cont.

	Sub-Tour Reversal Model						TOC Model			
	(1)	(2)	(3)	(4)	(5)	(6)	(7)	(8)	(9)	(10)
Runs	Routes	M_1	M_2	NV_m	Distance (km)	RLR_m (%)	Routes	CCR	Distance (km)	RLR_m (%)
9	7-10-9-8-13-11-15-7	32	12	2	41	30%	7-10-13-8-9-11-12-7	City 9	37	95%
10	7-11-10-8-13-14-15-7	37	8	2	32	20%	7-14-10-8-13-11-12-7	City 8	37	47.5%
11	7-10-11-14-13-12-8-7	44	10	3	32	16.67%	7-10-8-13-14-12-11-7	City 8	31	95%
12	7-14-9-10-13-8-12-7	39	10	2	43	16.67%	7-14-8-10-13-11-12-7	City 8	35	47.5%
13	7-8-15-14-13-12-9-7	42	9	3	31	15%	7-8-10-13-14-12-11-7	City 8	33	95%
14	7-15-14-11-13-9-12-7	49	9	3	38	15%	7-15-9-13-14-12-11-7	City 9	43	47.5%
15	7-12-8-10-13-15-11-7	40	12	2	39	30%	7-8-13-10-15-11-12-7	City 8	33	95%
16	7-10-9-14-13-15-12-7	39	11	2	36	27.5%	7-10-9-8-13-11-12-7	City 9	34	47.5%
17	7-8-11-10-13-14-15-7	37	8	2	31	20%	7-8-13-10-15-14-11-7	City 8	31	80%
18	7-15-12-8-13-11-9-7	45	10	3	39	16.67%	7-15-8-9-13-12-11-7	City 9	48	47.5%
19	7-11-9-10-13-12-8-7	39	10	2	41	16.67%	7-8-13-10-9-12-11-7	City 9	32	95%
20	7-14-15-10-13-8-11-7	42	11	3	33	18.33%	7-10-8-13-14-11-15-7	City 8	36	60%
21	7-12-14-8-13-10-11-7	47	14	3	32	23.33%	7-10-13-8-14-12-11-7	City 8	32	95%
22	7-9-15-14-13-11-8-7	44	11	3	29	18.33%	7-9-15-13-14-11-12-7	City 9	30	47.5%
23	7-11-8-14-13-9-12-7	44	9	3	30	15%	7-9-8-14-13-11-12-7	City 9	27	47.5%
24	7-9-14-10-13-15-12-7	39	11	2	43	27.5%	7-9-14-10-13-11-12-7	City 9	41	47.5%
25	7-8-9-10-13-15-11-7	32	12	2	38	30%	7-8-13-10-9-15-11-7	City 9	40	60%
26	7-15-8-11-13-10-12-7	42	13	3	38	21.67%	7-15-8-10-13-11-12-7	City 8	42	47.5%
27	7-12-9-8-13-15-11-7	42	12	3	33	20%	7-9-15-8-13-11-12-7	City 9	36	47.5%
28	7-10-12-15-13-11-9-7	45	10	3	33	16.67%	7-10-9-15-13-11-12-7	City 9	36	47.5%
29	7-15-12-9-13-10-14-7	47	10	3	43	16.67%	7-15-10-9-13-12-14-7	City 9	50	37.5%
30	7-15-11-8-13-12-10-7	42	13	3	37	21.67%	7-15-10-8-13-12-11-7	City 8	39	47.5%
				Avg.	36.47km	20.59%		Avg.	37.23km	64.83%

Table 9. The pair t-test result.

	Pair Difference		t Statistics	p Value	df
	Mean	Std. Deviation			
(1) Long-Route scenario	0.4326	0.1294	18.3190	0.001 <	29
(2) Short-Route scenario	0.4425	0.2199	11.0203	0.001 <	29

5. Conclusions

This study intends to propose a solution procedure to reduce the total transportation distance and to improve the return transportation vehicle loading rate by more than 50% in northern China. Our research contributes to developing the model considering two factors: the total transportation distance and the return loading rates of transport vehicles, which are different from previous studies mainly focusing on minimizing total logistics costs. The TOC model further optimizes the results of the Sub-Tour reversal method to determine the degree of improvement resulted from the TOC model. Based on bottleneck orientation, this paper broadens the view of the existing research field. We integrated the Sub-Tour reversal model and the TOC methodology, as the two-phase solution procedure solves the problem of the return loading rate. Usually, the Sub-Tour reversal model is applied to solve the minimization transportation distance problem. Our model adopts the deriving results of the Sub-Tour reversal model as the initial solution. Next, we applied the TOC model, employing the CCR concept to further optimize the current solution since the TOC model can quickly locate the CCR city (i.e., lowest return loading rate) and accurately find the limitations of the route. It should be noted that to find the optimal solution by maximizing the utility of the CCR in this study means replacing the CCR city with a non-CCR. In order to demonstrate our model capability, we further utilized two scenarios and employed the Monte Carlo simulation. In the northwest logistics network design, as presented in

Tables 7 and 8, the return loading rate is significantly improved by our proposed solution procedure, and this shows that our proposed model outperforms the conventional Sub-Tour reversal method. Note that our solution procedure can be implemented in real-world situations in a simple manner. The limitations of the study are calculating the total logistics costs of the optimized transportation distance. For future work, we suggest a number of issues for future researchers. (1) Our research investigated the deterministic manner. For future investigations, we may take uncertainty factors into account. (2) Our future model would further consider other factors, such as environmental protection, carbon pricing or emission. (3) In this research, we integrated the Sub-Tour reversal model with TOC. We can employ other approaches in future research.

Author Contributions: W.-T.H. contributed to manuscript preparation, experiment planning, and experiment measurements. C.-C.L. contributed to literature review and reviewer comments review. J.-F.D. contributed to data analysis, review and editing. All authors have read and agreed to the published version of the manuscript.

Funding: This research was funded by the Ministry of Science and Technology of Taiwan under Grant 109-2222-E-035 -007-.

Institutional Review Board Statement: Not applicable.

Informed Consent Statement: Not applicable.

Conflicts of Interest: The authors declare no conflict of interest.

References

1. Subulan, K.; Baykasoğlu, A.; Saltabas, A. An improved decoding procedure and seeker optimization algorithm for reverse logistics network design problem. *J. Intell. Fuzzy Syst.* **2014**, *27*, 2703–2714. [CrossRef]
2. Soysal, M.; Bloemhof-Ruwaard, J.M.; van der Vorst, J.G.A.J. Modelling food logistics networks with emission considerations: The case of an international beef supply chain. *Int. J. Prod. Econ.* **2014**, *152*, 57–70. [CrossRef]
3. Kim, J.S.; Lee, D.L. An integrated approach for collection network design, capacity planning and vehicle routing in reverse logistics. *J. Oper. Res. Soc.* **2015**, *66*, 76–85. [CrossRef]
4. Konstantakopoulos, G.D.; Gayialis, S.P.; Kechagias, E.P. Vehicle routing problem and related algorithms for logistics distribution: A literature review and classification. *Oper. Res.* **2020**, 1–30. [CrossRef]
5. Hillier, F.S.; Lieberman, G.J. *Introduction to Operations Research*; McGraw-Hill, Inc.: New York, NY, USA, 2006.
6. Moroko, W.; Caflisch, R. Quasi-Monte Carlo Simulation of Random Walks in Finance. In *Athens Conference on Applied Probability and Time Series Analysis*; Niederreiter, H., Hellekalek, P., Larcher, G., Zinterhof, P., Eds.; Springer: New York, NY, USA, 1998; Volume 127, pp. 340–352.
7. Huang, W.-T.; Chen, P.-S.; Liu, J.J.; Chen, Y.-R.; Chen, Y.-H. Dynamic configuration scheduling problem for stochastic medical resources. *J. Biomed. Inform.* **2018**, *80*, 96–105. [CrossRef]
8. Daganzo, C.F. The distance traveled to visit n points with a maximum of c stops per vehicle: An analytic model and an application. *Transp. Sci.* **1984**, *18*, 331–350. [CrossRef]
9. Dobos, I. Optimal production–inventory strategies for a HMMS-type reverse logistics system. *Int. J. Prod. Econ.* **2003**, *81-82*, 351–360. [CrossRef]
10. Ljungberg, D.; Gebresenbet, G. Mapping out the potential for coordinated goods distribution in city centres: The case of Uppsala. *Int. J. Transp. Manag.* **2004**, *2*, 161–172. [CrossRef]
11. Guo, R.; Guan, W.; Zhang, W. Route design problem of customized buses: Mixed integer programming model and case study. *J. Transp. Eng. Part A Syst.* **2018**, *144*, 04018069. [CrossRef]
12. Wang, C.-N.; Nguyen, N.-A.-T.; Dang, T.-T.; Lu, C.-M. A Compromised Decision-Making Approach to Third-Party Logistics Selection in Sustainable Supply Chain Using Fuzzy AHP and Fuzzy VIKOR Methods. *Mathematics* **2021**, *9*, 886. [CrossRef]
13. Duan, S.X.; Wibowo, S.; Chong, J. A multicriteria analysis approach for evaluating the performance of agriculture decision support systems for sustainable agribusiness. *Mathematics* **2021**, *9*, 884. [CrossRef]
14. Jiang, X.; Zhou, J. The Impact of Rebate Distribution on Fairness Concerns in Supply Chains. *Mathematics* **2021**, *9*, 778. [CrossRef]
15. Paksoy, T.; Bektaş, T.; Özceylan, E. Operational and environmental performance measures in a multi-product closed-loop supply chain. *Transp. Res. Part E: Logist. Transp. Rev.* **2011**, *47*, 532–546. [CrossRef]
16. Fahimnia, B.; Reisi, M.; Paksoy, T.; Özceylan, E. The implications of carbon pricing in Australia: An industrial logistics planning case study. *Transp. Res. Part D: Transp. Environ.* **2013**, *18*, 78–85. [CrossRef]
17. Özceylan, E.; Paksoy, T.; Bektas, T. Modeling and optimizing the integrated problem of closed-loop supply chain network design and disassembly line balancing. *Transp. Res. Part E Logist. Transp. Rev.* **2014**, *61*, 142–164. [CrossRef]

18. Özceylan, E.; Demirel, N.; Çetinkaya, C.; Demirel, E. A closed-loop supply chain network design for automotive industry in Turkey. *Comput. Ind. Eng.* **2017**, *113*, 727–745. [CrossRef]
19. Çil, Z.A.; Li, Z.; Mete, S.; Özceylan, E. Mathematical model and bee algorithms for mixed-model assembly line balancing problem with physical human–robot collaboration. *Appl. Soft Comput.* **2020**, *93*, 106394. [CrossRef]
20. Miraç, E.; Özceylan, E. P-median and maximum coverage models for optimization of distribution plans: A case of United Nations Humanitarian response depots. In *Smart and Sustainable Supply Chain and Logistics—Trends, Challenges, Methods and Best Practices*; Golinska-Dawson, P., Tsai, K.M., Kosacka-Olejnik, M., Eds.; EcoProduction (Environmental Issues in Logistics and Manufacturing); Springer: Cham, Switzerland, 2020.
21. Ryu, B.W.; Hyun, P.J. The study of logistics optimization model with empty transfer rate of reverse logistics. *J. Korea Saf. Manag. Sci.* **2006**, *8*, 125–141.
22. Lee, J.-H.; Chang, J.-G.; Tsai, C.-H.; Li, R.-K. Research on enhancement of TOC Simplified Drum-Buffer-Rope system using novel generic procedures. *Expert Syst. Appl.* **2010**, *37*, 3747–3754. [CrossRef]
23. Chang, Y.-C.; Huang, W.-T. An enhanced model for SDBR in a random reentrant flow shop environment. *Int. J. Prod. Res.* **2014**, *52*, 1808–1826. [CrossRef]
24. Schragenheim, E.; Dettmer, H.W. *Manufacturing at Warp Speed: Optimizing Supply Chain Financial Performance*; CRC Press: Boca Raton, FL, USA, 2000.
25. Benavides, M.B.; Landeghem, H.V. Implementation of S-DBR in four manufacturing SMEs: A research case study. *Prod. Plan. Control* **2015**, *26*, 1110–1127. [CrossRef]
26. Chakravorty, S.S.; Hales, D.N. Improving labor relations performance using a Simplified Drum Buffer Rope (S-DBR) technique. *Prod. Plan. Control* **2016**, *27*, 102–113. [CrossRef]

A Kronecker Algebra Formulation for Markov Activity Networks with Phase-Type Distributions

Alessio Angius [1], András Horváth [2,*] and Marcello Urgo [3]

[1] Enerbrain, 10132 Turin, Italy; alessio.angius.research@gmail.com
[2] Computer Science Department, University of Turin, 10149 Turin, Italy
[3] Mechanical Engineering Department, Polytechnic University of Milan, 20133 Milan, Italy; marcello.urgo@polimi.it
* Correspondence: horvath@di.unito.it

Abstract: The application of theoretical scheduling approaches to the real world quite often crashes into the need to cope with uncertain events and incomplete information. Stochastic scheduling approaches exploiting Markov models have been proposed for this class of problems with the limitation to exponential durations. Phase-type approximations provide a tool to overcome this limitation. This paper proposes a general approach for using phase-type distributions to model the execution of a network of activities with generally distributed durations through a Markov chain. An analytical representation of the infinitesimal generator of the Markov chain in terms of Kronecker algebra is proposed, providing a general formulation for this class of problems and supporting more efficient computation methods. This entails the capability to address stochastic scheduling in terms of the estimation of the distribution of common objective functions (i.e., makespan, lateness), enabling the use of risk measures to address robustness.

Keywords: stochastic makespan; markov activity network; phase-type distribution

1. Introduction

In the application of scheduling to real planning problems, such as industrial production, research and development, or software development, uncertainty or incomplete information are inevitably present. Deviations from what was planned can be due to a wide range of possible disturbances, both internal and external, affecting the execution of the scheduled activities. Among the most common source of disturbances are activities taking more or less than originally estimated, machine breakdowns, worker absenteeism, delayed supplies of materials and/or components, and so forth. A disrupted schedule quite often entails a cost due to missed due dates, resource idleness, a higher work-in-process inventory, or scarce utilization of the available resources.

Scheduling approaches should be able to cope with this uncertainty, such as exploiting stochastic models [1]. A specific class of approaches aims at providing a robust schedule, such as a schedule incorporating a protection, at least to a certain extent, against the possible occurrence of uncertain events. Robust scheduling approaches are based on a stochastic model for the objective function to optimize, and quite often require the calculation of the associated stochastic distribution. Nevertheless, this is a difficult problem to solve; hence, many approaches have been proposed to provide exact estimations, approximations, bounds or heuristic methods to cope with this problem. In this paper, grounded on the results in [2], we exploit a Markov chain to model the execution of a stochastic program evaluation and review technique (PERT) network; moreover, exploiting the approximation of phase-type (PH) distributions, we extend this model to generally distributed durations of the activity and general phase-type forms, thus generalizing the preliminary formalization described in [3] through the Kronecker algebra. This formal approach allows significant benefits in comparison with the existing approaches in the literature exploiting phase-type approximations [3–6]. In fact, it allows the embedding of PH distributions that are able to reach high/low coefficients of variations and higher-moment-matching with a number of phases that are smaller, compared to other phase-type subclasses. In addition, it provides a compact expression of the infinitesimal generator of the resulting continuous time Markov chain (CTMC), exploiting Kronecker algebra, without the need of explicitly enumerating all the possible states. Thank to this, it is possible to tackle the problem avoiding an uncontrollable increase of the dimension of the state space governing the Markov chain, which is typical when phase-type distributions are used to model multiple activities executed in parallel. Moreover, this also supports the definition of efficient and modular calculation strategies to solve the Markov chain and, hence, estimate the distribution of its time to absorption, that is, the makespan of the PERT network [7–9]. A test of the proposed approach and calculation method is provided, to assess the possibility to use phase-type distributions to approximate generally distributed stochastic durations with a reasonable number of phases.

The paper is organized as follows: Section 2 addresses the related literature, Section 3 describes Markov activity networks, while Sections 4 and 5 address phase-type distributions and their embedding in Markov activity networks, respectively. Section 6 shows how to deal with phase-type distributions in Markov activity networks using Kronecker algebra, and in Section 7 we report on several numerical experiments with the proposed approach. The paper is concluded in Section 8.

2. State of the Art

The literature related to stochastic scheduling mostly addresses scheduling problems where the duration of the jobs are modeled as random variables. A first class of approaches focuses on analytical approaches, with the aim of developing proper scheduling policies. To this aim, different classes of policies have been investigated together with their performance in the stochastic version of the considered scheduling problem. With the aim not to constrain the analysis to specific scheduling problems, we will address the project scheduling approaches as a generalization of any schedule problem. A network of activities with stochastic durations is often referred to as a stochastic PERT network. In this field, *pres-elective* policies have been proposed in [10,11] and further developed in [12]. Based on these theoretical results, specific dominance rules have been proposed by [12,13] and exploited by [13] in branch-and-bound algorithms for the optimization of the expected makespan.

Although being a reasonable objective function, the expected value of the makespan, as well as any other objective function addressed in terms of its expected value, does not entirely model the stochastic nature of the problem. In fact, minimizing the expected makespan aims at ensuring an average good performance, but does not protect against the worst case scenario if their probability is low [14–17]. A balanced compromise between values and the impact of rare but unfavourable events typically requires knowledge of the distribution of the objective function under study.

However, this problem has been demonstrated to be hard to solve in general [18]; hence, numerous approaches have been proposed to provide proper bounds to the distribution function [19,20]. For this reason, heuristics approaches for this class of problems have been proposed by many authors, where some examples are [21,22].

Under the hypothesis that the durations are independent and exponentially distributed, ref. [2] developed an exact approach for the calculation of the distribution of the makespan using a continuous-time Markov chain (CTMC) model. This approach has been a starting point for the work of many authors [23,24] exploiting the exponential distribution to support the analysis of the execution of the activities in the network in terms of net present value (NPV).

A different approach is to include in the model phase-type (PH) distributions which allow to approximate the behavior of general distributions by preserving the Markov property. Their use to provide an approximation to generic distributed activity durations in terms of a mixture of exponential distributions have been addressed by [25] and further developed in [3], where a preliminary analytical formulation of the problem has been provided for a generic network of activities. A further step to develop a framework based on PH distributions was taken in [26], which is the starting point of this article. Similar concepts have been used in [5,6] including resource constraints. Nevertheless, the author limits the analysis to phase-type subclasses (i.e., Cox and Erlang). Although these subclasses are capable of approximating a wide range of distributions, they cannot cope with high/low coefficients of variations and higher-moment-matching with a limited number of phases.

The use of PH distributions is interesting, since they allow the fitting of non-Markov distributions by means of Markov chains. Hence, PH distributions can be used as building blocks for the construction of Markov processes that embed generally distributed transitions instead of exponential only. Since PH distributions are fully described by a Markov chain, they preserve the Markov property, although the model dynamics are not memoryless anymore. As a consequence, the resulting model is able to match more realistic cases, but can still be modeled and solved using numerical methods designed for Markov chains whose performances are good from a computational point of view.

The class of PH distributions is dense in the field of distributions with a positive domain, that is, any distribution in this class can be approximated by PH distributions with any given accuracy, provided that a suitable number of phases is used. This fact, however, does not directly provide a practical method to fit distributions by PH distributions. Several authors proposed fitting methods, and most of these fall into two categories: maximum likelihood (ML)-based estimation of the parameters and moment-matching techniques.

One of the first works on ML estimation considered acyclic PH distributions [27], while an approach for the whole family, based on the expectation-maximization method, is proposed in [28]. Since these early papers, many new methods and improvements have been suggested for the whole PH family and for its sub-classes.

For what concerns moment-matching, for low-order (≤ 3) PH distributions, moment bounds and moment-matching formulas are either known in an explicit form, or there exist iterative numerical methods to check if given moments can be captured [29,30]. For higher-order ones, there exist matching algorithms, but these often result in improper density functions and the validity check is a non-trivial problem.

In [31], a simple method is provided that constructs a minimal-order acyclic PH distribution given three moments. Tool support is available for the construction of PH and ME distributions. Specifically, ML-based fitting is implemented in the software package PhFit [32], and a set of moment-matching functions is provided in the software package BuTools [33].

Although promising and useful in modeling general distributions for activity durations, the approximation through phase-types also has drawbacks. The main one is the considerable increase in terms of number of phases and, consequently, in terms of the states of the comprehensive Markov chain and the associated computational time. To cope with this issue, the use of Kronecker algebra to formalize the Markov model allows the exploitation of numerical methods able to operate the computation of the models without the need to explicitly build the whole infinitesimal generator [7–9].

3. Markov Activity Networks

The presented formulation of the scheduling problem grounds on the the formalization proposed in [2] to define a Markov activity network (MAN), that is, a Markov model describing the execution of a set of different activities whose durations follow an exponential distribution and are linked by precedence relations. Given that (i) the durations of the activities are mutually independent and (ii) exponentially distributed, the execution of the activity network can be represented through a CTMC.

The model is conveniently described by means of an Activity-on-Arc (AoA) network $(\mathcal{V}, \mathcal{E})$ where $\mathcal{V}$ is a set of vertices describing the precedences and $\mathcal{E}$ is a set of edges corresponding to the activities. The number of vertices is denoted with N, whereas K refers to the number of edges. Since each edge corresponds to an activity, the total number of activities composing the task is equal to K as well. Given a vertex v, $\mathcal{A}_v^+$ and $\mathcal{A}_v^-$ indicate the set of incoming and outgoing edges, respectively.

The graph also contains one root vertex without ingoing arcs ($\mathcal{A}_{v_1}^+ = \emptyset$) and one termination vertex without outgoing arcs ($\mathcal{A}_{v_N}^- = \emptyset$). These two vertices are connected by at least one path, while no cycle can exist in the whole network to guarantee the network to be acyclic. In the following, the root vertex will be referred to as v_1 and the termination vertex as v_N. The semantic of the model is such that:

1. $\mathcal{A}_{v_1}^-$ contains those activities without dependencies that can start as soon as the execution of the network begins;
2. The set of activities that departs from a vertex corresponds to $\mathcal{A}_v^-$, and they directly depend on $\mathcal{A}_v^+$;
3. Activities start as soon as all of the preceding activities are completed; this means that there is no time span between the end of an activity and the start of its successors;
4. Activities are never preempted and there is no limit to the number of activities that can be executed in parallel, that is, no resource constraint is enforced;
5. The duration of each activity i is modeled as a continuous distribution $\mathcal{X}_i$.

To provide an example, let us consider the activity network in Figure 1 with four vertices and five arcs, that is, modelling the processing of five activities. The execution of the network starts with the parallel processing of activities 1 and 2, since they belong to $\mathcal{A}_{v_1}^-$. As activity 1 is completed, activities 3 and 4 can start since vertex $\mathcal{A}_{v_2}^+$ only contains activity 1. Activity 5 can start only if both 2 and 3 have been completed, because $\mathcal{A}_{v_3}^+$ contains 2 and 3. Finally, when activities 4 and 5 are completed, the entire network is also completed because vertex v_4 does not have any outgoing arc.

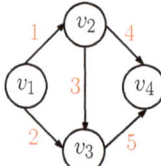

Figure 1. An activity network.

Because of the assumptions described above, the dynamics of an activity network can be described by a discrete-state stochastic process whose state corresponds to a vector $\mathbf{s} = |s_1, \ldots, s_K|$ where each entry s_i, $1 \leq i \leq K$ refers to an activity and can assume the values *Pending* (P), *Running* (R), and *Terminated* (T). Entry s_i is equal to P if activity i is waiting for its predecessors to complete and be allowed to start; it is equal to R if activity i is being executed; finally, it is equal to T if activity i has been completed.

The execution of the activities in the network can be modeled through a sequence of states whose transitions are triggered by the completion of a single activity and the number of transitions that departs from a state $\mathbf{s}$ is always equal to the number of activities that are running in $\mathbf{s}$. Let $f(\mathbf{s}, \mathbf{s}', i)$ be an indicator function that is equal to one when a transition between a state $\mathbf{s}$ and a state $\mathbf{s}'$ is possible by means of the termination of the ith activity. Then, we have:

$$f(\mathbf{s}, \mathbf{s}', i) = \begin{cases} 1 & (s_i = R \land s_i' = T) \land_{j \neq i} \left[(s_j = s_j') \lor (s_j = P \land s_j' = R \land d(\mathbf{s}, j) = \{i\}) \right] \\ 0 & \text{otherwise}; \end{cases} \quad (1)$$

where $d(\mathbf{s}, j)$ is a function that returns the set of activities that have to be completed to start activity j but are not yet finished in state $\mathbf{s}$. The condition in Equation (1) consists of two terms. The first term, $(s_i = R \land s_i' = T)$, refers to the activity that causes the transition, therefore activity i must be running in $\mathbf{s}$ and it must be finished in state $\mathbf{s}'$.

The second term refers to all the other activities defining two possible compatible scenarios. The first one, $(s_j = s'_j)$, refers to activities not changing their status moving from **s** to **s'**, that is, running activities keep running while completed activities remain completed. Nevertheless, the start of some activities could be triggered by the completion of activity i, hence, $(s_j = P \wedge s'_j = R) \wedge d(\mathbf{s}, j) = \{i\}$ states that there could be an activity j moving from state P to R, but this is possible only if activity i was the only one blocking the start of activity j.

The state transition graph of the stochastic process can be generated based on the conditions in (1). Specifically, starting from an initial state where all the activities are pending but those in $\mathcal{A}^+_{v_1}$, all the possible states can be generated enumerating the completion of each of the running activities, one at time. As all the states have been generated, the procedure necessarily reaches the ending state where all activities are terminated and stop.

Figure 2 provides an example of the state transition graph for the network in Figure 1. Within each node, the status of the system is defined, for example, the node on the right is the starting node where activities 1 and 2 are running, while the others are pending, that is, $|R, R, P, P, P|$. The labels on the arcs indicate the activity, hence, the arc labeled with 1 connects the state with the one representing the state of the system when activity 1 has been completed; that is, activity 4 and 5 can start and the new state is labeled as $|T, R, R, R, P|$. The last state on the right is the absorbing one, where all the activities are completed, $|T, T, T, T, T|$.

The state transition graph in Figure 2 has some properties: (i) it contains all the possible paths between the initial state and an absorbing one where all the activities are completed; (ii) the length of these paths is constant and equal to K because each of these paths represent a different ordering for the completion of the activities in the network; (iii) the graph does not contain cycles and all its states are transient, except for the absorbing one $|T, T, \ldots, T|$.

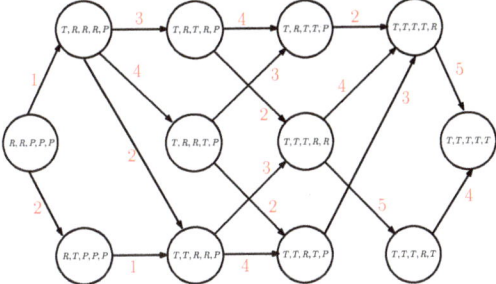

Figure 2. State transition graph of the stochastic process for the network of activities in Figure 1.

As claimed at the beginning of the paragraph, under the assumption that the durations of the activities are modelled with exponential distributions, that is, $\mathcal{X}_i \sim exp(\lambda_i)$, $1 \leq i \leq K$, where $1/\lambda_i$ is the average time of the activity and the activity network is a MAN. The state transition graph in Figure 2 is the support for the definition of a CTMC whose dynamics are described by a set of ordinary differential equations with cardinality equal to the number of states composing the process.

Assuming $\pi(\mathbf{s}, t)$ to be the probability to find the process in state **s** at time t, each equation has the following structure:

$$\frac{d\pi(\mathbf{s}, t)}{dt} = -\sum_{\forall i:\ s_i = R} \pi(\mathbf{s}, t)\lambda_i + \sum_{\forall \mathbf{s'}, i:\ f(\mathbf{s}, \mathbf{s'}, i) = 1} \pi(\mathbf{s'}, t)\lambda_i. \quad (2)$$

Assuming that all the state probabilities $\pi(\mathbf{s}, t)$ are collected in a vector $\pi(t)$ and all the transitions are grouped in an infinitesimal generator matrix Q, the solution of the ODE system described in Equation (2) is given by $\pi(t) = \pi(0) \exp(tQ)$ and the entry of

$\pi(t)$ that corresponds to $\pi(|T, T, \ldots, T|, t)$ contains the cumulative distribution of the time to absorption of the CTMC, that is, the distribution of the makespan of the network of activities. The term $\pi(0)$ is the initial probability vector whose entries are all equal to zero, but the one referring to the initial state whose probability is 1.0. The function $\exp(\bullet)$ is the matrix exponential, whose computation is a common problem whose solution can take advantage of a wide range of efficient and numerically safe methods (e.g., [34]).

As an example, Table 1 reports the infinitesimal generator for the model in Figure 1. The infinitesimal generator is a matrix describing the state transition graph in Figure 2. For each possible transition from a state to another, the infinitesimal generator contains the rate of the activity causing the transition. The values on the diagonal, on the contrary, correspond to the sum of all the rates going out from the state called intensity. The initial activities are 1 and 2, thus, the initial state is $|R, R, P, P, P|$, with $\pi(|R, R, P, P, P|, 0) = 1$, since it is the only nonzero entry of $\pi(0)$.

Table 1. Infinitesimal generator of the CTMC describing the execution of the activity network in Figure 1. Due to space constraints, row labels have not been included (the order of the rows is the same as that of the columns). For the same reason, commas have not been used to separate the entries of each state.

RRPPP	TRRRP	TRRTP	TRTTP	RTPPP	TRTRP	TTRRP	TTRTP	TTTTR	TTTRR	TTTRT	TTTTT
$-(\lambda_1+\lambda_2)$	λ_1	0	0	λ_2	0	0	0	0	0	0	0
0	$-(\lambda_2+\lambda_3+\lambda_4)$	λ_4	0	0	λ_3	λ_2	0	0	0	0	0
0	0	$-(\lambda_2+\lambda_3)$	λ_3	0	0	0	λ_2	0	0	0	0
0	0	0	$-\lambda_2$	0	0	0	0	λ_3	0	0	0
0	0	0	0	$-\lambda_1$	0	λ_1	0	0	0	0	0
0	0	0	λ_4	0	$-(\lambda_2+\lambda_4)$	0	0	0	λ_2	0	0
0	0	0	0	0	0	$-(\lambda_3+\lambda_4)$	λ_4	0	λ_3	0	0
0	0	0	0	0	0	0	$-\lambda_3 - \lambda_5$	λ_3	0	0	λ_5
0	0	0	0	0	0	0	0	$t_4 \otimes I_5$	$-(\lambda_4+\lambda_5)$	λ_5	0
0	0	0	0	0	0	0	0	0	0	$-\lambda_4$	λ_4
0	0	0	0	0	0	0	0	0	0	0	0

Although it provides a considerable advantage in terms of computation, the restriction to exponentially distributed activity durations represents a limiting hypothesis, specifically in the application to real industrial processes and, hence, to scheduling.

In the following, we present an extension of a classical MAN taking advantage of methods available for CTMCs by using a class of distributions able to approximate general distributions with arbitrary accuracy.

4. Phase-Type Distributions

In the field of Markov models, phase-type (PH) distributions are widely used to provide an approximation of a general distribution. Basically, a set of inter-related exponential delays are put together to approximate a general distribution. Formally, a continuous-time PH distribution is the distribution of the time to absorption of a CTMC, and its order is given by the number of the contained transient states. Consequently, the PH distribution is defined through a vector β, providing the initial probabilities of the transient states and a matrix T containing the intensities of the transitions among the transient states.

The cumulative distribution function is given by

$$F(x) = P\{X \leq x\} = 1 - \beta e^{Tx}\mathbb{1} \tag{3}$$

whereas the probability density function corresponds to

$$f(x) = \beta e^{Tx}(-T)\mathbb{1}. \tag{4}$$

The i-th moment of a PH distribution is equal to

$$m_i = i!\beta(-T)^{-i}\mathbb{1}, \tag{5}$$

where $\mathbb{1}$ denotes a column vector of ones of the same dimension of the T.

PH distributions provide a way to fit a general distribution with a certain degree of accuracy. It must be noted that a MAN can be seen as a PH distribution; in fact, the time to absorption in a MAN, usually not exponentially distributed, is represented through a structure of exponential delays (the execution of each activity) with mutual interactions (precedence relations among the activities).

As an example, let us consider the CTMC in Figure 3 with four states, three transient and one absorbing, whose initial probabilities and infinitesimal generator are:

$$\alpha = |0.5\ 0.2\ 0.3\ 0|, \quad Q = \begin{vmatrix} -0.7 & 0.2 & 0.5 & 0 \\ 1.2 & -2.95 & 1.5 & 0.25 \\ 2 & 1 & -3.5 & 0.5 \\ 0 & 0 & 0 & 0 \end{vmatrix}$$

The entries of α must sum up to one, while the sum of the elements in a row of Q must be equal to zero. Thus, the representation (α, Q) for this PH distribution is redundant. A non-redundant representation can be obtained by considering the transient states only, resulting in:

$$\beta = |0.5\ 0.2\ 0.3|, \quad T = \begin{vmatrix} -0.7 & 0.2 & 0.5 \\ 1.2 & -2.95 & 1.5 \\ 2 & 1 & -3.5 \end{vmatrix} \tag{6}$$

The structure of the generator T determines the class of the PH distribution and its capability to catch the moments of the distribution to approximate [35,36]. An exhaustive description of the possible classes of PH distributions is beyond the scope of this paper. Nevertheless, it is important to point out that the method presented in the next section supports the use of any class PH distribution without limitations.

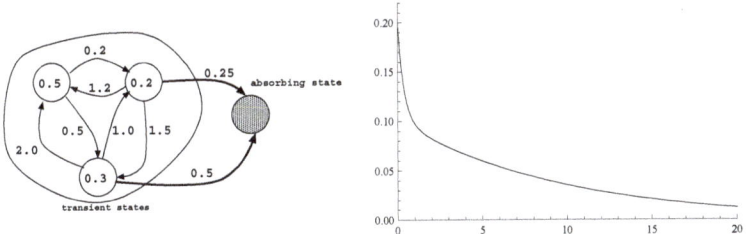

Figure 3. Graphical representation of an order of three PH distributions (**left**) and its probability density function (**right**).

5. Markov Activity Networks Enhanced with PH Distributions

In this section, we extend the formalism of MAN to embed the use of PH distributions and cope with non-exponential durations of the activities, while keeping the advantages provided by a CTMC model. We assume that the processing time of an activity i is distributed according to a PH distribution with representation $(\beta_i, T_i), 1 \leq i \leq K$. Additionally, in this case we assume that durations of the activities are mutually independent.

Due to this hypothesis, the structure of the CTMC modeling the execution of the activities is more complicated, because every state of the original model has to be expanded to consider the sub-states related to the PH model, that is, the aging of the activities in process. Thus, the execution of the activities progresses according to two levels: a macro level characterized by the completion of the activities, triggering the start of the execution of new ones; and the progress among the phases of the PH distribution of each activity during its execution.

The dynamics of the process modeling the execution of the MAN is still governed by an ODE system, but the probability of a single state evolves over time, according to the following differential equation:

$$\frac{\pi(\mathbf{s},t)}{dt} = \pi(\mathbf{s},t)D(\mathbf{s}) + \sum_{\forall \mathbf{s}',i:\ f(\mathbf{s},\mathbf{s}',i)=1} \pi(\mathbf{s}',t)O(\mathbf{s}',\mathbf{s},i). \qquad (7)$$

where $D(\bullet)$ and $O(\bullet,\bullet,\bullet)$ are matrices describing the transitions within a single state and between two states, respectively. In particular, $D(\mathbf{s})$ describes the aging of the PH distributions within state $\mathbf{s}$, that is, the parallel execution of activities that are being processed in $\mathbf{s}$. In more detail:

$$D(\mathbf{s}) = \bigoplus_{\forall i: s_i = R} T_i$$

where $\bigoplus$ is the Kronecker sum operator. On the contrary, matrix $O(\mathbf{s}',\mathbf{s},i)$ describes the dynamics triggered by the completion of an activity i and the consequent transition from state $\mathbf{s}'$ to state $\mathbf{s}$. Thus, the matrix embeds the dynamics related to:

1. The completion of the activity that caused the transition from $\mathbf{s}'$ to $\mathbf{s}$;
2. The memory of the states within the PH for the activities that were running in state $\mathbf{s}'$ and will continue to run in $\mathbf{s}$;
3. The start of activities triggered by the completion of activity i defined by precedence relations.

In formulas, this corresponds to:

$$O(\mathbf{s}',\mathbf{s},i) = \bigotimes_{\forall j \in \mathcal{V}} R_j \quad \text{with } R_j = \begin{cases} t_j & \text{if } j = i \\ \beta_j & \text{if } j \neq i \wedge s'_j = P \wedge s_j = R \\ I_j & \text{if } j \neq i \wedge s'_j = R \wedge s_j = R \\ 1 & \text{otherwise} \end{cases} \qquad (8)$$

In Equation (8), $\bigotimes$ denotes the Kronecker product operator iterated over all the activities in $\mathcal{V}$, and I_j is an identity matrix whose size is equal to the order of the PH distribution associated with activity j, and $t_j = -T_j \mathbf{1}$.

The value of R_j in Equation (8) is different according to the status of the considered activity j in $\mathbf{s}'$ and $\mathbf{s}$. If $j = i$, then activity j is the one that has been completed, and $R_j = t_j$ is a vector that contains the "finishing" intensities of the PH distribution associated to activity j.

If activity j is starting later due to the completion of activity i, then R_j coincides with the initial vector of the PH distribution (β_j).

If activity j is running in both $\mathbf{s}'$ and $\mathbf{s}$, R_j is equal to the identity matrix I_j.

Finally, if activity j is not active in both $\mathbf{s}'$ and $\mathbf{s}$, R_j is equal to 1 and has no impact in the Kronecker product.

The initial probability vector of the ODE system requires some reformulations. In fact, it is now composed of multiple blocks corresponding to phases of the PH distributions, and it must take into consideration the fact that each of them has its own initial vector.

Let us assume that $\mathbf{s}^*$ is the initial state; then $\pi(\mathbf{s}^*, 0)$ is equal to $\otimes_{j:s_j^*=R} \beta_j$ and coincides with the nonzero entries of the initial vector $\pi(0)$.

The infinitesimal generator for the network in Figure 1, where PH distributions are used for all the activities, is reported in Table 2. The structure of the matrix is the same as the exponential case (Table 1). However, the matrix contains sub-matrices instead of scalar values. Matrices D are placed on the diagonal, since they describe the dynamics within a given state, where one or more activities are in progress. Matrices O are placed out of the diagonal, describing the dynamics of the transitions between states.

If we consider state $|R,R,P,P,P|$, $D(|R,R,P,P,P|) = T_1 \oplus T_2$, modeling the superposition of the PH distributions active in that state, that is, the ones associated to activities 1 and 2. Additionally, if we consider the transition between state $|R,R,P,P,P|$ and $|T,R,R,R,P|$, this is caused by the completion of activity 1, but it must also keep the memory of the PH distribution of activity 2 (the associated sub-states remain the same during the transition). Moreover, the completion of activity i also triggers the start of activities 3 and 4. Thus, $O(|R,R,P,P,P|, |T,R,R,R,P|, 1) = t_1 \otimes I_2 \otimes \beta_3 \otimes \beta_4$, with t_1 containing the rates characterizing the completion of activity 1, I_2 keeps the memory of activity 2; β_3 and β_4 initialize the PH distributions of activities 3 and 4. The initial activities for the whole network are 1 and 2, thus, $\pi(|R,R,P,P,P|, 0) = \beta_1 \otimes \beta_2$. Finally, it is important to point out that the Kronecker sum and the Kronecker product do not have the commutative property, hence, the order of the operations must be preserved.

Table 2. Infinitesimal generator of the CTMC enhanced with PH distributions describing the model in Figure 1.

RRPPP	TRRRP	TRRTP	TRTTP	RTPPP	TRTRP	TTRRP	TTRTP	TTTTR	TTTRR	TTTRT	TTTTT
$T_1 \oplus T_2$	$t_1 \otimes I_2 \otimes \beta_3 \otimes \beta_4$	0	0	$I_1 \otimes t_2$	0	0	0	0	0	0	0
0	$T_2 \oplus T_3 \oplus T_4$	$I_2 \otimes I_3 \otimes t_4$	0	0	$I_2 \otimes t_3 \otimes I_4$	$t_2 \otimes I_3 \otimes I_4$	0	0	0	0	0
0	0	$T_2 \oplus T_3$	$I_2 \otimes t_3$	0	0	0	$t_2 \otimes I_3$	0	0	0	0
0	0	0	T_2	0	0	0	0	$t_2 \otimes \beta_5$	0	0	0
0	0	0	0	T_1	0	$t_1 \otimes \beta_3 \otimes \beta_4$	0	0	0	0	0
0	0	0	$I_2 \otimes t_4$	0	$T_2 \oplus T_4$	0	0	0	$t_2 \otimes I_4 \otimes \beta_5$	0	0
0	0	0	0	0	0	$T_3 \oplus T_4$	$I_3 \otimes t_4$	0	$t_3 \otimes I_4 \otimes \beta_5$	0	0
0	0	0	0	0	0	0	T_3	$t_3 \otimes \beta_5$	0	0	0
0	0	0	0	0	0	0	0	T_5	0	0	t_5
0	0	0	0	0	0	0	0	$t_4 \otimes I_5$	$T_4 \oplus T_5$	$I_4 \otimes t_5$	0
0	0	0	0	0	0	0	0	0	0	T_4	t_4
0	0	0	0	0	0	0	0	0	0	0	0

6. A Kroncker Algebra Approach for a Markov Activity Network with PH Distributions

There are several reasons supporting the use of a structured approach based on Kronecker algebra for the formal description of a Markov activity network enhanced through PH distributions. The first one, as stated in [37] with respect to matrix-geometric solutions for stochastic models, is to ensure that the models are in the best and most natural form for numerical computation. For this reason, although unstructured approaches have been proposed to expand the state space of a CTMC with additional states to mimic the behaviour of PH distributions [4,5], these are practicable only for a subset of PH distributions having a simple structure, such as the Coxian, the Erlang, or the Hyper-Exponential.

These approaches also quickly become unfeasible for certain classes of acyclic phase-types as, for instance, the Hyper-Erlang, which corresponds to a mixture of Erlang distributions having a different number of phases. Due to their structure, the Hyper-Erlang is extremely sparse and the tracking of each chain of states is not straightforward, already with a small number of phases. Despite this, their use is fundamental for the fitting of empirical data which are often characterized by irregular shapes and multi-modalities. These situations are often troublesome for the approaches based on moment-matching, because: (i) the moments might not satisfy the bounds of the PH class; (ii) a proper characterization of the shape of the distribution might be preferable to an exact estimation of the first N moments (see [38] for a detailed description of these cases).

Although such cases may seem pure theoretical speculations, their relevance in the modeling of real industrial cases is rather common. An example is manually executed activities, whose duration, as long as the execution goes smoothly, can be easily modeled through simple distributions. On the contrary, the occurrence of possible problems in their execution causes the duration to increase, and the probability distribution fitting the processing times is likely to be multi-modal. In such cases, simply fitting the first two (or three) moments does not provide a reasonable approximation, and different classes of PH distributions could be needed.

Figure 4 provides a graphical example of a case where moment-based approaches exploiting an acyclic PH are outperformed by a different fitting approach, namely a Hyper-Erlang distribution obtained through Hyper-* [39], a tool that couples clustering approaches with methods for the fitting of phase-type distributions. Figure 4 shows how the moment-based approaches fail in representing the bi-modality of the distribution, whereas the Hyper-Erlang is able to approximate the original distribution with good accuracy. Moment-based methods perform slightly better on the tail of the distribution where they are able to outperform the Hyper-Erlang between 10 and 12. The matching of three moments generates much smaller phase-types than the one provided by Hyper-* (in the case presented, 2 against 29 phases), but does not provide an adequate approximation. Specifically, the phases of the Hyper-Erlang are organized in three Erlang distributions: the first is made up of three phases, with $\lambda = 2.301$ and a probability to be chosen equal to $\approx 45\%$; the second with a single phase, $\lambda = 1.44$ and probability $\approx 0.44\%$; and the last one composed of 25 phases, with $\lambda = 0.10$ and probability $\approx 10\%$. Since with this class of fitting methods (e.g., [39]) the structure of the Hyper-Erlang (in terms of number of branches and length of each path) cannot be foreseen, it is not straightforward to enhance the Markov activity network with these classes of PH distributions by simply expanding the state space of the CTMC through unstructured approaches.

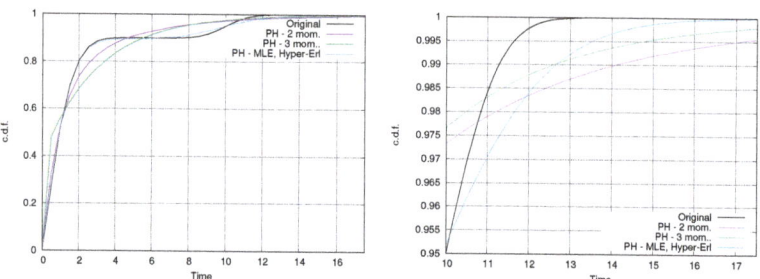

Figure 4. Comparison between an empirical distribution fitted by using both moment-based approaches and cluster-based fitting: cdf (**left**) and tail of the cdf (**right**).

In addition, unstructured approaches become even more impractical when coping with cyclic distributions, that is, creating loops within the states of the CTMC. The use of cyclic PH distributions in practical scheduling problems is justified by concrete requirements. The first one is that, although the class of acyclic phase types is able to represent any cyclic phase-type, there is no guarantee on the finiteness of the number of phases. Hence, a cyclic phase-type might be able to approximate a distribution in a more compact way. This applies for specific classes of distributions, for example, mixtures of monocyclic phase-types (see [40]). Algorithms for the fitting of cyclic phase-types exist and are currently being used; for example, the Butools library provides a method for the transformation of a matrix exponential distribution in a monocyclic phase-type [33].

Another notable reason is that cyclic phase-types are massively used in the context of fault tolerance systems, where mechanisms for restarting, rejuvenation, and check pointing are implemented to model the execution of specific classes of processes often restarted by a controller, when their duration exceeds a certain threshold, or when a risk of stalling [41] arises. These models also allow the use of Kronecker algebra to combine two or more simple dynamics to represent a more complex phenomenon, for example, the failure modes of a machine tool in isolation together with the degradation of a machine tool during the execution of a machining process [42].

Finally, the last argument supporting a structured approach based on Kronecker algebra is that it decouples the exploration of the state space from the PH distribution of each activity. Therefore, the selection of a proper PH approximation for certain activities does not require the redefinition of the whole state space, but only involves those states where the modifications have an impact.

In conclusion, the use of a structured approach to embed PH distributions into activity networks is preferable to unstructured approaches imposing limitations on the class of approximating distributions, as well as failing to provide support for efficient calculation.

7. Testing

This section provides a set of experiments to test the model described in Section 3, as well as an assessment of the degree of approximation entailed by the use of PH distributions with a limited number of phases. The goal is to demonstrate the accuracy achievable in the estimation of the distribution of the makespan with generally distributed durations of the activities modeled with PH distributions. In addition, results related to the required computation time are reported.

The testing phase is organized in three parts. The first one focuses on the evaluation of the impact of a single PH distribution in the activity network in Figure 1, by assessing the accuracy of the fitting for a small activity network. The second one takes into consideration a more general case, where all the activities are non-exponential and, thus, approximated through PH distributions. The aim is to show how the errors associated to each activity have an impact on the estimation of the makespan. Finally, the third part reports the results obtained by using the proposed approach on a set of 150 networks, randomly selected from the PSPLIB set of instances [43]. This part also reports a more detailed analysis of the performance of the approach in terms of computational effort, with respect to the number of states composing the underlying Markov chain. All the experiments have been carried out using a JAVA code implementing both the numerical methods for the analysis of the activity networks using PH approximations and a Monte Carlo simulator to provide an exact estimation of the makespan distribution. The fitting of PH distributions has been operated using the software packages *Butools 2.0* [33] and Mathematica 11.0 (other commercial tools like Matlab, and free software like Octave or Python could also be easily used to develop the approach).

7.1. Activity Network with a Single PH Distributed Activity Duration

To analyze the impact of a single PH distribution in a network, we consider the one in Figure 1. Three experiments have been carried out with different classes of distributions. Each experiment is organized in two steps:

1. A PH distribution is fitted to approximate the general distribution;
2. The resulting PH distribution is plugged into the activity network.

The first step takes advantage of two fitting methods, although any fitting method able to generate a PH distribution can be used. The first one, described in [36], is based on MLE and takes as input a set of data-points and the desired number of phases. The data-points are fitted into a PH distribution corresponding to a hyper-Erlang distribution having the specified number of phases. The second method, described in ([31]), is based on moment-matching. It considers up to three moments and returns a PH distribution with an arbitrary number of phases matching the moments. Both the methods are available in the *Butools* library. Alternative tools for the fitting of PH distributions are implemented in the software packages *PH-fit* [32] and *G-Fit* [36].

The second step addresses the activity network in Figure 1, assuming that all the activities but 3 are distributed according to the Erlang-5 distribution. Since this distribution belongs to the family of PH distributions, its PH representation is exact. On the contrary, the duration of activity 3 is assumed to be generally distributed and approximated through the fitting of a PH distribution according to what is described for the first step. Hence, the durations of activity 1, 2, 4 and 5 are modeled exactly, while an approximation can exist for activity 3. This approximation is the only source of error.

The decision to perform the testing on activity 3 is aimed at stressing the accuracy of the model. In fact, activity 3 connects the two paths 1–4 and 2–5 (1). To make the impact of the activities on the makespan uniform, their mean duration was set to 5.

The experiments have been carried out considering three different classes of distributions for activity 3:

Normal: It represents the ideal scenario for PH approximation because it is continuous and light-tailed on both sides;

Log-Normal: It is a more challenging scenario because of heavy tails;

Uniform: It represents the more difficult case, because PH distributions cannot model a distribution with finite support. Hence, a higher number of phases will be needed to reach a reasonable approximation.

Figure 5 provides a graphical representation of the differences between the distributions used in the three scenarios with each plot showing 1000 data-points. It is possible to notice that points related to a normal distribution are centered around 5 and their frequency decades with the distance from the average. On the contrary, points drawn from the log-normal are sparser and can be found at high values, whereas those drawn from the uniform distribution are equally distributed all along the interval.

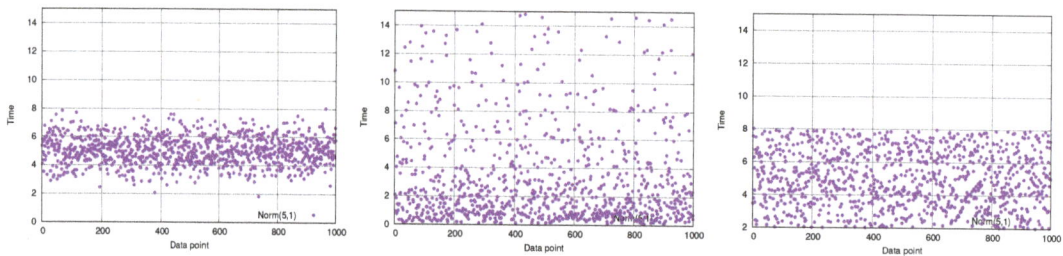

Figure 5. Scatter plots of the data-points used for the fitting of the Normal distribution (**left**), Log-Normal distribution (**center**), and Uniform distribution (**right**).

For this set of experiments, both the Monte Carlo simulations and the CTMC calculations required less than a second to be performed. Confidence intervals are not provided because more than 100,000 runs have been computed for the Monte Carlo approach; hence, the intervals are tight enough not to be visualized.

7.1.1. PH Approximation for a Normally Distributed Activity

The first set of experiments addresses the use of PH to fit a normal distribution, with a mean equal to 5 and a coefficient of variation equal to 0.2. A very small coefficient of variation has been used to mimic an industrial scenario where activities have a high level of formalization and, thus, are usually characterized by a small variation. This choice also guarantees the absence of negative values, impossible for a time span, during the fitting phase. Five different fitting approaches have been performed: the first three using the MLE method with a number of phases equal to 5, 10, and 15, respectively; the fourth and the fifth using the moment-based approach by matching the first two and the first three moments. The PHs generated through the moment-based approach had orders equal to 25 and 28, respectively.

Figure 6 reports the comparison between the original normal distribution and the fitted PH ones. It is possible to notice that only the distributions fitted using the moment-matching method have a Gaussian-like shape, whereas the other PH distributions under-estimate the peak and are far from being symmetric. This is a consequence of the small coefficient of variation used, requiring a high number of phases to be fitted. Despite this, observing the right tail of the distribution, it is possible to notice that the absolute error for the 95th and 99th percentile, obtained using the PH approximation with 15 phases, is rather small, that is, about 1 time unit.

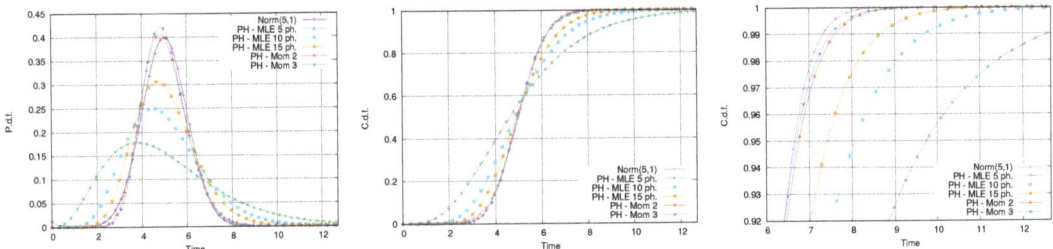

Figure 6. Comparison between the normal distribution with mean 5 and coefficient of variation equal to 0.2 and its fitting by means of PH distributions: pdf (**left**), cdf (**center**), and tail of the cdf (**right**).

Figure 7 reports the exact and PH-approximated distributions of the makespan for the network of activities in Figure 1. It can be observed that all the approximated makespans are able to provide a reasonable approximation. This is particularly evident by observing the right tail of the distribution, where the absolute error remains below 3% in all the cases.

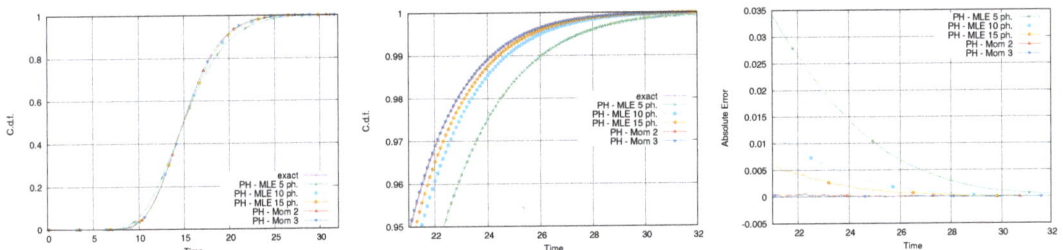

Figure 7. Makespan of the example activity network with the duration of activity 3 following a normal distribution with mean 5 and coefficient of variation equal to 0.2; full (**left**), only the tail (**center**), and the tail approximation error (**right**).

7.1.2. PH Approximation for a Log-Normal-Distributed Activity

The second set of experiments considers a log-normal distribution with parameters $a = -0.804$ and $b = 1.268$. This corresponds to a mean equal to 5 and coefficient of variation of 2. The same fitting approaches of the previous case have been used and, in this case, the moment-based approach was able to match the first three moments by using only two phases.

Figure 8 reports the result of the fitting. It is possible to notice that the worst approximation is provided by the PH distribution that matches three moments. This result is counter-intuitive because, at least in principle, the distribution matching the first three moments should provide a better approximation than the distribution that matches only the first two.

However, this phenomenon might occur when using a fitting method that only considers moment-matching, because it does not guarantee any accuracy with respect to the shape of the distribution. Therefore, for heavy-tailed distributions, matching additional moments is not likely to improve the accuracy of the fitting, particularly for small quantiles. In fact, the fitting matching three moments in this case provides an accurate approximation for extremely high values of the quantiles only.

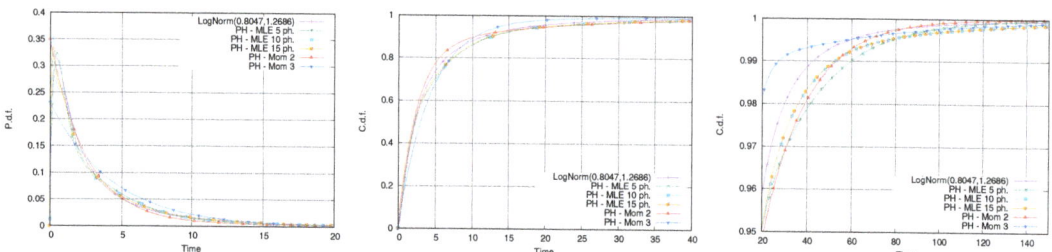

Figure 8. Comparison between the log-normal distribution with parameters $a = -0.804$ and $b = 1.268$ and its fitting by means of PH distributions: pdf (**left**), cdf (**center**), and tail of the cdf (**right**).

Figure 9 reports the exact and PH-approximated distributions of the makespan for the network of activities in Figure 1. The plots confirm the errors highlighted in Figure 8 for the fitting. Specifically, referring to the tail of the distribution of the makespan, this was partly expected. The Erlang-5, the PH distributions obtained using five phases, has a low coefficient of variation in comparison with the log-normal, therefore, the propagation of the error significantly affects the accuracy in the estimation of the distribution of the makespan. Despite this, the absolute error for all the approximated distributions of the makespans stays below 2.5%, with respect to the tail of the distribution (Figure 8, right).

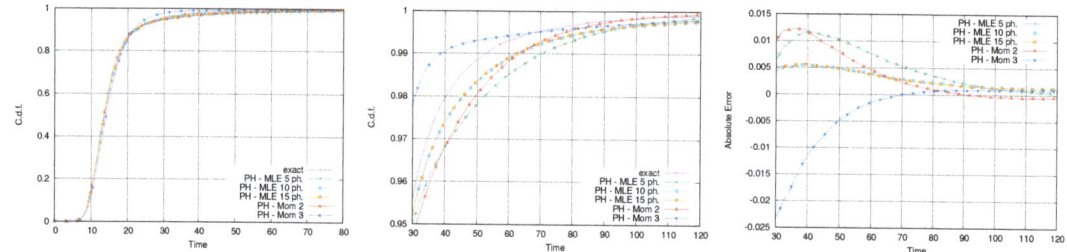

Figure 9. Makespan of the example activity network with the duration of activity 3 following a log-normal distribution with parameters $a = -0.804$ and $b = 1.268$; full (**left**), only the tail (**center**), and the tail approximation error (**right**).

7.1.3. PH Approximation for a Uniform Distribution

The third set of experiments considers a uniform distribution in the interval $[2, 8]$. In this case, there is no chance for a finite number of phases to be able to represent the finite support of the uniform distribution. The moment-based approach requires nine phases to match the first two moments and 12 to match the first three. The impact of this limitation for PH distributions is clear by observing Figure 10 reporting the result of the fitting. It is possible to notice that the pdf of the approximating PH distributions tends to be bell-shaped and trespasses the limited domain of the uniform distribution both in the right and left tails. Additionally, the shape of the tail is overestimated by all the approximating PH distributions (Figure 10, right).

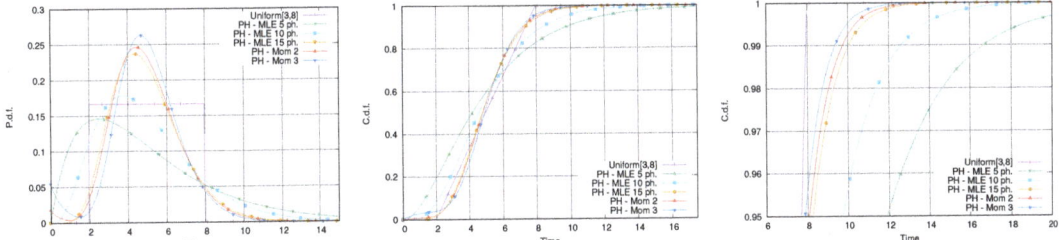

Figure 10. Comparison between the uniform distribution in the interval [2, 8] and its fitting by means of PH distributions: pdf (**left**), cdf (**center**), and tail of the cdf (**right**).

Figure 11 shows the comparison between the exact and approximated distribution of the makespan of the activity network in Figure 1 showing that the overall error reduces in comparison with the one related to activity 3 only. In particular, while the PH distribution of order 5 demonstrates an error greater than 4% in the tail of the distribution, the other approximating distributions have an error that stays below 2% and becomes almost zero near the 99th percentile.

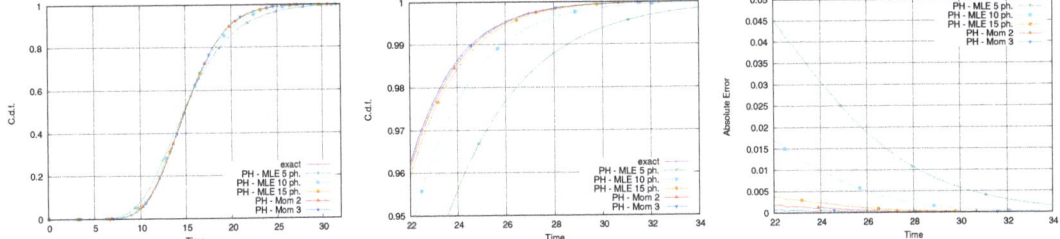

Figure 11. Makespan of the example activity network with the duration of activity 3 following a uniform distribution in the interval [2, 8]; full (**left**), only the tail (**center**), and the tail approximation error (**right**).

7.2. PH Approximation for All the Activities in the Network

In this section, we hypothesize a general distribution for all the activities in the network, with the aim to show that the accuracy in the estimation of the makespan is reasonable even with a large number of approximated distributions. For this test, we used a 30-job instance from the PSPLIB set of instances [43]. Since the repository only contains activity networks with deterministic durations, a distribution and a coefficient of variation have been randomly sampled for each activity, whereas the mean value has been set equal to the original deterministic duration. Two classes of distributions have been considered: log-normal and half-normal, with each activity having the same probability to be distributed as one or the other. The coefficients of variation have been assigned in a similar way, by sampling among three possible values: 0.5, 1, and 1.5.

The 32 activities in the network, together with their distributions, have been fitted by matching the first two moments. All the fittings required, at most, four phases to match the moments. Then, three sub-models have been considered in order to analyze the approximation error as the number of approximated distributions increases. Specifically, within the same network, subsets containing the first 5, 10, and 20 activities have been considered.

The structure of the network and its distributions are reported in Table 3. Horizontal lines delimit the considered subsets.

Table 3. Activities in the network with the associated distributions. Horizontal lines indicate the subsets that have been incrementally taken into consideration.

Activity	Distribution	Average	CV	Dependencies
1	L	1	0.5	-
2	H	8	1	1
3	H	9	1.5	1
4	H	1	1.5	1
5	L	4	0.5	3
6	H	4	1	2
7	L	8	1	3
8	L	3	0.5	7
9	H	3	1.5	6
10	L	9	0.5	8
11	H	6	1	9
12	H	3	1	5
13	H	5	1.5	9
14	H	4	1	12
15	H	9	1	8
16	H	5	1.5	7, 11, 13
17	H	9	1.5	6
18	H	9	1.5	4
19	H	7	1.5	12, 17
20	L	7	0.5	10, 14
21	L	8	1	16, 17, 20
22	H	6	1.5	10
23	H	10	1	13, 22
24	H	2	1.5	15, 21
25	H	1	1.5	13
26	L	9	0.5	19, 20, 23
27	H	3	1.5	14, 18
28	H	7	1.5	16
29	L	10	0.5	18
30	H	7	1.5	18, 24, 26
31	L	9	1	25, 27, 28
32	L	1	0.5	29, 30, 31

The four networks with an increasing number of activities have also been analyzed using a Monte Carlo simulation, to provide a reference distribution to be considered as the real one. Table 4 summarizes the experiments, showing that the state space of the CTMC grows very fast and the number of transitions increases even faster. Despite this, for a network consisting of up to 20 activities, the numerical solution requires less time than performing 100,000 samples for the Monte Carlo simulation.

Table 4. Summary of the experiments.

# Activities	# States	#Transitions	Time Numerical	Time Monte Carlo
5	13	77	0 s	15 s
10	73	1559	0 s	20 s
20	1351	60,710	6 s	22 s
32	6836	579,341	124 s	25 s

Figures 12–15 report the results of the calculation of the makespan for the four models considered. All the comparisons show good accuracy and an error on the percentiles that rarely exceeds 1%. Specifically, it is relevant to notice that the error does not increase with the number of activities. Although there is no guarantee for these results to represent a gen-

eral behavior, it is important to point out that all the fittings have been done automatically, without any specific selection or tuning of the fitting method.

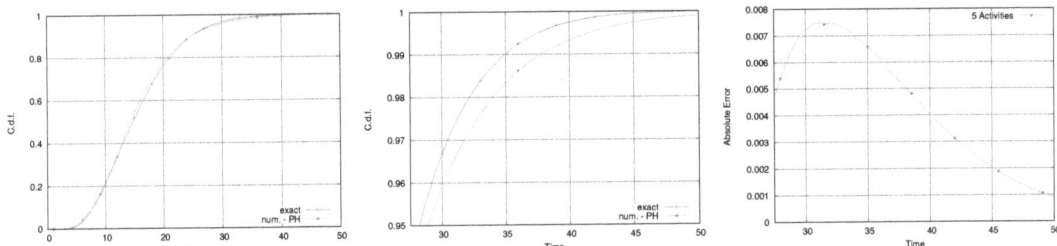

Figure 12. Makespan of the five-activity subset within the network; full makespan (**left**), tail (**center**), and approximation error on the tail (**right**).

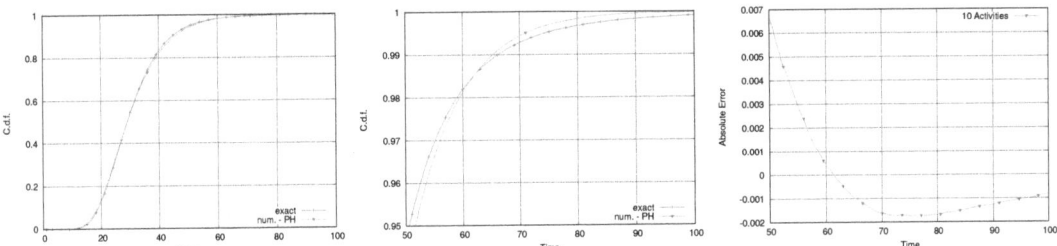

Figure 13. Makespan of the 10-activity subset within the network; full makespan (**left**), tail (**center**), and approximation error on the tail (**right**).

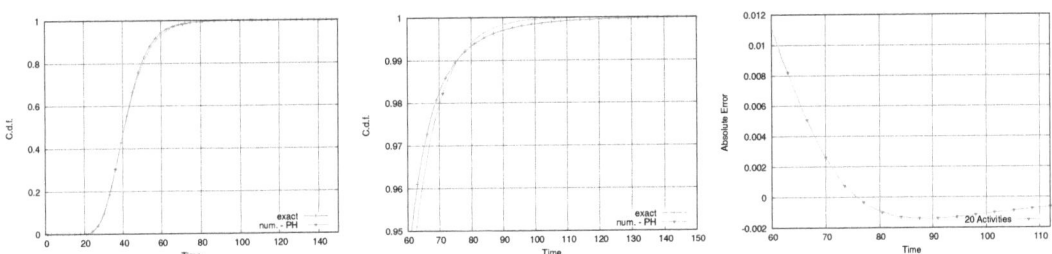

Figure 14. Makespan of the 20-activity subset within the network; full makespan (**left**), tail (**center**), and approximation error on the tail (**right**).

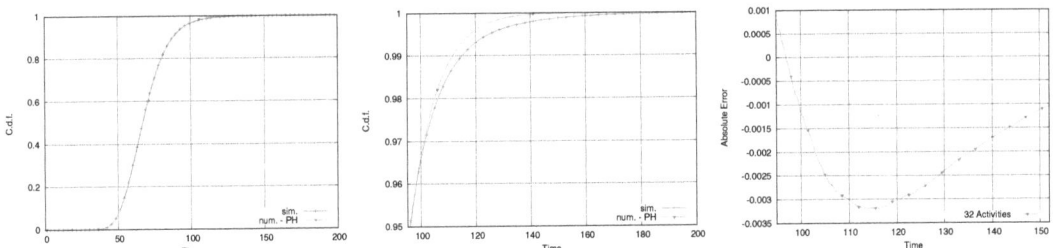

Figure 15. Makespan of the full 32-activity network; full makespan (**left**), tail (**center**), and approximation error on the tail (**right**).

7.3. Test on a Set of Activity Networks

The described approach has been tested on multiple networks of activities obtained by randomly selecting a set of 150 from the instances in the PSPLIB [43]. The structure of the networks has been considered as given, while the randomness in the duration of the activities has been introduced using the same approach described in Table 3. Hence, a PH distribution has been fitted for each activity in the network by matching the first two moments. The fitting of the activity distributions has been done in negligible time. In fact each activity required 0.017 s, on average, to find the approximating PH type.

Figure 16 provides a quick glimpse of the heterogeneity of the networks considered in the experiment, by showing a clear direct correlation between the overall number of transitions and the number of states for each network.

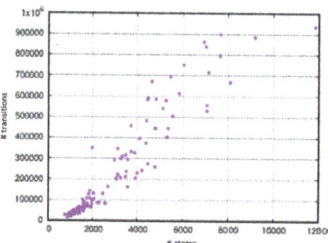

Figure 16. Scatter plot of the number of states against the number of transitions composing the CTMC of the networks.

However, the higher the number of states composing the CTMC, the more variable the number of transitions in the model. This is due to the fact that the number of transitions heavily depends on the level of concurrency in the network, that is, the number of activities that are in execution in each state. A network having many activities and, consequently, many states, has a higher probability of having multiple activities in execution at the same time. If this number is high, the number of transitions is high as well. As a consequence, the number of PH distributions enabled at the same time in the states, and the consequent number of transitions, is less variable among the different instances. A summary of the networks used for the experiments is reported in Table 5, providing information in terms of the number of states composing the CTMC and the number of transitions. It is possible to observe that the average number of states is equal to 2763.73, with a minimum of 779 states and a maximum of 11,858. On the contrary, the average number of transitions is equal to 207,343.59, with a maximum of 935,440 and a minimum of 17,298.

Table 5. Summary of the networks under investigation.

	Average	St. Dev.	Conf. Interval	Min	Max
States	2763.73	2037.00	(2335.28, 3192.16)	779	11,858
Transitions	207,343.59	237,726.36	(157,342.71, 257,344.45)	17,298	935,440
Entropy	0.0359	0.001	(0.03575, 0.0362)	0.0321	0.037

In order to quantify the error introduced by the approximation using PH distributions, we consider the average Kullback–Leibler divergence (KL) of the network, also called average entropy, defined as

$$\overline{KL} = \frac{\sum_{i=1}^{K} \int_0^\infty p_i(x) \log\left(\frac{p_i(x)}{q_i(x)}\right) dx}{K}, \qquad (9)$$

where $p_i(x)$ is the probability density function (pdf) of the exact distribution for activity i, $q_i(x)$ is the pdf of the approximated PH distribution, and K is the total number of activities in the network and the integral is operated on the duration of each activity. If the durations of the activities in the network are well-approximated by the PH distributions, $\overline{KL}$ is close to zero.

Table 5 also reports the value of the average entropy corresponding to 0.0359. Being the value rather close to zero, we can conclude that the PH distributions approximate the general distributions with good accuracy, despite the fact that only the first two moments are matched.

Table 6 compares the time required to compute the 99% quantile of the distribution of the time to absorption of the CTMC with PH distributions, that is, the distribution of the makespan for the network of activities. This performance is compared with the one obtained executing a Monte Carlo simulation with one million samples from the original distributions.

Table 6. Computation time for the 99% quantile of the distribution of the makespan for the experiments.

Approach	# Average	St. Dev.	Conf. Interval	Min	Max
PH distributions	85.10	138.035	(56.07, 114.135)	2	610
Monte Carlo	20.81	4.0	(27.2, 28.9)	20	30

It is possible to observe that the Monte Carlo simulation is, on average, faster than the numerical solution. The good performance of Monte Carlo simulation is due to:

- Large networks requiring a considerable amount of data to be stored in the RAM and slowing down the computation due to the swapping between primary and secondary memory;
- The calculation of the desired percentiles (performed with the bisection method) that increases the computational effort.

To provide a clearer picture of the overall performance of the approach, Figure 17 shows the computation times for all the experiments in relation to the number of states of the CTMC and the number of transitions. It is clear that the computation time increases exponentially with the size of the problem. Indeed, the Monte Carlo simulation does not suffer from this problem, since the overall complexity is bounded by $(K \times R)$, where R is the number of simulation runs. Nevertheless, it is important to point out that:

- The CTMC can be solved using more advanced techniques able to reduce the computation time;
- The CTMC defines a model for the execution of the network of activities that can support decomposition approaches, for example, calculating subnets and incorporating the obtained solution or estimation in the comprehensive network.

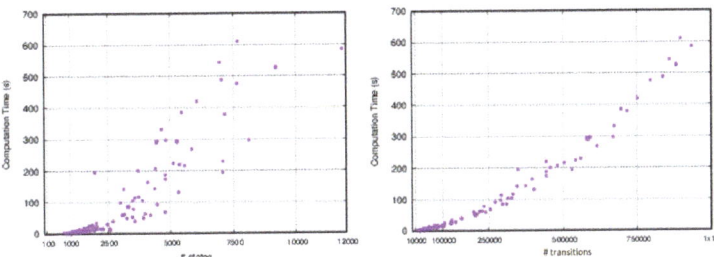

Figure 17. Time required to solve the networks with PH distributions as a function of the number of states (**left**) and the number of transitions (**right**).

Finally, Table 7 reports the error in the estimation of the distribution of the makespan using PH distributions for different percentiles. It is possible to notice that the error increases with the considered percentile. At the 90th percentile, the error is, on average

1%, and the confidence interval is (1.47, 1.96), with a minimum of 0.017% and a maximum of 3.46%, whereas at the 99th percentile, the error is, on average, 2.72% with a confidence interval equal to (2.28, 3.15), a minimum of 0.04% and a maximum of 9.6%.

Table 7. Accuracy of the estimation of the makespan distribution for different percentiles.

Percentile	# Average	St. Dev.	Conf. Interval	Min	Max
90	1.16	0.80	(0.99,1.33)	0.017	3.46
95	1.72	1.16	(1.47,1.96)	0.006	4.55
97	2.49	1.49	(2.17,2.79)	0.08	5.66
99	2.72	2.09	(2.28,3.15)	0.04	9.60

8. Conclusions

In this paper, a general approach has been proposed to exploit Markov chains and phase-type distributions to model the execution of a stochastic PERT network with general distributed processing times. A general and concise formulation has been proposed, based on Kronecker algebra, to support the application without any constraint on the class of phase-type distributions allowed.

The motivation for the approach stems from the need to fit real data and use this information in the planning and scheduling of manufacturing operations, overcoming the limitation of a large portion of the literature related to stochastic scheduling, requiring strong hypotheses on the underlying distributions. Coping with real data, characterized by asymmetry, multi-modality, and so forth, PH distributions are a very effective approximation approach also entailing the capability of using Markov activity networks and the associated corpus of approaches.

The proposed formulation also allows to easily change and/or refine the fitting according to the accuracy requirements to keep the computation time in line with the specific needs and without the need of recomputing the whole model from zero.

An analytical description has been provided, as well as experiments on a wide range of test instances. The testing demonstrated good performance both in terms of computation time and accuracy of the approximation, although the computational effort rapidly increases with the dimension of the network and the number of phases used.

Moreover, the algebraic formulation also allows the use of more advanced numerical methods able to evaluate the Markov activity network without the need to explicitly build the whole infinitesimal generator, thus paving the way to more efficient calculation approaches.

Hence, modeling the execution of an activity network through a Markov process and phase-type distributions provides a promising path towards embedding the proposed estimation approach in scheduling algorithms to optimize a function of the makespan distribution, such as an associated risk measure.

Further developments will address advanced calculation methods to speed up the computation time, as well as the exploitation of this method to support stochastic scheduling approaches.

Author Contributions: Conceptualization, M.U.; methodology, A.A., A.H. and M.U.; software, A.A.; writing, A.A., A.H. and M.U. All authors have read and agreed to the published version of the manuscript.

Funding: This research received no external funding.

Institutional Review Board Statement: Not applicable.

Informed Consent Statement: Not applicable.

Data Availability Statement: Not applicable.

Conflicts of Interest: The authors declare no conflict of interest.

References

1. Malcolm, D.; Rosenbloom, J.; Clark, C.; Fazar, W. Application of a technique for research and development program evaluation. *Oper. Res.* **1959**, *7*, 646–669. [CrossRef]
2. Kulkarni, V.; Adlakha, V. Markov and Markov-regenerative pert networks. *Oper. Res.* **1986**, *34*, 769–781. [CrossRef]
3. Urgo, M. Stochastic Scheduling with General Distributed Activity Durations Using Markov Activity Networks and Phase-Type Distributions. In *Sequencing and Scheduling with Inaccurate Data*; Nova Publisher: Hauppauge, NY, USA, 2014.
4. Elmaghraby, S.; Benmansour, R.; Artiba, A.; Allaoui, H. On The Approximation of Arbitrary Distributions by Phase-Type Distributions. In Proceedings of the 3rd International Conference on Information Systems, Logistics and Supply Chain, Casablanca, Morocco, 13 April 2010.
5. Creemers, S. Minimizing the expected makespan of a project with stochastic activity durations under resource constraints. *J. Sched.* **2015**, *18*, 263–273. [CrossRef]
6. Creemers, S. Maximizing the expected net present value of a project with phase-type distributed activity durations: An efficient globally optimal solution procedure. *Eur. J. Oper. Res.* **2018**, *267*, 16–22. [CrossRef]
7. Buchholz, P. Structured analysis approaches for large Markov chains. *Appl. Numer. Math.* **1999**, *31*, 375–404. [CrossRef]
8. Buchholz, P.; Kemper, P. Kronecker Based Matrix Representations for Large Markov Models. In *Validation of Stochastic Systems: A Guide to Current Research*; Baier, C., Haverkort, B.R., Hermanns, H., Katoen, J.P., Siegle, M., Eds.; Springer: Berlin/Heidelberg, Germany, 2004; pp. 256–295.
9. Ballarini, P.; Horváth, A. Memory Efficient Analysis for a Class of Large Structured Markov Chains: Work in Progress. In Proceedings of the Fourth International ICST Conference on Performance Evaluation Methodologies and Tools, Pisa, Italy, 20–22 October 2009; pp. 21:1–21:4.
10. Igelmund, G.; Radermacher, F.J. Preselective Strategies for the Optimization of Stochastic Project Networks under Resource Constraints. *Networks* **1983**, *13*, 1–28. [CrossRef]
11. Igelmund, G.; Radermacher, F.J. Algorithmic Approaches to Preselective Strategies for Stochastic Scheduling Problems. *Networks* **1983**, *13*, 29–48. [CrossRef]
12. Radermacher, F.J. Scheduling of Project Networks. *Ann. Oper. Res.* **1985**, *4*, 227–252. [CrossRef]
13. Stork, F. *Branch-and-Bound Algorithms for Stochastic Resource-Constrained Project Scheduling*; Technical Report; Research Report No. 702/2000; Technische Universität Berlin: Berlin, Germany, 2000.
14. Tolio, T.; Urgo, M.; Váncza, J. Robust production control against propagation of disruptions. *CIRP Ann. Manuf. Technol.* **2011**, *1*, 489–492. [CrossRef]
15. Radke, A.M.; Tolio, T.; Tseng, M.M.; Urgo, M. A risk management-based evaluation of inventory allocations for make-to-order production. *CIRP Ann. Manuf. Technol.* **2013**, *1*, 459–462. [CrossRef]
16. Urgo, M.; Buergin, J.; Tolio, T.; Lanza, G. Order allocation and sequencing with variable degree of uncertainty in aircraft manufacturing. *CIRP Ann. Manuf. Technol.* **2018**, *67*, 431–436. [CrossRef]
17. Urgo, M.; Váncza, J. A branch-and-bound approach for the single machine maximum lateness stochastic scheduling problem to minimize the value-at-risk. *Flex. Serv. Manuf. J.* **2018**, *31*, 472–496. [CrossRef]
18. Hagstrom, J.N. Computational complexity of PERT problems. *Networks* **1988**, *18*, 139–147. [CrossRef]
19. Kleindorfer, G. Bounding distributions for a stochastic acyclic network. *Oper. Res.* **1971**, *19*, 586–601. [CrossRef]
20. Moehring, R. Scheduling under Uncertainty: Bounding the Makespan Distribution. In *Computational Discrete Mathematics*; Springer: Berlin/Heidelberg, Germany, 2001; pp. 79–97.
21. Golenko-Ginsburg, D.; Gonik, A. Stochastic Network Project Scheduling with Non-Consumable Limited Resources. *Int. J. Prod. Econ.* **1997**, *48*, 29–37. [CrossRef]
22. Tsai, Y.W.; Gemmill, D.D. Using Tabu Search to Schedule Activities of Stochastic Resource-Constrained Projects. *Eur. J. Oper. Res.* **1998**, *111*, 129–141. [CrossRef]
23. Buss, A.H.; Rosenblatt, M.J. Activity delay in stochastic project networks. *Oper. Res.* **1997**, *45*, 661–677. [CrossRef]
24. Sobel, M.; Szmerekovsky, J.; Tilson, V. Scheduling projects with stochastic activity duration to maximize expected net present value. *Eur. J. Oper. Res.* **2009**, *198*, 697–705. [CrossRef]
25. Elmaghraby, S.E.; Ramachandra, G. *Optimal Resource Allocation in Activity Networks: II. The Stochastic Case*; Research Report; NCSU: Raleigh, NC, USA, 2012.
26. Angius, A.; Horváth, A.; Urgo, M. Analysis of activity networks with phase type distributions by Kronecker algebra. In Proceedings of the 14th International Conference on Project Management and Scheduling (PMS'14), Munich, Germany, 30 March–2 April 2014; pp. 1–5.
27. Bobbio, A.; Cumani, A. ML estimation of the parameters of a PH distribution in triangular canonical form. *Comput. Perform. Eval.* **1992**, *22*, 33–46.
28. Asmussen, S.; Nerman, O.; Olsson, M. Fitting Phase-Type Distributions via the EM Algorithm. *Scand. J. Stat.* **1996**, *23*, 419–441.
29. Telek, M.; Heindl, A. Matching moments for acyclic discrete and continuous phase-type distributions of second order. *Int. J. Simul. Syst. Sci. Technol.* **2002**, *3*, 47–57.
30. Horváth, G.; Telek, M. A canonical representation of order 3 phase type distributions. *Lect. Notes Theor. Comput. Sci.* **2007**, *4748*, 48–62.

31. Bobbio, A.; Horváth, A.; Telek, M. Matching Three Moments with Minimal Acyclic Phase Type Distributions. *Stoch. Model.* **2005**, *21*, 303–326. [CrossRef]
32. Horváth, A.; Telek, M. PhFit: A General Phase-Type Fitting Tool. In Proceedings of the 12th Performance TOOLS, London, UK, 14–17 April 2002; Volume 2324.
33. BuTools 2.0. 2018. Website online: http://webspn.hit.bme.hu/~telek/tools/butools (accessed on 1 April 2021).
34. Moler, C.; Loan, C.V. Nineteen Dubious Ways to Compute the Exponential of a Matrix, Twenty-Five Years Later. *SIAM Rev.* **2003**, *45*, 3–49. [CrossRef]
35. Horváth, A. Approximating Non-Markovian Behavior by Markovian Models. Ph.D. Thesis, Department of Telecommunications, Budapest University of Technology and Economics, Budapest, Hungary, 2002.
36. Thummler, A.; Buchholz, P.; Telek, M. A Novel Approach for Phase-Type Fitting with the EM Algorithm. *IEEE Trans. Dependable Secur. Comput.* **2006**, *3*, 245–258. [CrossRef]
37. Neuts, M.F. *Matrix-Geometric Solutions in Stochastic Models: An Algorithmic Approachg*; Dover: Mineola, NY, USA, 1981.
38. Angius, A.; Horvath, A.; Halawani, S.M.; Barukab, O.; Ahmad, A.R.; Balbo, G. Constructing Matrix Exponential Distributions by Moments and Behavior around Zero. *Math. Probl. Eng.* **2014**, *2014*, 610907. [CrossRef]
39. Reinecke, P.; Krauß, T.; Wolter, K. Cluster-based fitting of phase-type distributions to empirical data. *Comput. Math. Appl.* **2012**, *64*, 3840–3851. [CrossRef]
40. Mocanu, S.; Commault, C. Sparse representations of phase-type distributions. *Commun. Stat. Stoch. Model.* **1999**, *15*, 759–778. [CrossRef]
41. Wolter, K. *Stochastic Models for Fault Tolerance-Restart, Rejuvenation and Checkpointing*; Springer: Berlin/Heidelberg, Germany, 2010. [CrossRef]
42. Angius, A.; Colledani, M.; Yemane, A. Impact of condition based maintenance policies on the service level of multi-stage manufacturing systems. *Control Eng. Pract.* **2018**, *76*, 65–78. [CrossRef]
43. PSPLIB. 2018. Available online: http://www.om-db.wi.tum.de/psplib (accessed on 1 April 2021).

Article

Relocation Scheduling in a Two-Machine Flow Shop with Resource Recycling Operations

Ting-Chun Lo and Bertrand M. T. Lin *

Institute of Information Management, National Yang Ming Chiao Tung University, Hsinchu 300, Taiwan; chunchun.mg08@nycu.edu.tw
* Correspondence: bmtlin@mail.nctu.edu.tw

Abstract: This paper considers a variant of the relocation problem, which is formulated from an urban renewal project. There is a set of jobs to be processed in a two-machine flow shop subject to a given initial resource level. Each job consumes some units of the resource to start its processing on machine 1 and will return some amount of the resource when it is completed on machine 2. The amount of resource released by a job is not necessarily equal to the amount of resource acquired by the job for starting the process. Subject to the resource constraint, the problem is to find a feasible schedule whose makespan is minimum. In this paper, we first prove the NP-hardness of two special cases. Two heuristic algorithms with different processing characteristics, permutation and non-permutation, are designed to construct feasible schedules. Ant colony optimization (ACO) algorithms are also proposed to produce approximate solutions. We design and conduct computational experiments to appraise the performances of the proposed algorithms.

Keywords: resource-constrained scheduling; relocation problem; flow shop; resource recycling; heuristic algorithms; ant colony optimization

---

Citation: Lo, T.-C.; Lin, B.M.T. Relocation Scheduling in a Two-Machine Flow Shop with Resource Recycling Operations. *Mathematics* **2021**, *9*, 1527. https://doi.org/10.3390/math9131527

Academic Editors: Chin-Chia Wu and Win-Chin Lin

Received: 23 May 2021
Accepted: 23 June 2021
Published: 29 June 2021

Publisher's Note: MDPI stays neutral with regard to jurisdictional claims in published maps and institutional affiliations.

Copyright: © 2021 by the authors. Licensee MDPI, Basel, Switzerland. This article is an open access article distributed under the terms and conditions of the Creative Commons Attribution (CC BY) license (https://creativecommons.org/licenses/by/4.0/).

1. Introduction

Scheduling is a decision-making process that allocates limited resources to tasks in a given time period to optimize certain objectives in manufacturing as well as service industries [1]. Usually, resources are considered as machines that process the assigned tasks in manufacturing industries. In some scheduling contexts, there could be different extra resources, like capital, crews and technicians, storage space, energy, computer memory, and so on, that are required to support the execution of the tasks. Such scheduling problems are known as resource-constrained scheduling. Resource-constrained project scheduling problems (RCPSP) have received considerable attention for decades. Please refer to Brucker, Drexl, Möhring, Neumann, and Pesch [2], Habibi, Barzinpour, and Sadjadi [3] Herroelen, De Reyck, and Demeulemeester [4], Issa and Tu [5] for comprehensive reviews on RCPSP. The resource constraint featured in the relocation problem is different traditional ones in the sense that the amount of resource released by an activity is not necessarily the same as that acquired for commencing the activity. This study investigates the relocation problem in a two-machine flow shop with the specific feature of a resource recycling mechanism.

The construction industry have various optimization decisions to address in the project course [6,7]. The relocation problem originated from the public house redevelopment project in Boston [8,9]. The project had a set of buildings to be torn down and erected for redevelopment. During the redevelopment process, current tenants of the buildings under reconstruction needed to be relocated to temporary housing units. They could be assigned to new housing units. It was not mandatory for tenants to reside at the same place they lived before. Therefore, the authority had to determine a minimum budget of temporary housing units such that all tenants could be successfully relocated. Kaplan [8] first formulated the relocation problem of determining a feasible redevelopment sequence of the buildings with the initial budget. In the view of optimization, this problem can also be described as finding a feasible sequence of the redevelopment buildings that reflects the minimum initial budget. Kaplan and Amir [10] showed that the relocation problem is mathematically equivalent to the two-machine flow shop scheduling problem for minimizing makespan, implying that the basic relocation problem can be solved by the classical Johnson's algorithm [11].

Lin and Huang [12] first introduced the recycling operations for yielding the resource into the study on the relocation problem. In previous studies on the relocation problem, the resource is consumed when a job starts to process and returns immediately when it is completed. However, the concept of the resource recycling is assumed that we need to have a mechanism or procedure to recycle the resource before the resource can be used for later jobs. Therefore, a job is divided into two separate parts on two dedicated machines: one processed on machine 1 and the other for recycling the resource on machine 2. The operations on machine 1 should have sufficient resource so that they can commence the processing, and the operations on machine 2 should wait for the completion of its corresponding counterpart operations jobs on machine 1. However, the job sequences on the two machines are not necessarily the same. Cheng, Lin and Huang [13] presented an integer linear program formulation for the permutation case, in which the job sequences on all machines are the same. They also investigated the non-permutation case. We continue to study this relocation problem and discuss more theoretical proofs. Then, we present heuristic algorithms and ant colony optimization (ACO) algorithms for both permutation and non-permutation sequences to find feasible schedules with resource constraints for minimum makespan.

The rest of this paper is organised as follows. In Section 2, we present problem statements and give a numerical example followed by a literature review. The complexity results of two special cases are discussed in Section 3. In Sections 4 and 5, we present heuristic algorithms and ant colony optimization algorithms for constructing approximate schedules. Computational experiments and performance statics of the algorithms are given in Section 6. We conclude this search and suggest research directions for future study in Section 7.

2. Problem Definition

In this section, we first introduce the notation that will be used throughout our research. Then, a formal problem formulation follows. An integer programming model is also proposed. The notations are listed below:

Notation:

$\mathcal{N} = \{1, 2, \ldots, n\}$	set of jobs to be processed;
$p_{1,j}$	processing time of job j on machine 1;
$p_{2,j}$	processing time of job j on machine 2;
α_j	resource requirement of job j;
β_j	amount of the resource returned by job j;
$\sigma = (\sigma_1, \sigma_2, \ldots, \sigma_n)$	a particular sequence of the jobs (assumed for the case of permutation schedules);
v_0	initial resource level;
v_t	resource level at time $t \geq 0$;
$C_{m,j}$	completion time of job j on machine m, $m = 1, 2$.

We formally state the problem as follows: From time zero onwards, a set of jobs $\mathcal{N} = \{1, 2, \ldots, n\}$ is available to be processed in a two-stage flow shop consisting of machine 1 and machine 2. Initially, the common resource pool contains v_0 units of a single type of resource. Job $j \in \mathcal{N}$ can start processing only if machine 1 is not occupied and the resource level is larger than or equal to α_j. When job j starts processing, it immediately consumes α_j units of the resource and takes p_{1j} units of time on machine 1. After the operation on machine 1 is completed, p_{2j} units of time are required to complete its resource recycling operation on machine 2. When job j completes on machine 2, it produces and returns β_j units of the resource back to the resource pool. No preemption on either machine is permitted. Note that there is no strict relation between α_j and β_j. That is, β_j could be smaller than, equal to, or larger than α_j. The goal is to minimize the makespan. In other words, we want to find a feasible schedule that completes all jobs in the shortest time.

To illustrate the problem definition, we consider an instance of four jobs with an initial resource level $v_0 = 6$. The parameters are shown below in Table 1. We construct two example schedules.

Table 1. An example of four jobs.

	$p_{1,j}$	$p_{2,j}$	α_j	β_j
job 1	3	3	8	9
job 2	1	5	3	10
job 3	5	6	3	8
job 4	6	1	11	4

Figure 1 shows an optimal permutation solution with $C_{max} = 22$, and Figure 2 shows an optimal non-permutation solution with $C_{max} = 19$. Both of them are feasible. In this example, it is clear that the non-permutation solution can attain a better makespan than the permutation one.

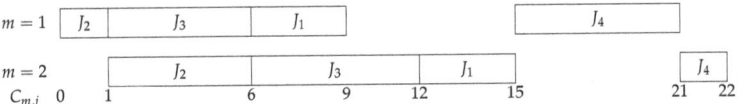

Figure 1. Permutation solution

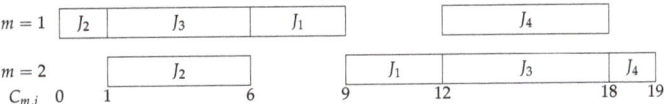

Figure 2. Non-permutation solution

Literature Review

To describe our problem, we denote use the standard three-field notation $F2|rp|C_{max}$, proposed by Graham et al. [14]. The first field indicates the machine environment of a two-machine flow shop, where the first machine is the operation of the building being torn down and the second machine is about re-constructing buildings corresponding to resource recycling operations. The second field indicates that the specific conditions for the job characteristics, i.e., the relocation problem. The last field specifies the objective function of makespan.

The study on the relocation problem was inaugurated by Kaplan [8] in 1986. The fundamental purpose of the basic relocation problem is to minimize the initial budget required for guaranteeing project feasibility. In Kaplan's study, multiple working crews were considered that if resources were sufficient, i.e., a number of buildings could be simultaneously developed. Kaplan and Amir [10] formulated the application of relocation problem as an integer program. They also noted the relationship between the minimum budget in relocation feasibility and the minimum makespan of two-machine flow shop scheduling, which is solvable in $O(n \log n)$ time [11]. To reflect real situations of the housing redevelopment project in East Boston, Kaplan, and Berman [15] refined the integer programming model and scheduling heuristics. Applications like the financial constraints on single machine scheduling problems [16] which can be reduced to the two-machine flow shop scheduling problem as a special case of the relocation problem. The relocation problem is also related to the memory management issue in database system in practical term [17]. Amir and Kaplan [17] showed that minimizing the makespan on parallel machines is NP-hard. Kononov and Lin [18] proved that parallel-machine setting is strongly NP-hard even if there are only two working crews and all jobs have the same processing time. They also designed approximation algorithms with performance ratio analysis for two special cases.

Cheng and Lin [19] presented more proofs and proposed the concept of composite jobs that can reduce the computational time for handling the relocation problem. Cheng and Lin [20] also demonstrated the concept of relocation scheduling to give an economic interpretation of Johnson's algorithm. The concept can also simplify proofs and reduce time complexity in some two-machine flow shop scheduling problems. There are other extensions from the relocation problem. Lin and Tseng [21] considered the problem with processing times and deadline constraints. Furthermore, they provided a complexity result and two polynomial algorithms to solve the restricted problems. Lin and Tseng [22] proposed a branch-and-bound algorithm to maximize the resource level under a specified due date and considered the precedence constraints [23] that is NP-hard even if the precedence constraints are specified by a bi-partite graph. Lin and Cheng [24] showed two relocation problems of minimizing the maximum tardiness is strongly NP-hard and of minimizing the number of tardy jobs under a due date is NP-hard even when all the jobs have an equal tardy weight and resource requirement. Based on the generalized due dates proposed by Hall [25], Lin and Liu [26] extended the scheduling problem and designed a branch-and-bound algorithm to reduce the computational time required. Sevastyanov, Lin and Huang [27] considered the relocation problem with arbitrary release dates. They developed a multi-parametric dynamic programming algorithm to solve the case with a fixed number if distinct due release dates and analyzed complexity of different problem settings. Kononov and Lin [28] considered minimizing the total weighted completion time and proved four special cases are strong NP-hardness. They established the equivalence between the UET (unit-execution-time) case and the unit-weighted case and presented a 2-approximation algorithm for the restricted special cases.

As per the feature of resource recycling in the relocation problem, there are some existing works. Lin and Huang [12] first introduced the concept of resource recycling. This operation can be processed on a secondary recycling machine that the whole procession can be described as a two-machine flow shop scheduling problem. In this paper, they showed that it is NP-hard and designed three heuristic algorithms to compose approximate schedules. Problem formulation and some complexity results were discussed in Cheng, Lin and Huang [13]. They presented integer linear programming models for finding the feasible permutation sequence and non-permutation sequence with minimum makespan. Lin [29] considered the setting where processing and recycling are carried out on the same single machine. The problem is a generalization of the knapsack problem. He designed a pseudo-polynomial time dynamic programming algorithm and formulated an integer program to solve this recycling problem that operations are processed on the same single machine.

3. Complexity Analysis

This section is dedicated to discussion of the complexity results of the problem of several special cases. First, let instance $\mathcal{I}$ contain n jobs with $p_{1,j}$, $p_{2,j}$, α_j, and β_j given for each job j and an initial resource level v_0. We create another instance $\bar{\mathcal{I}}$ having n jobs with $\bar{p}_{1,j} = p_{2,j}, \bar{p}_{2,j} = p_{1,j}, \bar{\alpha}_j = \beta_j,$ and $\bar{\beta}_j = \alpha_j$ that is symmetric to the instance $\mathcal{I}$. Set the initial resource level $\bar{v}_0 = v_0 - \sum_{j=1}^{4q+1}(\beta_j - \alpha_j)$. We claim that the two instances have the optimal makespan. The concept follows the results of Kononov and Lin [28].

Theorem 1. *(Mirror Property) Instance $\mathcal{I}$ and instance $\bar{\mathcal{I}}$ have the optimal makespan.*

Proof. Let $\sigma = (\sigma(1), \sigma(2), \ldots, \sigma(n))$ be a feasible permutation of jobs in the given instance $\mathcal{I}$ case. We show that $\bar{\sigma} = (\sigma(n), \sigma(n-1), \ldots, \sigma(1))$ is feasible for $\bar{\mathcal{I}}$ we created in the above. Assume $\overline{V}_k$ is the resource level after the jobs $\sigma(n), \ldots, \sigma(k+1)$ in $\bar{\sigma}$ complete. We have $\overline{V}_k - \overline{\alpha}_{\sigma(k)} = \overline{v}_0 + \sum_{j=k+1}^{n}(\overline{\beta}_{\sigma(j)} - \overline{\alpha}_{\sigma(j)}) - \overline{\alpha}_{\sigma(k)} = v_0 + \sum_{j=1}^{k}(\beta_{\sigma(j)} - \alpha_{\sigma(j)}) - \beta_{\sigma(k)} = v_0 + \sum_{j=1}^{k-1}(\beta_{\sigma(j)} - \alpha_{\sigma(j)}) - \alpha_{\sigma(k)} \geq 0$. The last inequality is feasible with the schedule σ. Therefore, we can get $\overline{V}_k \geq \overline{\alpha}_{\sigma(k)}$, which shows the schedule $\bar{\sigma}$ is feasible. As we know the schedule $\bar{\sigma}$ and σ are two-stage flow shop, if their sequences are reversed, they have same processing time. Therefore, we can construct another optimal schedule if we get an optimal one. □

For n jobs, there are $(n!)$ possible sequences when the permutation schedules are considered. If we consider the non-permutation variant, the number of schedules will become $O(n! \times n!)$ because the permutations on the two machines could be different. For technical constraints or dispatching fairness, say first comes first served, the processing sequence could be given and fixed [30,31]. In view of implementations, if an optimal schedule can be efficiently obtained from a given job sequence on either machine, we can then reduce the decision tree size from $O(n! \times n!)$ to $O(n!)$. This section will explore the complexity status of the setting with a fixed job sequence.

First, we prove the problems that when the sequence of the jobs on machine 2 is given and fixed, finding the optimal schedule is strongly NP-hard, even if all jobs have the same processing time on machine 2. On the other hand, if the given and fixed sequence of the jobs is on machine 1, we can get the same result that finding optimal schedules is also strongly NP-hard. The proof is given in the following:

3-Partition: Given an integer B and a set A of $3q$ elements $\{1, 2, \ldots, 3q\}$, each $j \in A$ has a size x_j, $B/4 < x_j < B/2$, such that $\sum_{j=1}^{3q} x_j = qB$, is there a partition $A_1, A_2, \ldots, A_q$ of the set A such that $\sum_{x_j \in A_l} x_j = B$, $1 \leq l \leq q$?

Theorem 2. *If a sequence of the jobs on machine 2 is given and fixed, then finding an optimal schedule is strongly NP-hard, even if all jobs have the same processing time on machine 2.*

Proof. Given an instance of 3-Partition, we create a corresponding set of $4q + 1$ jobs as follows:

Enforcer jobs: $p_{1,j} = 0, p_{2,j} = B, \alpha_j = 2B,$ and $\beta_j = 3B,$ $1 \leq j \leq q$;
Ordinary jobs: $p_{1,j} = x_i, p_{2,j} = B, \alpha_j = x_i,$ and $\beta_j = 0,$ $q+1 \leq j \leq 4q$;
Final job: $p_{1,j} = 3qB, p_{2,j} = B, \alpha_j = 0,$ and $\beta_j = 0,$ $j = 4q+1$.

The initial resource level $v_0 = 3B$. The jobs on machine 2 are sequenced in increasing order of their indices. We claim that there is a 3-Partition if there is a feasible schedule whose makespan is no greater than $(4q+1)B$.

Assume that there exists a desired partition $A_1, A_2, \ldots, A_q$ of 3-Partition. Because the total actual processing length on machine 2 is $(4q+1)B$, we know that no idle time on machine 2 is permitted. Then, we schedule the enforcer job 1 on machine one first followed by the the ordinary jobs corresponding to the three elements of A_1 and the resource level is brought back to $3B$. Repeat the dispatching pattern and then schedule the last job. It is to see that the schedule is feasible and the makespan is exactly $(4q+1)B$.

Assume that there is a feasible schedule whose makespan is no larger than $(4q+1)B$. The total actual processing length on machine 2 of all jobs is $(4q+1)B$. There is no idle time on machine 2. On the other hand, the total actual processing length on machine 1 is $4qB$. Considering the subsequent operations on machine 2, no idle time is allowed on machine 1. In other words, machine 1 and machine 2 cannot have any idle time in order to attain the makespan $(4q+1)B$. As given, machine 2 processes the enforcer jobs $1, 2, \ldots, q$ as in their indices. Since all enforcer jobs have the same parameter values, without loss of generality we assume that the enforcer jobs also follow the same processing order on machine 1.

We first note that job 1 should start first on machine 1 for otherwise non-zero idle time will be incurred on machine 2. After completing job 1 on machine 1, the resource level drops from $3B$ to B. To start the next enforcer job 2, machine 1 wait for the previous enforcer job 1 to be completed to accumulate sufficient resources. To avoid the idle time between the first and second enforcer jobs on machine 1, we assign ordinary jobs to fill up the idle period. Let A_1 be the set of elements defining these ordinary jobs. If $\sum_{x_j \in A_1} x_j < B$, then there is an idle time before job 2 on machine 1. On the other hand, if $\sum_{x_j \in A_1} x_j > B$, then the resource is insufficient and the completion time of some ordinary jobs are later than the second enforcer job, leading to the idle time on machine two. As the result, $\sum_{x_j \in A_1} x_j = B$ must hold. Continuing this process, we can find subsets $A_2, \ldots, A_n$ with $\sum_{x_j \in A_l} x_j = B$, $2 \leq l \leq q$, satisfied for the 3-Partition problem. Figure 3 shows the sequence of the optimal schedule: the purple blocks are enforcer jobs, the red ones are ordinary jobs, and the blue ones are final job. □

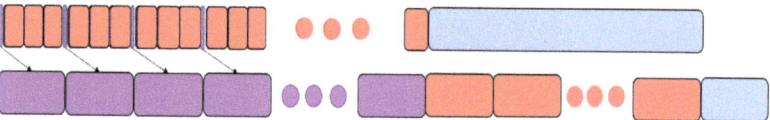

Figure 3. Given and fixed sequence on machine 2.

Theorems 1 and 2 together imply the following result.

Theorem 3. *If a sequence of the jobs on machine 1 is given and fixed, then finding an optimal schedule is strongly NP-hard, even if all jobs have the same processing time on machine 2.*

Proof. Owing to Theorems 1 and 2, we can get the feasibility of a given and fixed sequence of the jobs on machine 1 whose optimal schedule is strongly NP-hard. On the other hand, the idle time before job 1 on machine 2 is inevitable and the total actual processing length is $(4q + 1)B$. Furthermore, the sequence of Theorem 2 whose jobs on machine 2 is given and fixed, if we reverse this two-stage flow shop sequence, we can get the given and fixed sequence of the jobs on machine 1 which is equivalent to Theorem 3. It can seen that Figure 4 is derived form Figure 3 by reversing the Gantt chart from the right. As a result, the sequences of Theorem 2 and Theorem 3 have same total processing time. Then, we get the optimal schedule. □

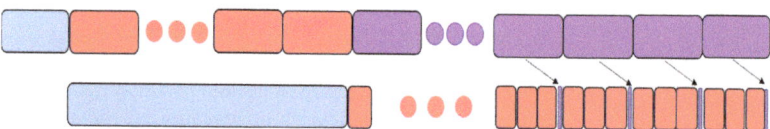

Figure 4. Given and fixed sequence on machine 1.

Theorem 4. *If sequences of the jobs on machine 1 and machine 2 are given and fixed, then the optimal schedule can be found in polynomial time.*

Proof. Assume that there is a feasible schedule whose sequences of the jobs on machine 1 and machine 2 are given and fixed. We schedule the first job on machine 1 followed by the first job on machine 2. If the resource of second job on machine 1 is insufficient, it should wait for previous job which on machine 2 to return the resource. Otherwise, it can be processed immediately when the previous job finished. Then, the job on machine 2 starts when the job on machine 1 completed. Continuing this process, we can schedule all the jobs and the makespan is minimum. On the other hand, if the job on machine 2 return resource is not enough for the next job to be processed, we can know that this sequence is not feasible. As a result, we can get an optimal schedule if sequences of the jobs on machine 1 and machine 2 are given and fixed. □

To simplify the problem, we consider the special case where processing sequences on both machines are given. By problem definition, the two sequences are not necessarily the same. For simplicity in presentation, we re-index the jobs to follow the natural sequence $\pi_1 = (\pi_{1,1}, \pi_{1,2}, \ldots, \pi_{1,n})$ on machine 1. Let $\pi_2 = (\pi_{2,1}, \pi_{2,2}, \ldots, \pi_{2,n})$ denote the sequence on machine 2 and $v(t)$ the resource level at a specific time point t. Notations t_1 and t_2 represent the current time points on machines 1 and 2. Note that if an operation whichever finished on machine 1 or machine 2 and another operation starts on next machine simultaneously at time t, we define $v(t)$ as the resource level after the finished operation on one machine and before the starting operation on the other machine. We will use this method, outlined in Algorithm 1, to calculate makespan for the problem.

Algorithm 1: Two Sequences

1 Let $\pi_1 = (\pi_{1,1}, \pi_{1,2}, \ldots, \pi_{1,n})$ and $\pi_2 = (\pi_{2,1}, \pi_{2,2}, \ldots, \pi_{2,n})$ be the given processing sequences;
2 $t_{1,\pi_{1,1}} = p_{1,\pi_{1,1}}; t_1 = t_{1,\pi_{1,1}}$;
3 $v(t_1) = v_0 - \alpha_{\pi_{1,1}}; t_2 = 0$;
4 $i = 2; j = 1$;
5 **while** $i \leq n$ and $j \leq n$ **do**
6   **if** $v(t_1) \geq \alpha_{\pi_{1,i}}$ **then**
7         $t_{1,\pi_{1,i}} = \max\{t_1, t_2\} + p_{1,\pi_{1,i}}$;
8         $t_1 = t_{1,\pi_{1,i}}$;
9         $v(t_1) = v(t_1) - \alpha_{\pi_{1,i}}$;
10         $i = i + 1$;
11   **else**
12   **if** job $\pi_{1,j}$ is not yet scheduled on machine 1 **then**
13   Report "No feasible solution!";
14   **else**
15             $t_2 = \max\{t_{1,\pi_{1,j}}, t_2\} + p_{2,\pi_{2,j}}$;
16   **if** $t_2 \geq t_1$ **then**
17                 $t_1 = t_2$;
18             $v(t_1) = v(t_1) + \beta_{\pi_{2,j}}$;
19             $j = j + 1$;
20 **while** $j \leq n$ **do**
21     $t_2 = \max\{t_{1,\pi_{2,j}}, t_2\} + p_{2,j}$;
22     $j = j + 1$;
23 **return** t_2.

In Algorithm 1, the first job of the sequence on machine 1 is processed first and there should be sufficient resource for it to start. Therefore, the first time point $t_{1,1}$ is the processing time of job 1 on machine 1, which is also t_1, and $v(t_1)$ is the resource level when job 1 finishes. In Line 6 we check the resource if we can process the job i on machine 1 or not. If resource is insufficient for job i on machine 1, we execute Line 11 to Line 19 for processing some job j on machine 2 to collect more resource. In Line 12, we need to check if job $\pi_{1,j}$ is scheduled first on machine 1 or not. Because we start to process jobs on machine 2 when the resource is not enough for machine 1, there may be several candidate jobs that can be processed on machine 2. Therefore, sometimes, t_2 is less than t_1 when the resource level is sufficient for the next job. In Lines 16 to 17, if t_2 is larger than t_1, we need to set t_1 equal to t_2, i.e., the next job on machine 1 should wait for the job on machine 2 to recycle its resource. When all the jobs on machine 1 finish, there are still some jobs on machine 2 not yet processed. As a result, we process the remaining jobs on machine 2 in the While loop of Line 20 to Line 22.

4. Heuristic Algorithms

Since the $F2|rp|C_{max}$ problem is computationally hard, it is hard to find optimal solutions when the problem size is large. We therefore design heuristic algorithms to produce approximate solutions in an acceptable time.

4.1. Permutation

We design two heuristic algorithms, using different sequences to construct feasible schedules for $F2|rp|C_{max}$ problem. If the job of the sequence does not violate the resource constraint, it must satisfy two conditions that the current resource level is sufficient for it, and that after its processing the resource level is sufficient for all remaining jobs. We denote the job sequence as σ and the remaining jobs sequenced by Johnson's rule using resource parameters α and β as σ_{JR}. Recall that v_0 is the initial resource requirement using Johnson's rule, and V_{needed} denotes the minimum resource requirement for the remaining jobs that the sequence is the same as σ_{JR} excluding job j. If the job violates the constraints, we will remove it. Then, we can get a set of feasible jobs which are the candidates to be processed next. Algorithm 2 examines each of the remaining jobs to determine if they are feasible candidates for the next position.

Algorithm 2: Check Resource

1 **Function** CheckResource($v_{now}, \sigma, \sigma_{JR}$):
2   **if** $\sigma.length = 1$ **then**
3   return σ;
4   **else**
5   **forall** $job\ j \in \sigma_{JR}$ **do**
6   **if** $v_{now} - \alpha_j + \beta_j < V_{needed}$ or $v_{now} \prec \alpha_j$ **then**
7   σ.remove($job\ j$);
8   return σ;

The first heuristic algorithm, JR-time Permutation Heuristic, is outlined in Algorithm 3. We define σ_{time} as the remaining jobs sequenced by Johnson's rule using $p_{1,i}$ and $p_{2,i}$ and σ_{JR} using α_i and β_i. Before a job is processed, we need to run CheckResource function for checking whether the job can be processed or not. Then, we append the job to the partial schedule σ. Repeat the same step until all the jobs are processed.

Algorithm 3: JR-time Permutation Heuristic

1 $\sigma = [\ ], v(t_i) = v_0$;
2 $i = 0$;
3 **while** $i \leq n$ **do**
4     $\tilde{\sigma} =$ CheckResource($v(t_{i+1}), \sigma_{time}, \sigma_{JR}$);
5     $v(t_{i+1}) = v(t_i) - \tilde{\sigma}(\alpha_1) + \tilde{\sigma}(\beta_1)$;
6   σ.append($\tilde{\sigma}(1)$);
7   σ_{JR}.remove($\tilde{\sigma}(1)$);
8     $\tilde{\sigma}$.remove($\tilde{\sigma}(1)$);
9 Stop.

The second heuristic, JR-resource Permutation Heuristic, is the same as the previous one except that the job sequence is ordered by Johnson's rule using α_i and β_i.

4.2. Non-Permutation

We design two heuristic algorithms that construct non-permutation schedules for $F2|rp|C_{max}$. Let σ_1 be the sequence of the jobs processed on machine 1, and σ_2 the job sequence on machine 2. Let $\hat{\sigma}_2$ contain the jobs eligible for processing on machine 2. Algorithm 4 processes the jobs on machine 1 first. If a job satisfies the two constraints, it will be appended to the schedule. Since we want to construct a non-permutation schedule, we create $\hat{\sigma}_2$ to collect the jobs which are finished on machine 1 but not yet on machine 2. When the resource level is insufficient for the candidate job, we need to process the jobs on machine 2 for acquiring more resource. Therefore, we choose the job that has the largest β in $\hat{\sigma}_2$ to be processed first. However, this strategy may lead to an idle time when we only process the selected job. To avoid this situation, we first find the arrival time of the selected job that can be processed and we call it *LargeBetaArrivalTime* here. Then, check if the completion time of any other job is earlier than *LargeBetaArrivalTime*. Furthermore, we change the *fraction* such that $1 - fraction$ is the acceptable time range that exceeds *LargeBetaArrivalTime*.

In Line 14, if the acceptable completion time is earlier than *LargeBetaArrivalTime*, then it is appended to σ_2 and removed form $\hat{\sigma}_2$. This process iterates until all the jobs of $\hat{\sigma}_2$ are checked. After that, the job having the largest β is appended to σ_2 and removed form $\hat{\sigma}_2$. We repeat the above steps until all the jobs be processed and then we get a feasible non-permutation schedule.

The second heuristic, JR-time Non-Permutation Heuristic, is similar to the first one except for using $p_{1,i}$ and $p_{2,i}$ to arrange the job sequence.

Algorithm 4: JR-resource Non-Permutation Heuristic

1　Order the jobs by Johnson's rule using α_i and β_i.;
2　$\sigma_1 = [\,], \sigma_2 = [\,], \hat{\sigma}_2 = \emptyset$;
3　$i = 0, j = 0, k = 0$;
4　$v(t_{1,i}) = v_0$;
5　**while** $i \leq n$ **do**
6　　$\bar{\sigma} = \text{CheckResource}(v(t_{1,i}), \sigma_{ori}, \sigma_{JR})$;
7　　**if** $v(t_{1,i}) \geq \bar{\sigma}(\alpha_1)$ **then**
8　　　$v(t_{1,i}) = v(t_{1,i}) - \bar{\sigma}(\alpha_1)$;
9　　　$\sigma_1.\text{append}(\bar{\sigma}(1))$;
10　　　$\hat{\sigma}_2.\text{append}(\bar{\sigma}(1))$;
11　　　$i = i + 1$;
12　　**else**
13　　　**forall** *job* $k \in \hat{\sigma}_2$ **do**
14　　　　**if** *LargeBetaArrivalTime* $\geq fraction * t_{2,k}$ **then**
15　　　　　$\sigma_2.\text{append}(job\ k)$;
16　　　　　$v(t_{2,j}) = v(t_{2,j}) + \beta_k$;
17　　　　　$\hat{\sigma}_2.\text{remove}(job\ k)$;
18　　　　　$j = j + 1$;
19　　　$v(t_{2,j}) = v(t_{2,j}) + \beta_{LargeBeta}$;
20　　　$\sigma_2.\text{append}(J_{2,LargeBeta})$;
21　　　$\hat{\sigma}_2.\text{remove}(J_{2,LargeBeta})$;
22　　　$j = j + 1$;
23　**while** $len(\hat{\sigma}_2) \leq n$ **do**
24　　$v(t_{2,j}) = v(t_{2,j}) + \beta_j$;
25　　$\sigma_2.\text{append}(job_{2,j})$;
26　　$\hat{\sigma}_2.\text{remove}(job_{2,j})$;
27　　$j = j + 1$;
28　Stop.

5. Ant Colony Optimization

In this section, we design an ACO algorithm to solve our problem. We will explain the framework and strategies of the algorithm for producing the approximate sequence.

State transition rule: In the ACO search process, each ant selects the next node to visit by calculating the preference for each path according to the pheromone intensity and heuristic visibility. In the proposed ACO algorithm, the preference P_{ij} of an ant, positioned at node i, for selecting node j is defined as:

$$P_{ij} = \begin{cases} \frac{\tau_{ij}^{w_\tau} \eta_{ij}^{w_\eta}}{\sum_{j \in I} \tau_{ij}^{w_\tau} \eta_{ij}^{w_\eta}}, & \text{if } j \in I; \\ 0, & \text{otherwise,} \end{cases} \tag{1}$$

where τ_{ij} is the pheromone intensity on the link from node i to node j, and η_{ij} the visibility value from node i to node j, and I the set of remaining admissible jobs to be processed. Parameters w_τ and w_η control the relative importance of τ_{ij} and η_{ij}. The greater a parameter is, the more influence of it to the preference value. In our design, the visibility value η_{ij} is based on a greedy strategy. We prefer less processing times on both machines, less resource requirement, and larger amount of the resource returned by job for priority selection. Visibility value is defined as:

$$\eta_{ij} = \frac{\beta_j}{\alpha_j + p_{1j} + p_{2j}}. \tag{2}$$

We use preference values P_{ij} for our exploration strategy. This method is just like the roulette wheel that every node, i.e., every job has their transition probability, based on which we select the next job randomly. Every node has a chance to be selected, even the probability is low.

Pheromone updating rule: After all the jobs are processed, we update the pheromone tails so that the ants can select their future paths according to previous experience. The trail intensity on link (i,j) is updated as below:

$$\tau_{ij} = (1 - \rho) \times \tau_{ij} + \Delta \tau_{ij}, \tag{3}$$

where ρ represents the pheromone evaporation rate, and $\Delta \tau_{ij}$ the incremental pheromone between nodes i and j given as:

$$\Delta \tau_{ij} = \begin{cases} \frac{Q}{C_k}, & \text{if } j \in I; \\ 0, & \text{otherwise.} \end{cases} \tag{4}$$

In the above definition, Q is an adjustable parameter and C_k the completion time of the last job on machine 2. This strategy is based on policy that the less C_k is, the more pheromone on the path enhanced.

Stopping criterion: The proposed ACO algorithm assigns a colony of ants to probe their own sequences and set a maximum number of iterations. When all the ants complete their routes in one iteration, we select the minimum makespan, i.e., the elite, to be our current best solution. Then, we iterate the process until reaching the maximum number of iterations. If there is a better solution in iterations, this new solution will replace the current best one.

Permutation: To take into account the resource constraints on two machines, we only choose the job that would return the sufficient resource for the remaining jobs to be successfully processed. Therefore, we use the function CheckResource before ants select the next job and then enter the ACO algorithm to get the permutation sequence. This method is similar to JR Permutation Heuristic except that we use ACO to choose the job sequence.

Non-Permutation: For non-permutation sequences, we divide the algorithm into two parts. In the first part, we use the ACO algorithm to obtain the sequence on machine 1, similar as in Permutation. In second part, the difference form Permutation is when the resource is insufficient for the next job, we use the same method *LargeBetaArrivalTime* in JR-resource Non-Permutation Heuristic to select jobs to process on machine 2. Then, we can get a complete sequence and we use it to update the ACO algorithm. As a result, we can get the non-permutation sequence by ACO combined with the heuristic method on machine 2.

6. Computational Experiments

In this section, we present computational experiments on the proposed methods through test data to compare and analyze the performance of these algorithms. The programs were coded in Python and executed on a personal computer with an Intel(R) Core(TM) i7-8700K CPU running at 3.70 GHz with 32.0 GB RAM. The operating system is Windows 10. We will describe how the test data sets were generated. Then, we present the related parameter settings and discuss the experimental results.

Data generation schemes: In our experiments, all parameters are integer. Processing times $p_{1,j}$ and $p_{2,j}$ of jobs on different machines were generated from the uniform distribution, $[1, 10]$. Resource parameters α_j and β_j were generated from the uniform distribution $[1, 20]$. The initial resource level was considered based on 1.1 and 1.4 times the minimum resource requirement that is at least how much the resource is needed for all the jobs of each data set. Test data sets are categorized into 8 different job numbers $n \in \{10, 20, 30, \ldots, 80\}$. For each job number, 5 independent sets were generated. Each set also has different uniform distributions for processing times $p_{1,j}$ and $p_{2,j}$, the resource parameters α_j and β_j, and the initial resource requirement. That means that we have 40 different data sets in all. On each data set, say 10 jobs, heuristic algorithms were run only once since they are deterministic. For a specific setting, the values were averaged over 5 independent sets of the same setting. The ACO algorithm, due to its randomness nature, was exercised 5 runs on each data set to get its average performance.

6.1. Results of Heuristic Algorithms

In this experiment, we apply the four heuristic algorithms on different data sets. We compare permutation solutions with non-permutation ones in two different methods, namely JR-resource and JR-time. The initial resource level in all experiment results set by multiplying the minimum resource requirement by 1.1. For each problem size, the average objective value of derived solution (minimum makespan) are reported. Since the elapsed execution times of four heuristic algorithms are almost negligible, we do not show the execution time in the following tables. All detail experiment results of different data sets are shown in Appendix A.

In Table 2, the makespan of permutation sequence of perm and non-permutation is maxβ. It can be seen that the JR-resource algorithm can get better makespan than the JR-time algorithm. Since the constraint is considered by resource, it is obviously that when the jobs sequenced by processing time, the resource would insufficient and the jobs should wait for the resource returned which lead to idle time. In most of the data sets, the makespans of permutation heuristics are less than non-permutation ones (maxβ in Table 2). However, sometimes, non-permutation can get a better solution that reported in JR-resource with 10 jobs. We speculate that some jobs on machine 2 can fill up the idle time and thus decrease the waiting time on machine 1.

In the experiment, there are different fractions of the bearable exceeding time, which is the acceptable time length that exceeds *LargeBetaArrivalTime*, used in JR-resource and JR-time Non-Permutation Heuristics. The fraction ranges from 5/10 to 10/10. Table 3 is for JR-time Non-Permutation Heuristics focused on processing times, and Table 4 for the heuristics focused on α_i and β_i. If we do not consider the bearable exceeding time, the makespan would be more than others that bear the exceed time because there is longer idle time in the sequences. It is clear that with the 10/10 fraction we get a longer makespan. In most cases, the makespan is the same regardless of the fraction. However, sometimes it is shown that if we bear too much exceeded time, we may get worse makespan with a 5/10 fraction of 20 jobs in JR-time and 5/10 to 6/10 fractions of 10 jobs in JR-resource. The results of fractions are not better than permutation ones, so we do not have further test for different fractions with 1.4 times the initial resource level.

Table 2. JR-resource and JR-time heuristics.

#	JR-Resource		JR-Time	
	perm	maxβ	perm	maxβ
		C_{max}		
10	80	78	87	92
20	137	142	166	177
30	199	202	251	256
40	258	262	332	345
50	307	308	425	439
60	351	357	474	494
70	426	431	596	605
80	467	472	636	642

Table 3. Different fractions in JR-time Non-Permutation Heuristics.

#	Fraction of Exceeding Time					
	10/10	9/10	8/10	7/10	6/10	5/10
			C_{max}			
10	92	88	87	87	86	86
20	177	169	166	166	166	167
30	256	252	252	252	252	252
40	345	332	332	333	333	332
50	439	424	424	425	425	425
60	494	475	475	475	475	475
70	605	595	597	597	597	597
80	642	638	638	637	637	637

6.2. Results of ACO Algorithms

We discuss the results of ACO algorithms with permutation and non-permutation options. We tuned several parameter values in preliminary tests to determine the setting for further experiments. We observed differences in the results, although not significant. The parameter values leading to better results were adopted as the base setting for the final computational tests. We set the two parameters $w_\tau = 2$ and $w_\eta = 3$ that could get better makespan in the experiment we tested before. The pheromone evaporation ρ is 0.95 to avoid early convergence. Parameter Q is defined as the number of jobs divided by 10 and multiplied by 50. Setting 100 epochs for a solution with the colony size the same as the number of jobs. Then, we have each data set run this process 5 times to get an average makespan and an average elapsed execution time. The time unit here is second and "-" means that no feasible solution was found. All complete experiment results are shown in Appendix A.

Table 4. Different fraction in JR-resource Non-Permutation Heuristics.

#	Fraction of Exceeding Time					
	10/10	9/10	8/10	7/10	6/10	5/10
	C_{max}					
10	78	78	78	79	80	80
20	142	138	137	137	137	137
30	202	199	199	199	199	199
40	262	258	258	258	258	258
50	308	307	307	307	307	307
60	357	351	351	351	351	351
70	431	428	426	426	426	426
80	472	467	467	467	467	467

In Table 5, perm indicates the permutation method, and maxβ the non-permutation method presented in Section 5. M2-enum is also non-permutation sequence that is different from the method we used in maxβ. The difference between them is that when the resource is insufficient on machine 1, M2-enum will enumerate all the possible sequences of candidate jobs on machine 2 to find the minimum makespan and return resource. This procedure is time-consuming because the number of possible sequences we need to compare is a factorial of the number of candidate jobs on machine 2. Therefore, when the job number is larger than 30, we cannot get a solution in 3600 s. We also experiment on the integer programming method (IP) proposed by Cheng et al. [13] to solve the problem. The IP can get the optimal solutions for data sets. However, when the number of jobs is over 10, it cannot find any feasible solution in 3600 s for most data sets. Therefore, it is regarded as no solution found. With the initial resource is multiplied by 1.1, it is clear that perm can obtain a better makespan and the required run time of perm is also less than maxβ.

Table 5. Results of ACO algorithms and IP with 1.1 × initial resource levels.

#	perm		maxβ		M2-enum		IP	
	C_{max}	Time	C_{max}	Time	C_{max}	Time	C_{max}	Time
10	69	0.41	70	0.80	69	0.84	68	446.81
20	125	2.33	128	4.54	123	27.47	-	-
30	184	8.24	187	13.46	184	3396.50	-	-
40	239	22.09	242	36.99	-	-	-	-
50	299	50.15	300	73.62	-	-	-	-
60	339	97.24	344	139.03	-	-	-	-
70	416	172.08	418	236.40	-	-	-	-
80	456	279.78	459	352.56	-	-	-	-

Table 6 indicates that with more initial resource a better makespan can be achieved, as comapred with those in Table 5. Similarly, M2-enum and IP cannot get any solution when job numbers are over 30 and 10, respectively, in 3600 s. It is also shown that when the initial resource level is larger, for like 30 jobs, M2-enum will waste more time so that it cannot find better solutions in 3600 s (Displayed in Table A18). Then, the permutation method can find an approximate solution with less time. However, results of 1.1 and 1.4 times are within spitting distance when the quantity of jobs increases. It is reckoned that the processing times of the subsequent jobs on machine 2 are larger and the resource is sufficient for them, so their starting times are later than their completion times on machine 1 and they keep processing continuous without idle time. Therefore, the makespan of larger data sets are not quite different when the initial resource increases.

Table 6. Results of ACO algorithms and IP with 1.4 × initial resource levels.

#	perm C_{max}	perm Time	maxβ C_{max}	maxβ Time	M2-enum C_{max}	M2-enum Time	IP C_{max}	IP Time
10	66	0.42	67	0.79	69	0.84	65	187.72
20	125	2.44	127	4.44	124	470.49	-	-
30	183	8.31	185	13.95	184	2967.66	-	-
40	239	21.59	241	34.04	-	-	-	-
50	299	49.88	300	69.53	-	-	-	-
60	339	95.94	343	129.10	-	-	-	-
70	416	160.34	418	226.25	-	-	-	-
80	456	263.11	458	364.20	-	-	-	-

Table 7 shows that different fractions of the bearable exceeding time used in ACO Non-Permutation algorithms with different times of the initial resource. In most cases, it would get less makespan with considering the bearable exceeding time. The makespan resulted from fractions between 5/10 and 9/10 are not quite different. It is clearly shown that the makespan of 10/10 fraction is longer than others. As a result, setting the bearable exceeding time can achieve a better performance.

Table 7. Results of different fraction in maxβ of ACO with different initial resource levels.

	Fraction of Exceeding Time with 1.1 × Initial Resource Levels									
#	5/10		6/10		7/10		8/10		9/10		10/10
	C_{max}	Time	C_{max}	Time	C_{max}	Time	C_{max}	Time	C_{max}	Time	C_{max}	Time
10	69	0.84	69	0.82	69	0.79	69	0.78	70	0.87	70	0.80
20	125	4.65	125	4.61	125	4.81	125	4.35	125	4.71	128	4.54
30	184	14.50	184	15.61	184	15.32	184	13.70	184	14.42	187	13.46
40	239	35.49	239	41.70	239	40.46	239	33.73	239	36.92	242	36.99
50	299	71.12	299	76.58	299	78.63	299	72.06	299	77.92	300	73.62
60	339	128.90	339	141.33	339	144.98	339	133.39	340	139.73	344	139.03
70	416	262.69	417	255.90	417	243.21	416	235.66	416	218.79	418	236.40
80	456	347.84	456	368.73	456	354.50	456	395.34	456	358.64	459	352.56
	Fraction of Exceeding Time with 1.4 × Initial Resource Levels									
#	5/10		6/10		7/10		8/10		9/10		10/10
	C_{max}	Time	C_{max}	Time	C_{max}	Time	C_{max}	Time	C_{max}	Time	C_{max}	Time
10	66	0.77	66	0.78	66	0.79	66	0.82	66	0.79	67	0.79
20	125	4.52	124	4.51	124	4.56	125	4.49	124	4.54	127	4.44
30	183	13.99	183	13.85	183	13.80	183	13.98	183	13.94	185	13.95
40	239	33.55	239	33.85	239	33.83	239	34.28	239	34.18	241	34.04
50	299	70.50	298	75.06	299	71.54	299	69.69	299	70.03	300	69.53
60	339	132.19	339	130.84	339	130.32	339	130.71	340	129.45	343	129.10
70	417	214.20	416	209.34	417	242.51	416	230.94	417	224.80	418	226.25
80	456	359.57	456	359.48	456	369.24	456	379.99	456	359.53	458	364.20

6.3. Comparison between Heuristics and ACO Algorithms

We discuss the experiment results of the heuristic and the ACO algorithms with special focus on permutation results of JR-time and JR-resource because the makespan of permutation cases are less than other non-permutation heuristic algorithms. ACO-10/10 is the 10/10 fraction of exceeding time bearable in the non-permutation ACO algorithm. We also choose the 5/10 fraction (ACO-5/10) and the permutaion of ACO (ACO-perm) as the control groups.

In Figures 5 and 6, both heuristic algorithms produced larger makespan no matter if the initial resource is multiplied by 1.1 or 1.4, especially for the JR-time. However, there is still some deviations in the ACO algorithms owing to the randomness nature. Except for the above situation, 1.4 times are still better than 1.1 times in most cases with permutation and non-permutation methods, especially JR-time. We reckon that with a higher initial resource level, more jobs on machine 1 can keep continuous processing, thus reducing the idle time waiting for resource return. Concerning the ACO algorithm, it is clear that that the 5/10 fraction is better than the 10/10 case.

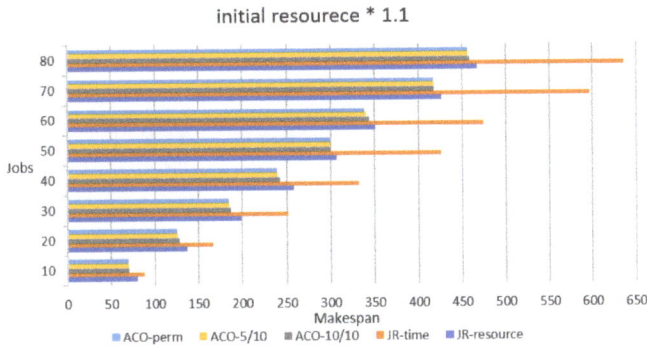

Figure 5. Bar chart of comparison with 1.1 × initial resource levels.

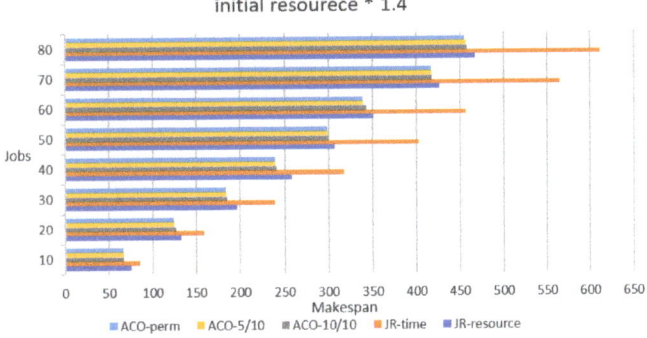

Figure 6. Bar chart of comparison with 1.4 × initial resource levels.

7. Conclusions and Future Works

In this paper, we considered the relocation problem in a two-machine flow shop scheduling problem with the second machine introduced for recycling the resource returned by jobs completed on the first machine. For this problem, we proved that given a sequence of jobs on whichever machine, the problem is still strongly NP-hard. The case with two fixed sequences of jobs on both machines can be solved in polynomial time. For the computationally hard problem, we proposed two heuristic algorithms to construct feasible schedules with permutation and non-permutation sequences. ACO algorithms were designed to find processing sequences on the two machines. Computational experiments indicates that JR-resource produced better makespan than JR-time, and both of their permutation algorithms are better than the non-permutation ones. This result is similar as in ACO algorithm that the makespan of permutation solutions are better. However, when we considered the bearable exceeding time, non-permutation sequences can get better solutions. Between heuristics and ACO algorithms, it is shown that ACO algorithms yielded schedules with less makespan.

For further studies, there is room for further improvements of our proposed solution methods by developing effective methods to arrange the job sequence on machine 2 to mitigate the incurred idle time. Furthermore, we can also combine processing time and resources as a control factor in the heuristic algorithms. It would be also interesting to deploy machine learning and reinforcement learning approaches for finding better parameter settings and updating strategies. We also note that to optimally solve the problem is still limited to small-scale instances. For larger instances, we need tighter lower bounds to facilitate the development of exact methods and provide a tight comparison base for approximation methods. Another direction is identifying application contexts in which the unique type of resource constraints, the amount of consumed resource and the amount of returned resource could be different, is applicable.

Author Contributions: Conceptualization, T.-C.L. and B.M.T.L.; methodology, T.-C.L. and B.M.T.L.; software, T.-C.L.; formal analysis, T.-C.L. and B.M.T.L.; writing—Original draft preparation, T.-C.L. and B.M.T.L.; writing, T.-C.L. and B.M.T.L.; supervision, T.-C.L. and B.M.T.L.; project administration, B.M.T.L.; funding acquisition, B.M.T.L. All authors have read and agreed to the published version of the manuscript.

Funding: Lo and Lin were partially supported by the Ministry of Science and Technology of Taiwan under the grant MOST-109-2410-H-009-029.

Institutional Review Board Statement: Not applicable.

Informed Consent Statement: Not applicable.

Data Availability Statement: Datasets analyzed in this study can be found at http://cpanel-199-19.nctu.edu.tw/~bmtlin/F2RP.zip (accessed on 29 June 2021).

Conflicts of Interest: The authors declare no conflict of interest.

Appendix A. Detail Experiment Results

All of the experiment results mentioned in Section 6 are summarized from the detail information shown below. We show the makespan (C_{max}) and execution time for each data set followed by the average over five data sets in different job numbers. In the tables, "perm" means permutation method, and "maxβ" non-permutation with a 10/10 fraction. We also set the execution time to stop the program if it exceeds 3600 s. Therefore, if we cannot obtain any solution in 3600 s, we indicate the situation by an "-" entry. In ACO, owing to its randomness property, we run each data sets for five times which is "Run." in the tables. In Tables A18 and A19, when the numbers of jobs are over 30 and 10, we cannot obtain solutions for most cases so that we do not experiment on the cases with more jobs. It is clear that most results are associated with an "-", indicating that no solutions were found within 3600 s.

Table A1. JR-resource with 1.1 × initial resource levels.

Sets	perm		maxβ		9/10		8/10		7/10		6/10		5/10	
	C_{max}	Time	C_{max}	Time	C_{max}	Time	C_{max}	Time	C_{max}	Time	C_{max}	Time	C_{max}	Time
10														
1	78	0.00	78	0.00	78	0.00	78	0.00	78	0.00	78	0.00	78	0.00
2	95	0.00	90	0.00	90	0.00	90	0.00	90	0.00	95	0.00	95	0.00
3	82	0.00	81	0.00	81	0.00	81	0.00	82	0.00	82	0.00	82	0.00
4	72	0.00	72	0.00	72	0.00	72	0.00	72	0.00	72	0.00	72	0.00
5	71	0.00	71	0.00	71	0.00	71	0.00	71	0.00	71	0.00	71	0.00
Avg.	80	0.00	78	0.00	78	0.00	78	0.00	79	0.00	80	0.00	80	0.00
20														
1	144	0.00	154	0.00	144	0.00	144	0.00	144	0.00	144	0.00	144	0.00
2	148	0.00	148	0.00	148	0.00	148	0.00	148	0.00	148	0.00	148	0.00
3	117	0.00	126	0.00	118	0.00	117	0.00	117	0.00	117	0.00	117	0.00
4	160	0.00	165	0.00	165	0.00	160	0.00	160	0.00	160	0.00	160	0.00
5	115	0.00	115	0.00	115	0.00	115	0.00	115	0.00	115	0.00	115	0.00
Avg.	137	0.00	142	0.00	138	0.00	137	0.00	137	0.00	137	0.00	137	0.00
30														
1	196	0.00	196	0.00	196	0.00	196	0.00	196	0.00	196	0.00	196	0.00
2	209	0.00	209	0.00	209	0.00	209	0.00	209	0.00	209	0.00	209	0.00
3	194	0.00	194	0.00	194	0.00	194	0.00	194	0.00	194	0.00	194	0.00
4	201	0.00	211	0.00	201	0.00	201	0.00	201	0.00	201	0.00	201	0.00
5	193	0.00	200	0.00	193	0.02	193	0.00	193	0.00	193	0.00	193	0.00
Avg.	199	0.00	202	0.00	199	0.00	199	0.00	199	0.00	199	0.00	199	0.00
40														
1	267	0.00	270	0.00	267	0.00	267	0.00	267	0.00	267	0.00	267	0.00
2	229	0.00	231	0.00	229	0.00	229	0.01	229	0.00	229	0.00	229	0.00
3	260	0.00	268	0.00	260	0.00	260	0.00	260	0.02	260	0.00	260	0.00
4	278	0.00	286	0.00	278	0.00	278	0.00	278	0.00	278	0.00	278	0.00
5	256	0.00	256	0.02	256	0.00	256	0.00	256	0.02	256	0.00	256	0.00
Avg.	258	0.00	262	0.00	258	0.00	258	0.00	258	0.01	258	0.00	258	0.00
50														
1	310	0.00	311	0.00	310	0.00	310	0.00	310	0.00	310	0.00	310	0.01
2	324	0.00	324	0.02	324	0.00	324	0.00	324	0.02	324	0.00	324	0.00
3	285	0.00	288	0.00	285	0.00	285	0.00	285	0.00	285	0.00	285	0.00
4	307	0.00	307	0.02	307	0.00	307	0.00	307	0.02	307	0.00	307	0.00
5	309	0.00	309	0.00	309	0.02	309	0.00	309	0.00	309	0.01	309	0.00
Avg.	307	0.00	308	0.01	307	0.00	307	0.00	307	0.01	307	0.00	307	0.00
60														
1	340	0.00	350	0.02	340	0.00	340	0.00	340	0.00	340	0.01	340	0.01
2	359	0.00	362	0.00	359	0.02	359	0.00	359	0.02	359	0.01	359	0.01
3	347	0.00	353	0.00	347	0.02	347	0.02	347	0.02	347	0.01	347	0.01
4	348	0.00	348	0.02	348	0.00	348	0.00	348	0.00	348	0.01	348	0.01
5	359	0.00	370	0.00	360	0.00	359	0.00	359	0.02	359	0.01	359	0.01
Avg.	351	0.00	357	0.01	351	0.01	351	0.00	351	0.01	351	0.01	351	0.01

Table A1. Cont.

Sets	perm		maxβ		9/10		8/10		7/10		6/10		5/10	
	C_{max}	Time	C_{max}	Time	C_{max}	Time	C_{max}	Time	C_{max}	Time	C_{max}	Time	C_{max}	Time
							70							
1	441	0.00	441	0.00	441	0.02	441	0.02	441	0.02	441	0.01	441	0.01
2	392	0.00	392	0.00	392	0.02	392	0.02	392	0.02	392	0.01	392	0.01
3	448	0.00	459	0.02	448	0.02	448	0.02	448	0.02	448	0.01	448	0.01
4	455	0.00	455	0.02	455	0.02	455	0.02	455	0.02	455	0.01	455	0.01
5	395	0.00	407	0.02	402	0.02	395	0.00	395	0.00	395	0.01	395	0.01
Avg.	426	0.00	431	0.01	428	0.02	426	0.01	426	0.01	426	0.01	426	0.01
							80							
1	505	0.00	505	0.02	505	0.02	505	0.02	505	0.02	505	0.01	505	0.01
2	466	0.00	473	0.02	466	0.02	466	0.02	466	0.02	466	0.01	466	0.01
3	479	0.00	483	0.02	479	0.02	479	0.02	479	0.02	479	0.01	479	0.01
4	438	0.00	441	0.02	438	0.02	438	0.02	438	0.02	438	0.01	438	0.01
5	449	0.00	457	0.02	449	0.02	449	0.02	449	0.02	449	0.01	449	0.01
Avg.	467	0.00	472	0.02	467	0.02	467	0.02	467	0.02	467	0.01	467	0.01

Table A2. JR-time with 1.1 × initial resource levels.

Sets	perm		maxβ		9/10		8/10		7/10		6/10		5/10	
	C_{max}	Time	C_{max}	Time	C_{max}	Time	C_{max}	Time	C_{max}	Time	C_{max}	Time	C_{max}	Time
							10							
1	89	0.00	89	0.00	89	0.00	89	0.00	89	0.00	89	0.00	89	0.00
2	101	0.00	109	0.00	101	0.00	101	0.00	101	0.00	101	0.00	101	0.00
3	77	0.00	86	0.00	84	0.00	77	0.00	77	0.00	77	0.00	77	0.00
4	78	0.00	84	0.00	78	0.00	78	0.00	79	0.00	74	0.00	74	0.00
5	90	0.00	90	0.00	90	0.00	90	0.00	90	0.00	90	0.00	90	0.00
Avg.	87	0.00	92	0.00	88	0.00	87	0.00	87	0.00	86	0.00	86	0.00
							20							
1	151	0.00	154	0.00	157	0.00	152	0.00	152	0.00	152	0.00	152	0.00
2	211	0.00	236	0.00	216	0.00	212	0.00	212	0.00	212	0.00	212	0.00
3	158	0.00	163	0.02	158	0.00	158	0.02	158	0.02	158	0.00	158	0.00
4	197	0.00	219	0.00	203	0.00	197	0.00	197	0.00	197	0.00	197	0.00
5	115	0.00	113	0.00	110	0.00	110	0.00	111	0.00	111	0.00	115	0.00
Avg.	166	0.00	177	0.00	169	0.00	166	0.00	166	0.00	166	0.00	167	0.00
							30							
1	317	0.02	316	0.00	317	0.00	317	0.00	317	0.00	317	0.01	317	0.00
2	239	0.00	231	0.02	234	0.00	240	0.00	240	0.00	240	0.00	240	0.00
3	220	0.00	229	0.00	222	0.00	222	0.00	222	0.00	222	0.00	222	0.00
4	272	0.00	289	0.02	278	0.00	272	0.02	272	0.02	272	0.00	272	0.00
5	208	0.00	213	0.00	209	0.00	208	0.00	208	0.00	208	0.00	208	0.00
Avg.	251	0.00	256	0.01	252	0.00	252	0.00	252	0.00	252	0.00	252	0.00

Table A2. Cont.

Sets	perm		maxβ		9/10		8/10		7/10		6/10		5/10	
	C_{max}	Time	C_{max}	Time	C_{max}	Time	C_{max}	Time	C_{max}	Time	C_{max}	Time	C_{max}	Time
40														
1	318	0.02	334	0.00	317	0.00	318	0.00	318	0.02	318	0.01	318	0.01
2	272	0.02	276	0.00	271	0.00	272	0.00	272	0.02	272	0.01	272	0.01
3	333	0.02	356	0.00	333	0.00	333	0.00	336	0.02	336	0.01	333	0.01
4	401	0.02	415	0.00	403	0.00	401	0.01	401	0.02	401	0.01	401	0.01
5	337	0.02	343	0.00	334	0.00	337	0.01	337	0.02	337	0.01	337	0.01
Avg.	332	0.02	345	0.00	332	0.00	332	0.00	333	0.02	333	0.01	332	0.01
50														
1	443	0.00	446	0.02	442	0.02	443	0.02	443	0.02	443	0.02	443	0.02
2	400	0.02	404	0.02	398	0.00	400	0.02	400	0.02	400	0.01	400	0.01
3	416	0.00	441	0.02	417	0.02	414	0.02	418	0.02	418	0.02	418	0.01
4	464	0.00	475	0.02	462	0.02	465	0.02	464	0.02	464	0.02	464	0.02
5	402	0.00	431	0.02	402	0.02	400	0.02	400	0.02	402	0.02	402	0.02
Avg.	425	0.00	439	0.02	424	0.01	424	0.02	425	0.02	425	0.02	425	0.02
60														
1	583	0.02	596	0.03	588	0.03	585	0.03	585	0.02	583	0.03	583	0.03
2	510	0.02	564	0.03	514	0.03	512	0.03	512	0.03	512	0.03	512	0.02
3	448	0.02	455	0.02	450	0.04	448	0.02	448	0.03	448	0.03	448	0.03
4	391	0.00	385	0.03	382	0.02	392	0.02	392	0.03	392	0.02	391	0.02
5	440	0.02	471	0.03	440	0.02	440	0.03	440	0.02	440	0.02	440	0.03
Avg.	474	0.01	494	0.03	475	0.03	475	0.02	475	0.02	475	0.03	475	0.03
70														
1	490	0.02	501	0.05	490	0.03	490	0.03	490	0.03	490	0.03	490	0.03
2	614	0.02	625	0.03	615	0.04	615	0.03	615	0.05	615	0.03	615	0.04
3	741	0.03	748	0.05	741	0.05	741	0.05	741	0.03	741	0.04	741	0.04
4	506	0.03	498	0.05	495	0.03	505	0.03	506	0.02	506	0.03	506	0.03
5	631	0.02	654	0.06	634	0.03	634	0.05	633	0.05	632	0.04	631	0.04
Avg.	596	0.02	605	0.05	595	0.04	597	0.04	597	0.03	597	0.04	597	0.04
80														
1	578	0.02	583	0.06	578	0.05	578	0.06	579	0.05	578	0.05	578	0.05
2	615	0.03	639	0.06	615	0.06	615	0.06	615	0.06	615	0.06	615	0.06
3	824	0.03	831	0.06	827	0.08	827	0.06	824	0.05	824	0.06	824	0.06
4	514	0.03	496	0.05	517	0.03	517	0.06	517	0.03	517	0.04	517	0.04
5	649	0.02	659	0.06	652	0.06	651	0.05	651	0.05	651	0.05	649	0.05
Avg.	636	0.02	642	0.06	638	0.06	638	0.06	637	0.05	637	0.05	637	0.05

Table A3. JR-resource and JR-time with 1.4 × initial resource levels.

Sets	JR-Resource				JR-Time			
	perm		maxβ		perm		maxβ	
	C_{max}	Time	C_{max}	Time	C_{max}	Time	C_{max}	Time
10								
1	69	0.00	69	0.00	91	0.00	91	0.00
2	95	0.00	90	0.00	101	0.00	109	0.00
3	73	0.00	75	0.00	77	0.00	85	0.00
4	70	0.00	70	0.00	63	0.00	64	0.00
5	71	0.00	71	0.00	90	0.00	90	0.00
Avg.	76	0.00	75	0.00	84	0.00	88	0.00
20								
1	136	0.00	145	0.00	141	0.00	149	0.00
2	141	0.00	141	0.00	189	0.00	189	0.00
3	117	0.00	126	0.00	158	0.00	163	0.00
4	160	0.00	165	0.00	197	0.00	219	0.00
5	111	0.00	111	0.00	107	0.00	107	0.00
Avg.	133	0.00	137.6	0.00	158	0.00	165	0.00
30								
1	191	0.00	191	0.00	316	0.00	321	0.01
2	209	0.00	209	0.00	216	0.00	213	0.00
3	194	0.00	194	0.00	197	0.00	207	0.01
4	201	0.00	211	0.00	272	0.00	289	0.00
5	187	0.00	194	0.00	188	0.00	186	0.00
Avg.	196	0.00	199.8	0.00	238	0.00	243	0.00
40								
1	267	0.00	270	0.00	318	0.00	334	0.01
2	229	0.00	231	0.00	238	0.00	240	0.01
3	260	0.00	268	0.00	299	0.00	307	0.01
4	278	0.00	286	0.00	401	0.00	415	0.01
5	256	0.00	256	0.00	327	0.00	332	0.01
average	258	0.00	262.2	0.00	317	0.00	326	0.01
50								
1	310	0.00	311	0.00	443	0.01	446	0.02
2	324	0.00	324	0.00	344	0.01	337	0.01
3	285	0.00	288	0.00	358	0.01	371	0.01
4	307	0.00	307	0.00	462	0.01	464	0.02
5	309	0.00	309	0.00	402	0.01	431	0.02
Avg.	307	0.00	307.8	0.00	402	0.01	410	0.02
60								
1	340	0.00	350	0.01	589	0.01	604	0.03
2	359	0.00	362	0.01	502	0.01	524	0.03
3	347	0.00	353	0.01	448	0.01	455	0.03
4	348	0.00	348	0.01	347	0.01	352	0.02
5	359	0.00	370	0.01	440	0.01	471	0.03
Avg.	351	0.00	356.6	0.01	465	0.01	481	0.03

Table A3. Cont.

Sets	JR-Resource				JR-Time			
	perm		maxβ		perm		maxβ	
	C_{max}	Time	C_{max}	Time	C_{max}	Time	C_{max}	Time
70								
1	441	0.00	441	0.01	490	0.02	501	0.03
2	392	0.00	392	0.01	569	0.02	587	0.04
3	448	0.00	459	0.01	680	0.02	699	0.04
4	455	0.00	455	0.01	479	0.02	459	0.03
5	395	0.00	407	0.01	603	0.02	621	0.04
Avg.	426	0.00	430.8	0.01	564	0.02	573	0.04
80								
1	505	0.00	505	0.01	550	0.03	521	0.05
2	466	0.00	473	0.01	615	0.02	639	0.06
3	478	0.00	482	0.01	817	0.02	824	0.06
4	438	0.00	441	0.01	481	0.03	446	0.04
5	449	0.00	457	0.01	591	0.03	610	0.05
Avg.	467	0.00	471.6	0.01	611	0.03	608	0.05

Table A4. ACO Permutation with $1.1 \times$ initial resource levels.

	Sets of Job									
Run	1	2	3	4	5	1	2	3	4	5
	C_{max} Time	C_{max} Time	C_{max} Time	C_{max} Time	C_{max} Time	C_{max} Time	C_{max} Time	C_{max} Time	C_{max} Time	C_{max} Time
	10					50				
1	66 0.41	82 0.41	73 0.41	61 0.39	64 0.40	303 48.86	321 50.82	277 49.20	294 51.35	300 50.37
2	66 0.41	82 0.41	73 0.41	61 0.41	64 0.41	303 49.31	321 50.27	277 49.09	294 50.77	300 50.55
3	68 0.41	83 0.41	73 0.41	61 0.41	64 0.41	303 49.18	321 50.91	279 50.17	294 50.99	300 50.57
4	66 0.41	82 0.41	73 0.41	60 0.41	64 0.41	303 48.76	321 49.94	277 50.67	294 50.63	298 51.03
5	66 0.41	82 0.42	73 0.39	61 0.41	64 0.39	303 48.91	320 49.75	278 50.36	294 50.90	301 50.48
Avg.	66 0.41	82 0.41	73 0.40	61 0.40	64 0.40	303 49.00	321 50.34	278 49.90	294 50.93	300 50.60
	20					60				
1	122 2.37	134 2.32	109 2.31	155 2.34	108 2.31	325 91.75	337 92.80	339 103.47	346 97.93	349 97.29
2	123 2.34	134 2.32	109 2.32	153 2.39	108 2.33	325 92.25	337 93.09	339 100.06	346 101.74	349 98.19
3	122 2.31	134 2.30	111 2.31	154 2.36	108 2.31	325 92.80	337 101.40	339 95.92	343 98.36	349 99.48
4	122 2.30	134 2.32	109 2.31	153 2.38	108 2.31	324 93.05	337 102.11	339 96.97	346 95.87	349 98.85
5	122 2.30	134 2.33	108 2.33	155 2.38	108 2.31	325 93.06	337 101.48	339 98.08	346 95.88	349 99.10
Avg.	122 2.33	134 2.32	109 2.32	154 2.37	108 2.32	325 92.58	337 98.18	339 98.90	345 97.96	349 98.58
	30					70				
1	181 8.22	199 8.24	187 8.08	183 8.26	172 8.28	427 237.11	382 232.36	437 243.98	446 232.11	396 244.93
2	178 8.32	199 8.23	185 8.30	187 8.19	172 8.33	426 231.01	383 231.93	436 240.15	446 233.42	396 236.99
3	179 8.34	199 8.02	187 8.15	186 8.34	172 8.19	426 231.26	383 232.20	437 242.17	446 235.37	397 240.01
4	178 8.27	199 8.31	187 8.16	186 8.19	172 8.19	428 233.16	384 235.12	438 243.92	446 228.57	396 235.52
5	180 8.35	199 8.07	187 8.26	185 8.35	172 8.24	426 231.87	382 234.44	437 242.01	446 235.86	395 244.66
Avg.	179 8.30	199 8.18	187 8.19	185 8.27	172 8.25	427 232.88	383 233.21	437 242.45	446 233.06	396 240.42

Table A4. Cont.

Run	Sets of Job																			
	1		2		3		4		5		1		2		3		4		5	
	C_{max}	Time	C_{max}	Time	C_{max}	Time	C_{max}	Time	C_{max}	Time	C_{max}	Time	C_{max}	Time	C_{max}	Time	C_{max}	Time	C_{max}	Time
	40										80									
1	234	22.94	226	22.26	237	22.21	244	21.88	255	21.37	493	262.88	459	272.06	474	280.10	429	286.30	426	290.69
2	234	22.34	225	22.15	237	22.17	244	22.18	255	21.70	494	264.76	457	268.81	476	301.40	429	282.64	426	279.41
3	234	22.39	225	22.12	237	22.22	244	21.97	255	21.79	491	269.19	459	274.39	476	289.93	429	286.31	427	279.23
4	234	22.23	226	22.42	237	23.49	244	21.46	255	21.92	491	267.44	459	274.37	476	290.52	429	286.06	426	283.96
5	234	22.20	226	22.37	237	21.27	244	21.66	255	21.58	491	261.35	460	274.82	476	287.02	429	292.30	428	288.61
Avg.	234	22.42	226	22.26	237	22.27	244	21.83	255	21.67	492	265.12	459	272.89	476	289.80	429	286.72	427	284.38

Table A5. 10/10 fraction in $\max\beta$ of ACO with $1.1 \times$ initial resource levels.

Run	Sets of Job																			
	1		2		3		4		5		1		2		3		4		5	
	C_{max}	Time	C_{max}	Time	C_{max}	Time	C_{max}	Time	C_{max}	Time	C_{max}	Time	C_{max}	Time	C_{max}	Time	C_{max}	Time	C_{max}	Time
	10										50									
1	66	0.80	83	0.82	76	0.80	61	0.77	64	0.78	304	73.59	322	72.11	281	74.46	295	76.75	298	75.98
2	66	0.81	83	0.84	76	0.81	61	0.77	64	0.78	304	73.20	323	71.61	284	74.40	295	76.59	300	74.93
3	66	0.80	83	0.82	76	0.81	61	0.77	64	0.77	305	72.03	320	72.18	281	74.08	294	76.54	301	73.19
4	66	0.80	83	0.81	76	0.81	61	0.77	64	0.77	305	70.09	321	72.05	281	73.93	294	75.74	300	71.40
5	66	0.81	82	0.83	76	0.81	62	0.78	64	0.77	303	72.64	321	72.12	282	73.60	294	76.02	300	71.38
Avg.	66	0.80	83	0.82	76	0.81	61	0.77	64	0.77	304	72.31	321	72.02	282	74.09	294	76.33	300	73.38
	20										60									
1	125	4.45	134	4.35	109	4.52	165	4.84	108	4.34	335	133.58	342	139.84	345	137.31	346	137.39	353	138.52
2	125	4.47	137	4.33	110	4.50	159	4.92	108	4.36	335	141.68	343	140.46	343	137.01	346	140.27	353	140.43
3	124	4.56	137	4.48	111	4.58	162	4.95	108	4.38	334	139.79	341	138.67	343	140.84	346	136.81	353	142.69
4	125	4.58	135	4.49	110	4.51	162	4.84	108	4.35	333	141.93	341	138.21	345	140.37	346	136.16	352	138.97
5	124	4.44	137	4.59	110	4.56	163	4.86	108	4.34	332	143.57	341	134.99	344	139.70	346	136.96	354	139.60
Avg.	125	4.50	136	4.45	110	4.53	162	4.88	108	4.36	334	140.11	342	138.43	344	139.05	346	137.52	353	140.04
	30										70									
1	183	13.86	199	12.84	187	13.31	189	13.47	173	13.65	427	237.11	382	232.36	437	243.98	446	232.11	396	244.93
2	184	13.72	199	12.80	188	13.28	190	13.20	172	13.71	426	231.01	383	231.93	436	240.15	446	233.42	396	236.99
3	185	13.77	199	12.80	187	13.64	189	13.39	173	13.54	426	231.26	383	232.20	437	242.17	446	235.37	397	240.01
4	184	13.71	199	12.85	187	13.79	190	13.70	174	13.81	428	233.16	384	235.12	438	243.92	446	228.57	396	235.52
5	184	13.82	199	12.84	187	13.63	189	13.80	174	13.64	426	231.87	382	234.44	437	242.01	446	235.86	395	244.66
Avg.	184	13.77	199	12.83	187	13.53	189	13.51	173	13.67	427	232.88	383	233.21	437	242.45	446	233.06	396	240.42
	40										80									
1	240	38.52	226	38.32	242	36.42	249	35.61	255	36.54	492	375.78	462	349.85	477	360.96	431	352.80	432	347.36
2	238	38.52	226	38.49	241	36.89	251	35.49	255	36.70	491	370.97	461	351.44	476	354.39	432	341.28	432	343.65
3	239	37.96	226	39.25	243	37.06	250	35.57	255	36.49	491	346.62	461	353.05	476	371.67	432	346.74	432	346.67
4	237	38.03	227	38.29	241	35.97	253	35.50	255	36.04	491	346.77	462	370.16	476	355.03	432	347.73	433	343.81
5	238	39.12	226	36.22	242	35.79	249	35.57	255	36.28	492	338.27	462	349.98	477	352.28	432	350.91	432	345.74
Avg.	238	38.43	226	38.11	242	36.43	250	35.55	255	36.41	491	355.68	462	354.90	476	358.87	432	347.89	432	345.45

Table A6. 9/10 fraction in maxβ of ACO with 1.1 × initial resource levels.

Run	Sets of Job									
	1	2	3	4	5	1	2	3	4	5
	C_{max} Time	C_{max} Time	C_{max} Time	C_{max} Time	C_{max} Time	C_{max} Time	C_{max} Time	C_{max} Time	C_{max} Time	C_{max} Time
	10					50				
1	66 0.86	82 0.90	75 0.88	61 0.83	64 0.85	303 77.82	321 74.82	278 80.54	294 81.34	300 75.74
2	66 0.91	82 0.91	75 0.88	61 0.83	64 0.85	303 77.56	320 71.95	279 82.37	294 80.57	300 74.94
3	64 0.89	82 0.90	76 0.88	61 0.85	64 0.85	303 76.57	320 74.97	277 80.30	294 80.34	296 75.27
4	67 0.91	82 0.91	75 0.89	62 0.83	64 0.86	303 76.99	321 77.19	279 79.90	294 80.85	300 75.89
5	64 0.89	82 0.91	74 0.89	61 0.84	64 0.85	303 77.53	321 78.59	278 76.85	294 83.66	300 75.41
Avg.	65 0.89	82 0.91	75 0.88	61 0.84	64 0.85	303 77.30	321 75.50	278 79.99	294 81.35	299 75.45
	20					60				
1	123 4.71	134 4.64	109 4.68	153 4.95	108 4.51	327 141.19	337 136.65	339 138.21	346 138.00	349 141.32
2	122 4.72	134 4.63	108 4.66	154 5.01	108 4.50	328 139.83	337 137.01	339 138.86	346 141.57	349 140.31
3	122 4.70	134 4.65	107 4.67	152 4.96	108 4.54	325 135.63	337 136.45	339 138.58	346 141.70	349 143.52
4	123 4.72	134 4.65	110 4.70	155 4.98	108 4.53	326 133.17	337 137.07	339 142.03	346 140.06	349 145.28
5	122 4.72	134 4.64	110 4.70	153 4.95	108 4.52	328 138.14	338 138.98	339 143.39	346 140.57	349 145.65
Avg.	122 4.71	134 4.64	109 4.68	153 4.97	108 4.52	327 137.59	337 137.23	339 140.21	346 140.38	349 143.21
	30					70				
1	180 14.65	199 13.82	187 14.55	183 14.10	172 14.36	426 226.28	382 221.02	435 219.65	445 221.61	394 223.12
2	179 14.66	199 14.17	187 14.45	185 14.62	172 14.47	426 222.66	382 212.25	435 220.95	446 215.60	394 217.91
3	179 14.91	199 14.14	187 14.30	185 14.25	172 14.38	426 217.61	382 213.16	435 220.95	446 217.37	394 218.23
4	178 14.69	199 13.99	187 14.70	185 14.13	172 14.42	426 218.11	382 215.98	435 219.74	445 218.73	394 218.14
5	178 15.52	199 14.08	187 14.17	184 14.47	172 14.43	426 221.69	382 215.92	435 219.88	445 215.06	394 218.20
Avg.	179 14.89	199 14.04	187 14.43	184 14.31	172 14.41	426 221.27	382 215.66	435 220.23	445 217.67	394 219.12
	40					80				
1	234 35.61	226 34.97	237 37.93	245 38.02	255 36.98	492 375.78	462 349.85	477 360.96	431 352.80	432 347.36
2	234 35.83	226 35.40	237 37.82	245 38.44	253 37.02	491 370.97	461 351.44	476 354.39	432 341.28	432 343.65
3	234 35.44	225 34.85	237 37.90	246 38.10	253 36.78	491 346.62	461 353.05	476 371.67	432 346.74	432 346.67
4	234 36.39	226 35.38	237 38.30	245 38.07	255 36.94	491 346.77	462 370.16	476 355.03	432 347.73	433 343.81
5	234 36.14	226 36.89	237 38.16	244 38.38	254 37.28	492 338.27	462 349.98	477 352.28	432 350.91	432 345.74
Avg.	234 35.88	226 35.50	237 38.02	245 38.20	254 37.00	491 355.68	462 354.90	476 358.87	432 347.89	432 345.45

Table A7. 8/10 fraction in maxβ of ACO with 1.1 × initial resource levels.

Run	Sets of Job									
	1	2	3	4	5	1	2	3	4	5
	C_{max} Time	C_{max} Time	C_{max} Time	C_{max} Time	C_{max} Time	C_{max} Time	C_{max} Time	C_{max} Time	C_{max} Time	C_{max} Time
	10					50				
1	66 0.83	82 0.80	74 0.78	60 0.74	64 0.77	303 75.91	320 68.71	278 71.71	294 73.90	300 72.48
2	64 0.81	82 0.80	75 0.78	61 0.73	64 0.75	303 71.79	320 70.20	278 71.97	294 73.30	300 71.27
3	66 0.80	82 0.80	73 0.77	61 0.73	64 0.77	303 71.89	320 71.23	277 71.16	294 75.10	300 71.72
4	66 0.81	82 0.80	74 0.78	61 0.75	64 0.75	303 73.03	320 69.73	277 71.86	294 73.61	300 71.68
5	64 0.80	82 0.80	75 0.78	61 0.73	64 0.75	303 71.47	320 69.86	277 72.09	294 73.92	300 71.85
Avg.	65 0.81	82 0.80	74.2 0.78	61 0.74	64 0.76	303 72.82	320 69.95	277 71.76	294 73.97	300 71.80

Table A7. Cont.

Run	Sets of Job									
	1	2	3	4	5	1	2	3	4	5
	C_{max} Time	C_{max} Time	C_{max} Time	C_{max} Time	C_{max} Time	C_{max} Time	C_{max} Time	C_{max} Time	C_{max} Time	C_{max} Time
	20					60				
1	121 4.43	134 4.24	110 4.31	151 4.61	108 4.14	326 139.02	337 135.00	339 135.43	346 133.79	349 130.65
2	121 4.34	134 4.30	109 4.44	153 4.59	108 4.14	325 138.47	337 136.73	339 129.84	346 130.84	349 128.93
3	121 4.39	134 4.33	109 4.30	154 4.58	108 4.14	327 138.06	337 137.37	339 129.56	343 130.46	349 127.88
4	122 4.36	134 4.27	109 4.38	154 4.56	108 4.09	325 139.67	337 134.98	339 131.12	346 133.02	349 128.04
5	123 4.39	134 4.28	109 4.37	155 4.53	108 4.17	327 135.08	337 134.00	339 138.65	343 130.87	349 127.35
Avg.	122 4.38	134 4.28	109 4.36	153 4.57	108 4.14	326 138.06	337 135.62	339 132.92	345 131.80	349 128.57
	30					70				
1	180 14.04	199 13.41	187 13.70	184 13.45	172 13.86	426 241.06	382 221.76	435 234.29	446 231.74	394 231.07
2	178 13.89	199 13.31	187 13.64	184 13.43	172 14.02	426 250.99	382 228.99	435 238.28	446 231.08	394 246.31
3	179 14.05	198 13.30	187 13.73	184 13.71	172 14.00	426 252.71	382 230.86	435 239.93	445 221.51	394 238.70
4	180 14.11	199 13.36	187 13.56	186 13.86	172 13.94	426 232.75	381 228.95	435 239.59	445 226.95	394 239.24
5	179 14.10	199 12.98	187 13.58	184 13.51	172 13.93	426 225.05	382 248.09	435 240.17	445 235.00	394 236.33
Avg.	179 14.04	199 13.27	187 13.64	184 13.59	172 13.95	426 240.51	382 231.73	435 238.45	445 229.25	394 238.33
	40					80				
1	234 33.57	226 33.11	237 33.34	244 33.83	255 34.57	492 390.40	456 392.66	476 397.85	429 389.19	428 388.73
2	234 33.64	226 33.13	237 33.66	244 34.47	255 34.15	491 401.52	459 392.01	476 410.50	429 395.19	427 395.45
3	234 33.27	226 33.00	237 33.73	244 34.18	253 34.07	492 386.09	459 403.42	476 408.59	429 386.62	426 391.20
4	234 33.92	226 32.90	237 33.51	244 34.04	255 34.22	491 392.32	461 393.11	476 406.42	429 388.55	426 390.01
5	234 33.55	226 32.66	237 33.83	244 34.17	253 34.72	491 394.83	459 398.15	476 406.28	429 383.86	425 400.62
Avg.	234 33.59	226 32.96	237 33.61	244 34.14	254.2 34.35	491 393.03	459 395.87	476 405.93	429 388.68	426 393.20

Table A8. 7/10 fraction in $\max\beta$ of ACO with $1.1 \times$ initial resource levels.

Run	Sets of Job									
	1	2	3	4	5	1	2	3	4	5
	C_{max} Time	C_{max} Time	C_{max} Time	C_{max} Time	C_{max} Time	C_{max} Time	C_{max} Time	C_{max} Time	C_{max} Time	C_{max} Time
	10					50				
1	66 0.80	82 0.81	76 0.78	61 0.75	64 0.78	303 81.13	320 76.29	279 78.45	294 80.38	300 78.42
2	64 0.82	82 0.81	73 0.80	61 0.75	64 0.77	303 81.72	319 77.53	278 76.26	294 81.02	300 78.29
3	64 0.80	82 0.83	74 0.78	61 0.78	64 0.77	303 82.78	320 75.95	278 76.19	294 81.27	300 77.83
4	66 0.81	82 0.83	74 0.80	61 0.77	64 0.77	303 80.47	319 75.51	278 76.16	294 81.51	300 77.96
5	66 0.80	82 0.80	74 0.78	60 0.77	64 0.78	303 79.49	320 74.80	277 76.11	294 81.57	300 78.58
Avg.	65 0.81	82 0.82	74.2 0.79	61 0.76	64 0.77	303 81.12	320 76.02	278 76.64	294 81.15	300 78.22
	20					60				
1	122 4.75	134 4.78	109 4.94	155 5.00	108 4.66	325 145.04	337 144.45	339 145.83	346 144.10	349 146.28
2	122 4.76	134 4.73	110 4.94	153 4.98	108 4.58	326 147.64	337 142.39	339 144.15	346 140.60	349 144.70
3	123 4.75	134 4.70	110 4.94	152 5.11	108 4.58	325 151.54	337 144.42	339 146.35	346 140.53	349 145.21
4	123 4.67	134 4.84	110 4.89	153 5.20	108 4.64	325 149.09	337 145.36	339 145.84	344 140.23	349 144.35
5	121 4.72	134 4.88	110 4.75	152 5.00	108 4.59	324 148.09	337 145.79	339 146.22	346 141.77	349 144.56
Avg.	122 4.73	134 4.78	110 4.89	153 5.06	108 4.61	325 148.28	337 144.48	339 145.68	346 141.44	349 145.02

Table A8. Cont.

Run	Sets of Job									
	1	2	3	4	5	1	2	3	4	5
	C_{max} Time	C_{max} Time	C_{max} Time	C_{max} Time	C_{max} Time	C_{max} Time	C_{max} Time	C_{max} Time	C_{max} Time	C_{max} Time
	30					70				
1	178 15.69	199 14.88	187 15.30	184 15.28	172 15.81	427 239.67	382 226.74	435 240.19	446 243.32	394 253.61
2	179 15.70	199 14.66	187 15.30	185 15.49	172 15.47	426 243.98	382 231.35	435 247.64	445 261.67	394 252.07
3	178 15.63	199 14.66	187 15.20	183 15.28	172 15.72	426 241.49	382 232.59	435 250.78	446 256.78	394 249.48
4	178 15.54	199 14.82	187 15.15	185 15.36	172 15.56	426 234.57	381 234.01	435 241.50	446 250.07	394 254.81
5	177 15.86	199 14.87	187 15.11	185 15.11	172 15.59	427 226.02	382 238.63	435 229.16	445 246.84	394 253.42
Avg.	178 15.68	199 14.78	187 15.21	184 15.30	172 15.63	426 237.15	382 232.66	435 241.85	446 251.74	394 252.68
	40					80				
1	234 36.13	226 40.48	237 41.07	244 40.77	255 39.08	491 352.66	457 372.15	476 364.67	429 347.80	425 350.14
2	234 41.42	226 40.43	237 41.15	244 40.58	253 39.08	491 362.31	460 359.62	476 356.04	429 353.10	426 350.78
3	234 41.59	226 40.43	237 40.66	244 40.61	255 40.77	491 357.62	457 356.03	476 362.06	429 340.50	427 352.36
4	234 41.07	226 41.04	237 40.04	244 40.48	254 41.65	492 349.59	460 355.83	476 362.85	429 341.16	426 351.01
5	234 41.35	226 40.95	237 40.07	244 38.95	255 41.56	491 350.00	459 363.36	476 358.65	429 340.25	428 352.00
Avg.	234 40.32	226 40.67	237 40.60	244 40.28	254.4 40.43	491 354.44	459 361.40	476 360.85	429 344.56	426 351.26

Table A9. 6/10 fraction in maxβ of ACO with 1.1 × initial resource levels.

Run	Sets of Job									
	1	2	3	4	5	1	2	3	4	5
	C_{max} Time	C_{max} Time	C_{max} Time	C_{max} Time	C_{max} Time	C_{max} Time	C_{max} Time	C_{max} Time	C_{max} Time	C_{max} Time
	10					50				
1	66 0.82	82 0.87	75 0.84	61 0.77	64 0.80	303 84.10	320 71.40	277 76.13	294 79.02	300 75.71
2	64 0.82	82 0.85	74 0.80	60 0.77	64 0.79	303 78.00	321 73.81	277 75.54	294 80.19	299 76.62
3	66 0.80	82 0.87	74 0.82	61 0.78	64 0.80	303 74.56	321 76.09	277 77.72	294 78.72	300 76.84
4	66 0.81	82 0.86	74 0.83	61 0.80	64 0.78	303 73.70	319 74.11	278 76.30	294 79.37	300 77.50
5	66 0.82	82 0.86	75 0.85	61 0.78	64 0.79	303 73.27	321 74.28	278 74.55	294 79.70	300 77.19
Avg.	66 0.82	82 0.86	74.4 0.83	61 0.78	64 0.79	303 76.73	320 73.94	277 76.05	294 79.40	300 76.77
	20					60				
1	121 4.60	134 4.50	109 4.50	153 4.85	108 4.56	327 141.96	337 142.32	339 142.07	346 138.16	349 141.01
2	121 4.66	134 4.48	107 4.60	153 4.88	108 4.56	326 141.70	337 140.15	339 145.36	346 138.41	349 138.29
3	123 4.65	134 4.55	110 4.62	153 4.83	108 4.46	326 144.69	337 137.77	339 144.49	342 138.68	349 140.56
4	122 4.59	134 4.56	107 4.60	154 4.82	108 4.47	326 148.28	337 138.50	339 144.32	346 139.33	349 141.35
5	122 4.63	134 4.50	109 4.57	154 4.87	108 4.38	326 144.58	337 138.47	339 144.73	346 137.57	349 140.42
Avg.	122 4.63	134 4.52	108 4.58	153 4.85	108 4.49	326 144.24	337 139.44	339 144.20	345 138.43	349 140.33
	30					70				
1	179 14.79	199 15.13	187 15.93	185 15.89	172 16.42	426 250.03	382 255.22	435 253.63	446 249.17	394 263.45
2	178 14.66	199 14.95	187 15.75	185 15.90	172 16.94	427 248.29	382 252.99	435 264.60	445 248.31	394 258.80
3	179 14.74	199 14.91	187 15.78	185 16.66	172 14.83	426 260.56	382 253.93	435 266.52	445 244.39	394 255.52
4	178 15.47	199 15.02	187 15.88	185 16.33	172 15.14	426 257.73	382 259.06	435 258.85	446 250.25	394 262.16
5	178 16.22	199 15.01	187 15.75	183 16.55	172 15.51	426 261.93	382 251.41	435 260.53	446 248.11	394 262.06
Avg.	178 15.18	199 15.01	187 15.82	185 16.26	172 15.77	426 255.71	382 254.52	435 260.82	446 248.05	394 260.40

Table A9. Cont.

Run	\multicolumn{10}{c}{Sets of Job}									
	1		2		3		4		5		1		2		3		4		5	
	C_{max}	Time	C_{max}	Time	C_{max}	Time	C_{max}	Time	C_{max}	Time	C_{max}	Time	C_{max}	Time	C_{max}	Time	C_{max}	Time	C_{max}	Time
	\multicolumn{10}{c}{40}									\multicolumn{10}{c}{80}								
1	234	40.62	226	39.70	237	42.47	244	42.91	254	43.84	491	345.15	459	363.19	476	405.04	429	371.25	427	395.82
2	234	40.83	226	40.25	237	41.81	244	43.00	255	42.47	491	349.86	459	362.35	476	401.10	429	408.69	427	400.23
3	235	40.70	224	40.67	237	42.36	244	42.46	253	43.08	491	363.31	459	380.68	476	406.06	429	396.13	427	398.01
4	234	40.81	226	40.62	237	42.19	244	42.81	254	41.81	491	355.77	459	410.13	476	407.89	429	399.04	426	400.03
5	234	40.64	226	41.40	237	42.97	245	43.13	254	39.00	491	353.87	460	400.75	476	409.60	429	396.60	426	396.07
Avg.	234	40.72	226	40.53	237	42.36	244	42.86	254	42.04	491	353.59	459	383.42	476	405.94	429	394.34	427	398.03

Table A10. 5/10 fraction in maxβ of ACO with 1.1 × initial resource levels.

Run	\multicolumn{10}{c}{Sets of Job}									
	1		2		3		4		5		1		2		3		4		5	
	C_{max}	Time	C_{max}	Time	C_{max}	Time	C_{max}	Time	C_{max}	Time	C_{max}	Time	C_{max}	Time	C_{max}	Time	C_{max}	Time	C_{max}	Time
	\multicolumn{10}{c}{10}									\multicolumn{10}{c}{50}								
1	66	0.81	82	0.87	73	0.80	61	0.80	64	0.89	303	72.07	320	69.82	277	70.59	294	73.93	300	70.89
2	66	0.80	82	0.84	74	0.84	60	0.80	64	0.97	303	71.26	321	68.68	279	69.98	294	73.50	300	70.51
3	66	0.81	82	0.84	73	0.84	60	0.86	64	0.85	303	71.84	321	68.00	277	70.00	294	73.57	300	71.20
4	66	0.82	82	0.83	73	0.83	60	0.80	64	0.89	303	71.96	321	69.43	277	70.30	294	73.22	300	70.50
5	66	0.81	82	0.83	74	0.82	60	0.84	64	0.84	303	71.82	320	69.12	279	70.78	294	73.73	300	71.41
Avg.	66	0.81	82	0.85	73.4	0.83	60	0.82	64	0.89	303	71.79	321	69.01	278	70.33	294	73.59	300	70.90
	\multicolumn{10}{c}{20}									\multicolumn{10}{c}{60}								
1	122	4.50	134	4.48	109	4.71	154	4.92	108	4.42	326	129.69	337	134.31	339	128.27	346	125.95	349	127.23
2	122	4.55	134	4.51	111	4.76	153	4.94	108	4.43	326	128.17	337	128.05	339	128.40	346	125.94	349	130.87
3	122	4.51	134	4.55	108	4.86	155	4.94	108	4.52	326	131.61	337	127.31	339	128.30	346	125.48	349	130.67
4	121	4.47	134	4.57	110	4.70	153	5.12	108	4.45	325	133.03	337	126.95	339	128.40	346	125.64	349	131.79
5	122	4.49	134	4.66	108	4.56	153	4.90	108	4.66	325	132.79	337	127.01	339	128.45	346	125.79	349	132.32
Avg.	122	4.50	134	4.55	109	4.72	154	4.97	108	4.49	326	131.06	337	128.73	339	128.36	346	125.76	349	130.58
	\multicolumn{10}{c}{30}									\multicolumn{10}{c}{70}								
1	179	14.93	199	14.20	187	14.44	185	14.93	172	14.59	426	238.87	382	257.53	435	271.36	445	269.61	394	276.76
2	179	14.57	199	13.86	187	14.38	185	14.70	172	14.62	426	223.68	382	270.23	435	279.23	445	267.17	394	254.83
3	178	14.58	198	13.98	187	14.72	185	14.65	172	14.50	426	244.73	382	277.17	435	269.71	446	268.80	394	267.50
4	179	14.68	199	13.87	187	14.32	185	14.69	172	14.59	426	242.79	382	270.89	435	270.17	445	268.87	394	276.46
5	179	14.61	199	14.01	187	14.87	185	14.52	172	14.56	426	234.45	382	269.20	435	274.05	446	267.15	394	256.08
Avg.	179	14.68	199	13.98	187	14.55	185	14.70	172	14.57	426	236.90	382	269.00	435	272.90	445	268.32	394	266.33
	\multicolumn{10}{c}{40}									\multicolumn{10}{c}{80}								
1	234	35.81	224	35.41	237	35.25	244	35.76	253	36.26	491	349.43	458	347.66	476	355.28	429	336.65	427	345.07
2	234	35.10	226	35.36	237	35.29	244	35.63	253	37.32	492	348.48	460	346.87	476	356.84	429	340.46	426	347.93
3	234	35.50	226	34.47	237	35.36	244	35.61	255	35.52	493	360.13	460	345.98	476	358.21	429	341.88	426	344.11
4	234	35.01	226	34.38	237	35.33	244	35.16	255	35.23	491	348.64	459	350.96	474	352.55	429	340.05	425	344.19
5	234	36.18	226	35.07	237	35.31	244	35.82	254	36.18	491	346.19	460	346.46	476	353.20	429	342.86	427	345.99
Avg.	234	35.52	226	34.94	237	35.31	244	35.60	254	36.10	492	350.57	459	347.59	476	355.22	429	340.38	426	345.46

Table A11. ACO Permutation with 1.4 × initial resource levels.

Run	Sets of Job																			
	1		2		3		4		5		1		2		3		4		5	
	C_{max}	Time	C_{max}	Time	C_{max}	Time	C_{max}	Time	C_{max}	Time	C_{max}	Time	C_{max}	Time	C_{max}	Time	C_{max}	Time	C_{max}	Time
	10										50									
1	60	0.40	82	0.42	65	0.43	59	0.42	64	0.47	303	44.12	319	52.14	277	51.69	293	48.35	299	49.90
2	60	0.41	82	0.41	65	0.42	59	0.40	64	0.43	303	44.19	321	51.98	278	51.82	293	49.51	300	50.13
3	60	0.41	82	0.42	65	0.41	59	0.41	64	0.43	303	46.80	321	51.32	277	51.86	293	48.41	300	47.95
4	59	0.41	82	0.42	65	0.42	59	0.45	64	0.44	303	48.62	321	54.77	277	51.95	293	48.42	300	48.78
5	59	0.42	83	0.41	65	0.41	59	0.46	64	0.44	303	50.69	321	53.74	279	50.94	292	48.92	300	49.98
Avg.	60	0.41	82	0.42	65	0.42	59	0.43	64	0.44	303	46.88	321	52.79	278	51.65	293	48.72	300	49.35
	20										60									
1	119	2.36	132	2.37	111	2.75	154	2.43	108	2.38	325	94.56	337	94.70	339	96.27	346	98.59	349	96.67
2	119	2.44	132	2.35	111	2.38	156	2.43	108	2.39	326	93.79	337	94.69	339	95.27	346	101.80	349	96.86
3	119	2.40	132	2.33	108	2.39	152	2.44	108	2.37	325	94.10	337	95.00	339	90.30	344	99.10	349	102.86
4	119	2.70	132	2.41	110	2.42	154	2.45	108	2.37	325	94.11	337	95.32	339	97.20	346	96.52	349	96.35
5	119	2.64	132	2.62	110	2.38	154	2.44	108	2.41	324	95.23	337	95.11	339	98.55	341	98.13	349	87.34
Avg.	119	2.51	132	2.41	110	2.46	154	2.44	108	2.38	325	94.36	337	94.96	339	95.52	345	98.83	349	96.02
	30										70									
1	175	8.19	199	8.52	187	8.52	186	8.18	171	8.23	426	151.96	382	155.61	435	161.65	445	164.42	394	166.58
2	174	8.66	199	8.36	187	8.57	187	8.11	171	8.22	427	153.84	382	158.69	435	161.05	446	160.26	394	159.41
3	176	8.29	199	8.24	187	8.20	185	8.37	171	8.09	427	161.72	382	160.61	435	162.51	446	158.98	394	165.49
4	174	8.06	199	8.62	187	8.38	184	8.42	171	8.04	426	156.67	381	156.82	435	161.68	445	166.21	394	165.23
5	173	8.33	199	8.25	187	8.48	187	8.20	171	8.10	426	155.25	380	154.27	435	159.76	446	167.91	394	161.85
Avg.	174	8.31	199	8.40	187	8.43	186	8.26	171	8.14	426	155.89	381	157.20	435	161.33	446	163.56	394	163.72
	40										80									
1	234	22.11	226	21.04	237	21.19	244	21.40	252	21.31	491	267.29	459	262.07	475	257.31	429	264.95	425	257.91
2	234	22.67	226	22.56	237	21.92	244	21.51	252	21.08	491	271.28	458	261.66	474	260.63	429	271.69	425	251.82
3	234	21.59	226	22.20	237	21.26	244	21.36	252	21.49	491	266.73	459	259.88	474	262.55	429	267.61	425	269.71
4	234	21.46	226	21.55	237	21.93	244	22.19	252	20.70	491	261.10	459	258.10	474	264.23	429	269.63	425	255.35
5	234	21.87	226	21.41	237	21.49	244	21.70	252	20.79	491	267.99	458	255.27	474	270.05	429	263.06	425	259.92
Avg.	234	21.94	226	21.75	237	21.56	244	21.63	252	21.07	491	266.88	459	259.39	474	262.96	429	267.39	425	258.94

Table A12. 10/10 fraction in maxβ of ACO with 1.4 × initial resource levels.

Run	Sets of Job																			
	1		2		3		4		5		1		2		3		4		5	
	C_{max}	Time	C_{max}	Time	C_{max}	Time	C_{max}	Time	C_{max}	Time	C_{max}	Time	C_{max}	Time	C_{max}	Time	C_{max}	Time	C_{max}	Time
	10										50									
1	60	0.82	83	0.83	69	0.79	59	0.75	64	0.80	304	70.10	321	69.38	281	68.83	294	71.00	300	71.04
2	61	0.81	83	0.81	65	0.79	59	0.74	64	0.80	305	70.57	321	66.71	281	67.84	294	70.75	300	71.30
3	59	0.80	83	0.83	68	0.78	59	0.72	64	0.81	304	70.14	321	66.19	280	67.47	293	70.70	300	71.48
4	60	0.78	83	0.85	68	0.78	59	0.74	64	0.78	303	70.10	321	67.39	281	68.16	292	71.29	300	70.54
5	59	0.80	83	0.82	68	0.79	59	0.73	64	0.81	306	70.25	321	66.72	281	68.49	293	70.75	300	70.94
Avg.	60	0.80	83	0.83	68	0.79	59	0.74	64	0.80	304	70.23	321	67.28	281	68.16	293	70.90	300	71.06

Table A12. Cont.

Run	Sets of Job									
	1	2	3	4	5	1	2	3	4	5
	C_{max} Time	C_{max} Time	C_{max} Time	C_{max} Time	C_{max} Time	C_{max} Time	C_{max} Time	C_{max} Time	C_{max} Time	C_{max} Time
	20					60				
1	119 4.38	134 4.40	111 4.49	160 4.71	108 4.12	331 132.65	341 133.61	343 130.06	346 125.04	351 131.37
2	119 4.53	133 4.51	111 4.48	157 4.76	108 4.01	333 132.66	339 126.78	340 127.96	346 130.12	353 130.25
3	119 4.51	134 4.34	111 4.46	162 4.72	108 4.13	333 129.97	339 127.06	345 127.95	346 128.06	354 130.49
4	119 4.62	133 4.29	111 4.45	163 4.78	108 4.05	334 128.33	338 126.99	344 127.21	346 127.61	353 131.08
5	119 4.39	134 4.38	112 4.47	162 4.78	108 4.22	335 128.70	341 127.05	342 127.76	346 126.90	351 131.90
Avg.	119 4.49	134 4.38	111 4.47	161 4.75	108 4.11	333 130.46	340 128.30	343 128.19	346 127.54	352.4 131.02
	30					70				
1	180 14.53	199 13.45	187 13.63	190 14.18	172 13.97	426 223.65	383 224.15	438 224.10	446 225.36	396 228.87
2	181 14.39	199 13.33	187 13.95	189 14.19	172 13.97	427 221.73	383 225.46	437 224.98	446 218.05	395 234.39
3	180 14.48	199 13.20	187 13.84	189 14.35	171 14.08	427 221.17	382 220.01	437 235.77	445 219.83	394 240.44
4	180 14.65	199 13.10	187 13.77	190 14.22	172 14.04	427 222.83	383 219.55	438 235.66	446 219.19	396 234.23
5	179 14.58	199 12.99	187 13.62	188 14.24	172 14.06	428 230.13	383 218.47	436 242.94	446 212.65	395 232.63
Avg.	180 14.52	199 13.21	187 13.76	189 14.23	171.8 14.02	427 223.90	383 221.53	437 232.69	446 219.01	395.2 234.11
	40					80				
1	238 34.79	226 32.29	241 33.85	248 34.13	252 34.58	491 361.10	461 365.65	476 367.03	432 356.03	432 374.91
2	237 33.80	226 32.30	240 34.02	250 34.71	252 35.05	491 351.14	460 379.02	476 384.25	432 358.12	432 361.36
3	238 34.36	226 33.10	241 33.96	250 34.79	252 35.29	492 351.12	461 365.51	476 389.75	431 361.65	431 364.59
4	239 33.40	226 32.46	241 33.98	248 34.44	252 35.94	493 352.39	461 365.27	476 375.20	430 354.07	432 362.76
5	241 33.58	226 32.41	242 33.93	249 34.47	252 35.30	492 349.35	461 358.78	476 370.01	430 359.80	428 366.10
Avg.	239 33.99	226 32.51	241 33.95	249 34.51	252 35.23	492 353.02	461 366.84	476 377.25	431 357.93	431 365.94

Table A13. 9/10 fraction in maxβ of ACO with 1.4 × initial resource levels.

Run	Sets of Job									
	1	2	3	4	5	1	2	3	4	5
	C_{max} Time	C_{max} Time	C_{max} Time	C_{max} Time	C_{max} Time	C_{max} Time	C_{max} Time	C_{max} Time	C_{max} Time	C_{max} Time
	10					50				
1	59 0.79	82 0.83	65 0.78	59 0.74	64 0.80	303 70.98	320 67.09	279 70.00	293 72.83	299 70.42
2	60 0.79	82 0.88	68 0.77	59 0.73	64 0.81	303 71.25	320 67.34	277 69.35	292 72.52	300 70.00
3	60 0.78	82 0.84	68 0.81	59 0.76	64 0.80	303 70.85	321 67.20	277 70.08	293 72.78	300 68.74
4	60 0.76	82 0.85	68 0.80	59 0.79	64 0.80	303 70.94	320 67.33	277 69.82	292 72.58	300 68.60
5	59 0.78	82 0.82	68 0.81	59 0.77	64 0.79	303 70.49	321 67.69	278 69.74	292 72.73	300 69.49
Avg.	60 0.78	82 0.85	67 0.79	59 0.76	64 0.80	303 70.90	320 67.33	278 69.80	292 72.69	300 69.45
	20					60				
1	119 4.43	132 4.53	109 4.58	152 4.81	108 4.27	327 129.64	337 130.01	339 132.17	346 125.70	349 129.38
2	119 4.60	132 4.57	110 4.57	152 4.88	108 4.42	325 129.85	337 127.65	339 131.66	346 125.47	349 130.15
3	119 4.58	132 4.58	107 4.58	155 4.85	108 4.23	328 131.25	337 130.42	339 131.22	346 127.41	349 131.58
4	119 4.47	132 4.51	110 4.62	155 4.80	108 4.16	326 130.20	337 128.24	339 130.18	346 126.64	349 130.51
5	119 4.63	132 4.48	108 4.54	155 4.84	108 4.06	327 130.41	337 129.65	339 131.22	346 125.51	349 130.05
Avg.	119 4.54	132 4.53	109 4.58	154 4.84	108 4.23	327 130.27	337 129.19	339 131.29	346 126.15	349 130.33

Table A13. Cont.

	Sets of Job									
Run	1	2	3	4	5	1	2	3	4	5
	C_{max} Time	C_{max} Time	C_{max} Time	C_{max} Time	C_{max} Time	C_{max} Time	C_{max} Time	C_{max} Time	C_{max} Time	C_{max} Time
	30					70				
1	175 14.43	199 13.19	187 13.89	185 14.45	171 13.45	427 244.28	382 224.02	435 226.72	445 227.96	394 226.79
2	175 14.16	199 12.87	187 13.80	184 14.65	171 13.96	427 228.50	382 223.32	435 225.61	445 227.60	394 222.79
3	177 14.13	199 12.90	187 13.97	184 14.19	171 14.14	428 223.29	382 222.66	435 227.02	445 216.39	394 221.29
4	175 14.98	199 12.96	187 14.08	185 13.91	171 13.99	426 222.01	382 219.02	435 228.56	446 215.56	394 221.98
5	173 14.69	199 13.36	187 13.87	185 14.24	171 14.31	427 222.75	382 223.12	435 236.94	445 218.18	394 223.71
Avg.	175 14.48	199 13.05	187 13.92	185 14.29	171 13.97	427 228.17	382 222.43	435 228.97	445 221.14	394 223.31
	40					80				
1	234 33.55	225 32.81	237 33.76	246 35.03	252 37.16	491 349.76	459 351.18	474 359.25	429 339.20	425 361.69
2	234 33.77	226 32.87	237 33.90	246 35.67	252 34.21	491 351.63	460 349.46	476 368.82	429 348.57	425 361.43
3	234 34.07	226 32.98	237 33.99	245 34.85	252 34.40	491 352.73	458 358.00	476 368.09	429 337.87	426 372.25
4	234 33.95	226 32.42	237 33.81	244 35.01	250 34.63	491 351.55	458 364.62	476 362.86	429 345.34	426 363.50
5	235 33.87	226 32.38	237 34.32	244 36.21	252 34.89	491 348.60	460 362.70	476 367.12	429 353.93	427 438.14
Avg.	234 33.84	226 32.69	237 33.95	245 35.36	252 35.06	491 350.85	459 357.19	476 365.23	429 344.98	426 379.40

Table A14. 8/10 fraction in $\max\beta$ of ACO with $1.4 \times$ initial resource levels.

	Sets of Job									
Run	1	2	3	4	5	1	2	3	4	5
	C_{max} Time	C_{max} Time	C_{max} Time	C_{max} Time	C_{max} Time	C_{max} Time	C_{max} Time	C_{max} Time	C_{max} Time	C_{max} Time
	10					50				
1	60 0.80	82 0.92	65 0.81	59 0.75	64 0.82	303 69.44	321 65.64	277 67.94	293 73.12	300 70.41
2	58 0.79	82 0.92	65 0.83	59 0.78	64 0.80	303 69.44	321 66.30	277 69.99	293 73.27	300 70.45
3	59 0.80	82 0.91	67 0.84	59 0.76	64 0.79	303 68.72	321 65.93	279 69.78	293 73.30	300 71.13
4	60 0.87	82 0.93	66 0.82	59 0.75	64 0.79	303 69.31	321 65.67	279 69.48	292 73.22	298 72.03
5	60 0.85	82 0.88	65 0.80	59 0.74	64 0.79	303 68.95	321 65.64	278 69.60	293 73.21	300 70.32
Avg.	59 0.82	82 0.91	66 0.82	59 0.76	64 0.80	303 69.17	321 65.84	278 69.36	293 73.22	300 70.87
	20					60				
1	119 4.37	132 4.36	109 4.37	154 4.97	108 4.06	326 129.48	337 128.02	339 132.21	346 127.52	349 131.45
2	119 4.39	132 4.56	111 4.39	155 5.05	108 4.08	326 130.63	337 132.50	339 135.24	346 132.26	349 131.68
3	119 4.40	132 4.60	110 4.60	156 4.96	108 4.23	328 128.73	337 131.23	339 133.96	346 128.72	349 131.43
4	119 4.27	132 4.54	110 4.64	153 4.82	108 4.24	326 130.76	337 130.80	339 132.43	346 127.90	349 131.03
5	119 4.33	132 4.45	110 4.70	153 4.72	108 4.21	326 130.31	337 127.97	339 131.18	343 127.68	349 132.76
Avg.	119 4.35	132 4.50	110 4.54	154 4.90	108 4.16	326 129.98	337 130.10	339 133.01	345 128.82	349 131.67
	30					70				
1	174 14.53	198 13.46	187 13.70	184 14.11	171 13.83	426 231.32	382 229.68	435 237.95	445 225.90	394 235.27
2	174 14.44	199 13.21	187 14.41	185 14.15	171 13.76	426 232.78	382 226.54	435 237.43	445 231.07	394 235.68
3	174 14.69	199 13.29	185 13.84	185 14.29	171 14.22	427 226.33	382 231.25	435 228.80	446 224.58	394 233.44
4	175 14.55	198 12.96	187 14.01	184 14.24	171 14.79	426 228.31	381 228.21	435 235.13	445 224.31	394 233.68
5	175 14.10	199 13.03	187 13.75	185 14.10	171 14.16	427 229.95	382 230.07	435 232.42	445 222.44	394 240.95
Avg.	174 14.46	199 13.19	187 13.94	185 14.18	171 14.15	426 229.74	382 229.15	435 234.35	445 225.66	394 235.81

Table A14. Cont.

Run	Sets of Job																			
	1		2		3		4		5		1		2		3		4		5	
	C_{max}	Time	C_{max}	Time	C_{max}	Time	C_{max}	Time	C_{max}	Time	C_{max}	Time	C_{max}	Time	C_{max}	Time	C_{max}	Time	C_{max}	Time
	40										80									
1	234	33.74	226	32.75	237	34.00	244	34.79	252	35.40	491	372.89	459	371.61	476	378.11	429	431.14	427	367.77
2	234	35.39	226	33.79	237	34.33	244	34.56	252	35.25	491	385.00	457	376.62	476	386.75	429	437.42	425	355.19
3	234	34.76	226	32.94	237	33.78	244	35.11	252	35.64	491	367.65	459	379.97	476	380.16	429	433.80	425	366.88
4	234	34.00	226	32.51	237	33.50	244	35.07	252	35.27	491	369.67	459	371.16	474	378.50	429	375.88	424	375.39
5	234	34.47	226	32.77	237	33.62	244	35.29	252	34.29	491	366.16	458	369.34	476	373.08	429	365.21	425	364.34
Avg.	234	34.47	226	32.95	237	33.85	244	34.96	252	35.17	491	372.27	458	373.74	476	379.32	429	408.69	425	365.92

Table A15. 7/10 fraction in maxβ of ACO with 1.4 × initial resource levels.

Run	Sets of Job																			
	1		2		3		4		5		1		2		3		4		5	
	C_{max}	Time	C_{max}	Time	C_{max}	Time	C_{max}	Time	C_{max}	Time	C_{max}	Time	C_{max}	Time	C_{max}	Time	C_{max}	Time	C_{max}	Time
	10										50									
1	60	0.76	82	0.84	65	0.76	59	0.72	64	0.79	303	80.16	321	69.24	277	70.24	292	74.02	300	71.51
2	60	0.78	82	0.81	65	0.77	59	0.74	64	0.82	303	73.05	321	65.99	277	70.62	293	73.80	300	71.51
3	60	0.80	82	0.81	65	0.84	59	0.75	64	0.83	303	72.41	321	66.72	279	70.26	292	74.34	300	72.53
4	61	0.79	82	0.81	65	0.83	59	0.75	64	0.87	302	73.04	321	67.55	277	70.30	293	74.22	300	71.41
5	60	0.77	82	0.86	65	0.78	59	0.73	64	0.83	303	71.77	321	67.53	277	71.02	292	73.75	300	71.42
Avg.	60	0.78	82	0.83	65	0.80	59	0.74	64	0.83	303	74.09	321	67.41	277	70.49	292	74.03	300	71.68
	20										60									
1	119	4.39	132	4.46	108	4.78	153	4.89	108	4.26	326	128.65	337	129.66	339	134.27	346	130.59	349	128.63
2	119	4.26	132	4.51	110	4.65	152	4.76	108	4.38	326	128.75	337	127.03	339	133.06	346	127.68	349	129.51
3	119	4.32	132	4.59	109	4.71	153	4.80	108	4.29	325	128.42	337	128.60	339	135.36	346	127.38	349	130.51
4	119	4.27	132	4.48	109	5.00	154	4.79	108	4.48	324	129.78	337	128.45	339	136.13	346	128.11	349	132.49
5	119	4.67	132	4.49	109	4.83	156	4.76	108	4.26	326	129.25	337	132.78	339	133.66	346	126.74	349	132.60
Avg.	119	4.38	132	4.51	109	4.79	154	4.80	108	4.33	325	128.97	337	129.30	339	134.50	346	128.10	349	130.75
	30										70									
1	174	14.30	199	13.06	187	13.64	186	13.73	171	13.66	427	239.39	382	235.44	435	246.05	445	236.55	394	255.11
2	177	14.71	199	13.04	187	13.63	184	13.88	171	14.31	426	242.99	382	234.77	435	238.37	445	243.76	394	245.57
3	174	14.45	199	12.97	187	13.72	185	13.93	171	13.86	426	240.07	382	232.38	435	253.19	446	241.68	394	244.85
4	174	14.69	199	13.12	187	13.55	183	14.08	171	13.99	428	236.13	382	238.29	435	251.76	445	237.30	394	248.74
5	174	14.42	199	13.07	185	13.60	185	13.97	171	13.73	427	237.71	382	239.42	435	256.39	445	243.75	394	243.09
Avg.	175	14.52	199	13.05	187	13.63	185	13.92	171	13.91	427	239.26	382	236.06	435	249.15	445	240.61	394	247.47
	40										80									
1	234	34.18	226	32.34	237	33.47	244	34.39	252	34.75	491	355.44	459	387.20	476	367.73	429	348.69	425	361.64
2	234	34.21	226	32.27	237	33.84	244	34.22	252	34.78	491	358.90	458	452.33	476	357.15	429	352.06	425	364.45
3	234	33.97	225	32.37	237	33.51	244	34.47	252	34.77	491	352.64	460	446.64	476	365.63	429	341.14	425	367.49
4	234	34.10	226	32.39	237	33.53	244	34.46	252	34.69	491	354.75	460	411.01	473	375.85	429	343.29	425	358.24
5	234	34.00	225	32.48	237	33.61	244	34.21	252	34.71	491	350.60	459	371.80	475	370.45	429	349.52	425	366.38
Avg.	234	34.09	226	32.37	237	33.59	244	34.35	252	34.74	491	354.47	459	413.79	475	367.36	429	346.94	425	363.64

Table A16. 6/10 fraction in maxβ of ACO with 1.4 × initial resource levels.

Run	Sets of Job																			
	1		2		3		4		5		1		2		3		4		5	
	C_{max}	Time	C_{max}	Time	C_{max}	Time	C_{max}	Time	C_{max}	Time	C_{max}	Time	C_{max}	Time	C_{max}	Time	C_{max}	Time	C_{max}	Time
	10										50									
1	59	0.77	82	0.84	65	0.77	59	0.71	64	0.76	303	70.78	320	67.86	278	69.21	293	88.11	300	77.73
2	60	0.75	82	0.85	65	0.76	59	0.73	64	0.80	303	70.28	321	69.51	277	69.30	292	87.96	298	77.74
3	60	0.75	82	0.85	65	0.77	59	0.74	64	0.79	303	70.48	320	68.20	277	69.85	292	92.54	300	75.20
4	59	0.78	82	0.85	65	0.76	59	0.80	64	0.78	303	71.06	319	67.40	277	69.75	293	90.99	300	74.56
5	60	0.77	82	0.85	65	0.78	59	0.78	64	0.80	303	71.99	320	69.42	277	75.94	293	83.33	300	77.39
Avg.	60	0.77	82	0.85	65	0.77	59	0.75	64	0.79	303	70.92	320	68.48	277	70.81	293	88.59	300	76.52
	20										60									
1	119	4.49	132	4.56	109	4.53	153	4.75	108	4.29	327	129.50	337	129.90	339	131.28	346	128.13	349	132.61
2	119	4.46	132	4.45	108	4.54	154	5.02	108	4.33	326	131.09	337	129.46	339	130.31	346	127.91	349	133.17
3	119	4.33	132	4.51	109	4.44	152	4.72	108	4.27	326	131.82	337	131.52	339	132.01	346	128.22	349	132.52
4	119	4.33	132	4.50	109	4.40	155	4.67	108	4.28	324	131.20	337	128.93	339	132.86	346	129.25	349	132.61
5	119	4.52	132	4.61	110	4.44	152	5.05	108	4.31	325	129.73	337	129.75	339	133.50	346	130.48	349	133.23
Avg.	119	4.43	132	4.53	109	4.47	153	4.84	108	4.29	326	130.67	337	129.91	339	131.99	346	128.80	349	132.83
	30										70									
1	173	14.08	199	12.86	187	13.64	185	13.96	171	14.39	426	215.26	381	218.51	435	216.39	446	198.86	394	206.93
2	176	14.11	199	12.90	187	13.64	185	14.32	171	14.08	426	215.64	382	207.97	435	208.71	445	197.88	394	207.29
3	175	14.23	199	12.98	187	13.72	185	14.33	171	13.91	427	208.16	382	212.68	435	212.47	446	197.33	394	206.72
4	173	14.35	199	13.02	187	13.68	185	14.70	171	14.08	426	226.05	382	202.80	435	209.57	445	197.59	394	207.20
5	174	14.38	199	12.88	187	13.64	185	14.43	171	13.96	426	230.31	382	216.97	435	205.56	445	198.29	394	208.30
Avg.	174	14.23	199	12.93	187	13.66	185	14.35	171	14.08	426	219.08	382	211.79	435	210.54	445	197.99	394	207.29
	40										80									
1	234	34.21	226	32.54	237	33.76	244	34.32	252	34.62	491	348.91	460	365.14	474	376.18	429	345.95	425	366.90
2	234	34.15	226	32.60	237	33.57	244	34.47	252	34.72	491	356.95	460	366.77	475	368.62	429	343.70	425	364.87
3	234	34.26	226	32.66	237	33.57	244	34.58	252	34.80	493	355.42	459	357.71	476	362.79	429	345.14	426	360.93
4	234	33.85	226	32.45	237	33.87	244	34.23	252	34.14	491	361.16	458	365.88	476	375.33	429	344.93	425	358.11
5	234	34.10	226	32.57	237	33.72	244	34.49	252	33.89	491	355.63	456	366.90	475	369.19	429	346.99	425	357.00
Avg.	234	34.11	226	32.57	237	33.70	244	34.42	252	34.43	491	355.61	459	364.48	475	370.42	429	345.34	425	361.56

Table A17. 5/10 fraction in $\max\beta$ of ACO with 1.4 × initial resource levels.

Run	Sets of Job																			
	1		2		3		4		5		1		2		3		4		5	
	C_{max}	Time	C_{max}	Time	C_{max}	Time	C_{max}	Time	C_{max}	Time	C_{max}	Time	C_{max}	Time	C_{max}	Time	C_{max}	Time	C_{max}	Time
	10										50									
1	58	0.80	82	0.80	65	0.75	59	0.72	64	0.77	303	70.50	319	68.01	279	69.68	292	74.77	300	71.71
2	60	0.79	82	0.82	65	0.74	59	0.71	64	0.78	303	70.92	321	66.38	277	69.63	292	74.27	300	72.12
3	60	0.79	82	0.83	65	0.76	59	0.70	64	0.77	303	70.80	320	65.30	278	69.15	293	74.18	300	71.46
4	60	0.79	82	0.82	65	0.76	59	0.70	64	0.78	303	70.82	320	65.58	277	70.80	292	73.60	298	70.41
5	60	0.76	82	0.82	65	0.76	59	0.70	64	0.77	302	71.10	320	65.66	279	70.27	292	74.84	300	70.52
Avg.	60	0.79	82	0.82	65	0.75	59	0.70	64	0.77	303	70.83	320	66.19	278	69.91	292	74.33	300	71.24
	20										60									
1	119	4.27	132	4.40	110	4.68	155	4.87	108	4.26	324	134.13	337	131.51	339	134.55	346	129.39	349	135.28
2	119	4.53	132	4.63	110	4.62	154	4.90	108	4.19	325	132.49	337	130.62	339	134.62	346	128.56	349	133.57
3	119	4.48	132	4.51	110	4.62	152	4.84	108	4.23	326	133.00	337	131.47	339	130.36	346	127.61	349	134.02
4	119	4.44	132	4.49	109	4.63	154	4.86	108	4.12	325	136.22	337	131.36	339	129.65	344	128.76	349	134.32
5	119	4.34	132	4.47	111	4.67	155	4.86	108	4.15	325	138.10	337	130.66	339	131.27	346	132.26	349	131.01
Avg.	119	4.41	132	4.50	110	4.64	154	4.87	108	4.19	325	134.79	337	131.12	339	132.09	346	129.31	349	133.64
	30										70									
1	176	14.57	199	13.36	187	13.96	185	14.06	171	13.99	426	213.89	382	210.60	435	212.37	446	203.49	394	219.23
2	175	14.69	199	13.09	187	13.81	185	14.10	171	14.01	426	212.78	382	211.25	435	213.80	446	204.17	394	228.61
3	175	14.95	199	13.29	187	13.72	186	14.26	171	14.13	427	212.19	382	211.88	435	214.04	446	208.09	394	239.13
4	173	14.50	199	13.33	187	13.84	185	14.12	171	13.87	426	211.30	382	208.96	435	212.23	446	208.38	394	230.80
5	174	14.70	199	13.21	187	14.10	185	14.25	171	13.75	427	215.19	382	209.14	435	212.17	445	208.73	394	222.70
Avg.	175	14.68	199	13.26	187	13.89	185	14.16	171	13.95	426	213.07	382	210.37	435	212.92	446	206.57	394	228.09
	40										80									
1	234	34.42	225	32.72	237	33.05	244	34.08	252	34.27	492	353.73	458	366.59	476	367.47	429	341.96	425	354.92
2	234	34.92	226	31.71	237	32.88	244	33.96	252	34.55	491	353.13	460	361.57	474	371.96	429	345.05	425	356.15
3	234	34.76	226	31.76	237	32.93	244	34.09	251	34.17	491	352.95	459	362.61	476	366.73	429	349.67	425	359.46
4	234	34.30	226	32.02	237	33.52	244	33.71	252	33.99	491	373.47	459	353.15	476	376.03	429	347.99	425	370.89
5	234	34.11	226	31.76	237	33.01	244	34.12	252	34.05	491	370.20	459	355.60	476	368.67	429	343.99	425	365.23
Avg.	234	34.50	226	31.99	237	33.08	244	33.99	252	34.21	491	360.70	459	359.90	476	370.17	429	345.73	425	361.33

Table A18. Experiment results of ACO M2-enum.

Sets of Job

10 jobs with 1.1 × initial resource

Run	1 C_{max}	1 Time	2 C_{max}	2 Time	3 C_{max}	3 Time	4 C_{max}	4 Time	5 C_{max}	5 Time
1	64	0.86	82	0.84	73	0.83	61	0.84	64	0.83
2	66	0.84	82	0.83	73	0.83	60	0.86	64	0.84
3	66	0.84	82	0.84	73	0.83	60	0.86	64	0.83
4	66	0.81	82	0.86	73	0.84	61	0.86	64	0.83
5	64	0.83	82	0.86	74	0.84	60	0.86	64	0.83
Avg.	65	0.84	82	0.85	73	0.83	60	0.86	64	0.83

20 jobs with 1.1 × initial resource

Run	1 C_{max}	1 Time	2 C_{max}	2 Time	3 C_{max}	3 Time	4 C_{max}	4 Time	5 C_{max}	5 Time
1	122	8.27	134	17.43	109	7.01	154	5.14	108	106.89
2	121	8.00	134	17.74	110	6.98	153	5.19	108	105.95
3	121	8.33	134	18.44	108	6.85	155	5.34	108	96.28
4	122	8.39	134	18.19	110	6.91	153	5.22	108	92.69
5	121	8.42	134	18.19	109	6.96	155	5.26	108	92.67
Avg.	121	8.28	134	18.00	109	6.94	154	5.23	108	98.90

30 jobs with 1.1 × initial resource

Run	1 C_{max}	1 Time	2 C_{max}	2 Time	3 C_{max}	3 Time	4 C_{max}	4 Time	5 C_{max}	5 Time
1	179	1363.26	200	-	187	-	183	2379.65	172	-
2	177	1350.69	200	-	187	-	184	3020.92	172	-
3	178	1649.96	200	-	187	-	184	2696.27	172	-
4	177	1282.78	200	-	187	-	184	-	172	-
5	180	1084.22	200	-	187	-	184	-	172	-
Avg.	178	1346.18	200	3600.00	187	3600.00	184	3059.37	172	3600.00

10 jobs with 1.4 × initial resource

Run	1 C_{max}	1 Time	2 C_{max}	2 Time	3 C_{max}	3 Time	4 C_{max}	4 Time	5 C_{max}	5 Time
1	64	0.87	82	0.87	65	0.86	59	1.12	64	0.86
2	64	0.85	82	0.87	65	0.87	59	1.22	64	0.87
3	64	0.88	82	0.87	65	0.86	59	1.16	64	0.86
4	64	0.92	82	0.87	65	0.88	59	1.17	64	0.86
5	64	0.86	82	0.86	65	0.85	59	1.20	64	0.86
Avg.	64	0.87	82	0.87	65	0.86	59	1.17	64	0.86

20 jobs with 1.4 × initial resource

Run	1 C_{max}	1 Time	2 C_{max}	2 Time	3 C_{max}	3 Time	4 C_{max}	4 Time	5 C_{max}	5 Time
1	119	14.59	132	35.62	109	6.65	153	5.09	108	1864.11
2	119	13.13	132	37.49	107	6.65	153	5.03	108	2062.52
3	119	15.01	132	39.95	108	6.69	154	5.08	108	2313.29
4	119	14.46	132	41.77	107	6.70	153	5.03	108	2376.40
5	119	14.22	132	40.31	110	6.66	153	5.05	108	2820.85
Avg.	119	14.28	132	39.03	108	6.67	153	5.05	108	2287.43

30 jobs with 1.4 × initial resource

Run	1 C_{max}	1 Time	2 C_{max}	2 Time	3 C_{max}	3 Time	4 C_{max}	4 Time	5 C_{max}	5 Time
1	174	1450.22	200	-	187	-	185	2586.26	172	-
2	174	1056.23	200	-	187	-	185	2700.09	172	-
3	173	1699.81	200	-	187	-	184	2049.86	172	-
4	175	1441.64	200	-	187	-	185	2385.19	172	-
5	175	1218.19	200	-	187	-	185	3604.12	172	-
Avg.	174	1373.22	200	3600.00	187	3600.00	185	2665.10	172	3600.00

Table A19. Experiment results of IP.

IP Sets	Times of Initial Resource			
	1.1		1.4	
	C_{max}	Time	C_{max}	Time
	10			
1	64	1518.71	56	344.81
2	81	59.93	81	62.71
3	73	110.96	65	23.72
4	59	74.42	59	15.25
5	64	470.02	64	492.10
Avg.	68	446.81	65	187.72
	20			
1	123	3600.11	119	2864.56
2	-	3600.00	132	1233.87
3	-	3600.27	-	3600.20
4	-	3600.32	x	3600.61
5	-	3600.47	107	366.97
Avg.	-	-	-	-

References

1. Pinedo, M. *Scheduling*; Springer: New York, NY, USA, 2016; Volume 29.
2. Brucker, P.; Drexl, A.; Möhring, R.; Neumann, K.; Pesch, E. Resource-constrained project scheduling: Notation, classification, models, and methods. *Eur. J. Oper. Res.* **1999**, *112*, 3–41. [CrossRef]
3. Habibi, F.; Barzinpour, F.; Sadjadi, S. Resource-constrained project scheduling problem: Review of past and recent developments. *J. Proj. Manag.* **2018**, *3*, 55–88. [CrossRef]
4. Herroelen, W.; Reyck, B.D.; Demeulemeester, E. Resource-constrained project scheduling: A survey of recent developments. *Comput. Oper. Res.* **1998**, *25*, 279–302. [CrossRef]
5. Issa, S.; Tu, Y. A survey in the resource-constrained project and multi-project scheduling problems. *J. Proj. Manag.* **2020**, *5*, 117–138. [CrossRef]
6. Ibadov, N.; Kulejewski, J.; Krzemiński, M. Fuzzy ordering of the factors affecting the implementation of construction projects in Poland. *AIP Conf. Proc.* **2013**, *1558*, 1298–1301.
7. Krzemiński, M. KASS v. 2.2. scheduling software for construction with optimization criteria description. *Acta Phys. Pol. A* **2016**, *130*, 1439–1442. [CrossRef]
8. Kaplan, E.H. Relocation models for public housing redevelopment programs. *Environ. Plan. B Plan. Des.* **1986**, *13*, 5–19. [CrossRef]
9. PHRG. *New Lives for Old Buildings: Revitalizing Public Housing Project*; Public Housing Group, Department of Urban Studies and Planning, MIT: Cambridge, MA, USA, 1986.
10. Kaplan, E.H.; Amir, A. A fast feasibility test for relocation problems. *Eur. J. Oper. Res.* **1988**, *35*, 201–206. [CrossRef]
11. Johnson, S.M. Optimal two- and three-stage production schedules with setup times included. *Nav. Res. Logist. Q.* **1954**, *1*, 61–68. [CrossRef]
12. Lin, B.M.T.; Huang, H.L. On the relocation problem with a second working crew for resource recycling. *Int. J. Syst. Sci.* **2006**, *37*, 27–34. [CrossRef]
13. Cheng, T.C.E.; Lin, B.M.T.; Huang, H.L. Resource-constrained flowshop scheduling with separate resource recycling operations. *Comput. Oper. Res.* **2012**, *39*, 1206–1212. [CrossRef]
14. Graham, R.L.; Lawler, E.L.; Lenstra, J.K.; Kan, A.R. Optimization and approximation in deterministic sequencing and scheduling: A survey. *Ann. Discret. Math.* **1979**, *5*, 287–326.
15. Kaplan, E.H.; Berman, O. OR hits the heights: Relocation planning at the orient heights housing project. *Interfaces* **1988**, *18*, 14–22. [CrossRef]
16. Xie, J. Polynomial algorithms for single machine scheduling problems with financial constraints. *Oper. Res. Lett.* **1997**, *21*, 39–42. [CrossRef]
17. Amir, A.; Kaplan, E.H. Relocation problems are hard. *Int. J. Comput. Math.* **1988**, *25*, 101–110. [CrossRef]
18. Kononov, A.V.; Lin, B.M.T. On relocation problems with multiple identical working crews. *Discret. Optim.* **2006**, *3*, 366–381. [CrossRef]

19. Cheng, T.C.E.; Lin, B.M.T. Johnson's rule, composite jobs and the relocation problem. *Eur. J. Oper. Res.* **2009**, *192*, 1008–1013. [CrossRef]
20. Cheng, T.C.E.; Lin, B.M.T. Demonstrating Johnson's algorithm via resource-constrained scheduling. *Int. J. Prod. Res.* **2017**, *55*, 3326–3330. [CrossRef]
21. Lin, B.M.T.; Tseng, S.S. Some results of the relocation problems with processing times and deadlines. *Int. J. Comput. Math.* **1991**, *40*, 1–15. [CrossRef]
22. Lin, B.M.T.; Tseng, S.S. Relocation problems of maximizing new capacities under a common due date. *Int. J. Syst. Sci.* **1992**, *23*, 1433–1448. [CrossRef]
23. Lin, B.M.T.; Tseng, S.S. Resource-requirement minimization in relocation problems with precedence constraints. In Proceedings of the ICCI '92: Fourth International Conference on Computing and Information, Toronto, ON, Canada, 28–30 May 1992; pp. 26–29.
24. Lin, B.M.T.; Cheng, T.C.E. Minimizing the weighted number of tardy jobs and maximum tardiness in relocation problem. *Eur. J. Oper. Res.* **1999**, *116*, 183–193. [CrossRef]
25. Hall, N.G. Scheduling problems with generalized due dates. *IIE Trans.* **1986**, *18*, 220–222. [CrossRef]
26. Lin, B.M.T.; Liu, S.T. Maximizing the reward in relocation problems with generalized due dates. *Int. J. Prod. Econ.* **2008**, *115*, 55–63. [CrossRef]
27. Sevastyanov, S.V.; Lin, B.M.T.; Huang, H.-L. Tight complexity analysis of the relocation problem with arbitrary release dates. *Theor. Comput. Sci.* **2011**, *412*, 4536–4544. [CrossRef]
28. Kononov, A.V.; Lin, B.M.T. Minimizing the total weighted completion time in the relocation problem. *J. Sched.* **2010**, *13*, 123–129. [CrossRef]
29. Lin, B.M.T. Resource-constrained scheduling with optional recycling operations. *Comput. Ind. Eng.* **2015**, *90*, 39–45. [CrossRef]
30. Shafransky, Y.M.; Strusevich, V.A. The open shop scheduling problem with a given sequence of jobs on one machine. *Nav. Res. Logist.* **1998** *45*, 705–773 . [CrossRef]
31. Lin, B.M.T.; Hwang, F.J.; Kononov, A.V. Relocation scheduling subject to fixed processing sequences. *J. Sched.* **2016**, *19*, 153–163. [CrossRef]

Article

An Imitation and Heuristic Method for Scheduling with Subcontracted Resources

Anna Antonova *, Konstantin Aksyonov and Olga Aksyonova

Faculty of Information Technology and Automatics, Ural Federal University, 620002 Ekaterinburg, Russia; k.a.aksyonov@urfu.ru (K.A.); wiper99@mail.ru (O.A.)
* Correspondence: a.s.antonova@urfu.ru

Abstract: A scheduling problem with subcontracted resources is widely spread and is associated with the distribution of limited renewable and non-renewable resources, both own and subcontracted ones based on the work's due dates and the earliest start time. Scheduling's goal is to reduce the cost of the subcontracted resources. In the paper, application of a few scheduling methods based on scheduling theory and the optimization algorithm is considered; limitations of these methods' application are highlighted. It is shown that the use of simulation modeling with heuristic rules for allocation of the renewable resources makes it possible to overcome the identified limitations. A new imitation and heuristic method for solving the assigned scheduling problem is proposed. The comparison of the new method with existing ones in terms of the quality of the found solution and performance of the methods is carried out. A case study is presented that allowed a four-fold reduction of the overall subcontracted resources cost in a real project portfolio.

Keywords: scheduling theory; operations research; subcontracted resources; scheduling on parallel machines; renewable and non-renewable resources; heuristic methods

Citation: Antonova, A.; Aksyonov, K.; Aksyonova, O. An Imitation and Heuristic Method for Scheduling with Subcontracted Resources. *Mathematics* **2021**, *9*, 2098. https://doi.org/10.3390/math9172098

Academic Editors: Chin-Chia Wu and Frank Werner

Received: 26 July 2021
Accepted: 27 August 2021
Published: 30 August 2021

Publisher's Note: MDPI stays neutral with regard to jurisdictional claims in published maps and institutional affiliations.

Copyright: © 2021 by the authors. Licensee MDPI, Basel, Switzerland. This article is an open access article distributed under the terms and conditions of the Creative Commons Attribution (CC BY) license (https://creativecommons.org/licenses/by/4.0/).

1. Introduction

The problem of business processes and project works scheduling is one of the key problems of organizational system control. Organizational systems are widely spread: examples are enterprises of various industries, multiservice communication networks, and project organizations.

To date, some methods to solve the scheduling problem with resource constraints have been developed depending on the problem statement, restrictions, and objective function. These are scheduling theory and network methods [1–5], simulation and multi-agent modelling methods [6,7], heuristic methods [8,9], and methods based on the application of commercially available solvers [10–14].

According to the machine environment, scheduling problems can be divided into the main following types with the type notation in brackets [15]: identical parallel machines (Pm) where work can be performed at any of the machines [1–5,8,9,12,14]; Job shop (Jm) where each work has to be processed on each of the machines and all works have different routes [6,7,11,13]; Open shop (Om) where each work has to be processed on each of the machines and some of this processing time may be zero [10].

According to the processing restrictions, there are several main constraints specified: release dates (r_j), where work j cannot start its processing before its release date [3–5]; precedence constraints ($prec$), where one or more works have to be completed before execution of a certain consequent work is allowed [1–9,12–14]; batch processing ($batch(b)$), where a machine may be able to process a number of b works simultaneously [11,12]; and breakdowns ($brkdwn$), where a machine may not be permanently available [3,5].

The main objective functions of the scheduling problem are: Makespan minimization (C_{max}) is a minimization of the last work's completion time to leave the system [1–4,6–9];

and the weighted number of tardy jobs minimization ($\sum \omega_j U_j$) [5,11–14], throughput maximization [10].

In general, the scheduling problem is a problem of work's sequence definition satisfying the given restrictions and minimizing makespan. A schedule, or calendar plan, is a set of work start dates. Commonly, researchers have investigated the renewable restricted resources that are occupied while a work is being performed and released when the work is over, as in [1,2,4–6,8–12,14]. Examples of the renewable resources are personnel, aggregates, vehicles, and other. At the same time, there is an urgent problem to consider non-renewable resources' distribution, as in [3,7,13]. The non-renewable resources are consumed at the work input and produced at the work output. Examples are raw materials, fuel, finance, and goods.

We consider the scheduling problem on parallel machines in the presence of the orders' earliest start times and due dates. We have extended the problem by using subcontracted resources' cost optimization and accounting for restricted non-renewable resources. The optimization of subcontracted resources is relevant either for enterprises with a relatively small set of their own resources, or for enterprises with resources of constrained competence. These enterprises flexibly respond to demand fluctuations in products manufactured or services provided and attract subcontracted resources when it is necessary. Examples are construction companies and project organizations. They need to reduce the engaged subcontractors' cost while keeping with existing time restrictions to reduce the company's waste.

The rest of the paper is organized as follows. In Section 2, a literature overview is presented. Notations used are presented in Section 3. In Section 4, a scheduling problem is formulated considering own and subcontracted renewable and non-renewable resources. Application of the scheduling methods to the trial problem and their shortcomings are given in Section 5. In Section 6, we propose an imitation and heuristic method for the problem considered. In Sections 7 and 8, a case study is presented providing a comparison of the application results of the methods considered. We conclude this research and propose directions for the further work in Section 9.

2. Literature Overview

Network methods for the scheduling problem are introduced in [1,2]. They are intended to determine a critical path and backup time for a work. A PERT method [1] is used when works have a probabilistic duration with due dates. A GERT method [2] extends the PERT method by considering the probabilities of individual works' implementation. The methods are very useful in the case of precedence relation occurrence. Network methods do not support non-renewable resources' implementation and restrictions on the works' earliest start time.

An approximate algorithm proposed in [3] is intended to solve the scheduling problem with restricted renewable and non-renewable resources in the presence of a deadline and restrictions on the earliest start time for a work. The algorithm has two stages. At the first stage, a feasible schedule is calculated based on an assumption that all resources are non-renewable. At the second stage, the renewable resources' constraints are added to the schedule found and then works are packed to satisfy all resource constraints and minimize the execution time. A disadvantage of the algorithm is that the arrangement solution is excluded from consideration if contradictions arise between the resource constraints and work deadlines. For example, excess availability of the own renewable resources for works on the critical path can exist. In this case, companies can use subcontracted renewable resources, the cost of which needs to be optimized.

For a scheduling optimization problem in a parallel system with identical non-renewable resources and orders' earliest start dates, two algorithms of the scheduling theory are presented in [4,5]. The algorithms are discussed in detail in Section 5.

A decentralized scheduling approach applied to a job-shop scheduling problem via agent-based simulation is presented in [6,7]. Multi-agent simulation is a popular technique intended to represent decision makers as a community of interacting agents [16]. In this case, each renewable resource and each single work is represented by an agent. The agents interact with each other and allocate the resources to the works by mean of negotiations. In [6], a new control algorithm based on dispatching rules is described. This algorithm is used to assign a score to each agent-resource proposal to agent-work, then the agent-work chooses the highest scored one. The resource score depends on the work's minimum processing time provided by the current resource compared to all the available alternatives, and the average cycle time of the system. An advantage of the algorithm is simultaneous multi-objective optimization of the system's throughput and cycle time, with weighted linear convolution of the two into a single objective function. In [7], the schedule is built via self-organization of software agents forming two networks of demands and resources, competing and cooperating on a virtual market. Agents of the basic types are the agents of orders (works), tasks (operations), resources (renewable resources), products (non-renewable resources), as well as the scene agent. Each agent has its own objective function named the satisfaction function, which is a weighted sum of components that meet various criteria. The value of the system's objective function is refined through the normalized sum of the agents' objective functions. The scheduling algorithm given in [7] includes a negotiation stage used by agents to build a set of conflicting orders for the resources and a conflict resolution stage, which recursively search for placement options taking into account existing limitations. The main advantages of the agent-based scheduling approach proposed in [7] are the agent's knowledge base of preferences when assigning resources to operations and the ability to reschedule in real time if the list of available resources or works has changed.

Heuristic scheduling algorithms based on a genetic algorithm are presented in [8,9]. The genetic algorithm (GA) is a popular optimization technique applicable to various application fields proposed by Goldberg [17]. In [8,9], a sequence of works is encoded into a chromosome, a population of the chromosomes is formed to represent a project, and then the population is transformed via genetic operators. Each of the investigations proposes a fitness function of the total project duration that is to be minimized. In [8], the authors introduce a dense gene concept, which is a fixed chromosome section that encodes a sequence of works that uses the available renewable resources in the most optimal way, providing the smallest remainder of free resources. The scarcity of the renewable resource is determined by solving the simplified scheduling problem with assumption that all the resources are non-renewable and calculating the remaining free resources. In [9], a scheduling problem of prefabricated buildings is considered. The authors consider both renewable resources and prefabricated blocks, or non-renewable resources, and their supplies. The GA consists of two stages: first, a schedule is formed with the assumption that the non-renewable resources are in shortage; second, the non-renewable resource supply's time constraints are added to the schedule and a new search is performed. The advantage of the algorithm presented in [9] is the use of heuristics to generate an initial population that allows a reduction in the number of unfeasible solutions produced by GA.

Commercial solvers were applied to the scheduling problem in [10–14].

OptQuest optimizer is a software that incorporates a combination of metaheuristics, such as scatter search, tabu search, and neural networks [18]. As a part of the AnyLogic modeling system [19], OptQuest is used to solve the multi-objective scheduling problem in [10,11]. A simulation model is used as an objective function for the optimization, which in return determines an optimal configuration of input parameters for the simulation model. Two algorithms of the renewable resources' allocation based on dispatching rules or on agent negotiation are compared using the OptQuest optimizer in [10]. Controlled variables are the assignment of resources to works, and the objective function is a multi-object function that estimates the total delay of the works and the total consumption of the renewable resources. The authors concluded that the use of the multi-agent approach decreases the resource waiting time for works but does not increase the throughput. The advantage of the study [11] is that energy consumption is considered as a part of the objective function. The author searches for an effective batch size using OptQuest optimizer and then improves the resources' utilization within the obtained schedule. It is concluded that batch size has a greater impact on the total delay rather than on the energy consumption.

A GA-based solver is embedded into the Tecnomatix Plant Simulation software package [20]. An application of this solver to the scheduling problem is presented in [12,13]. In [12], a flow shop problem is considered with a multi-objective function minimizing the mean flow time and total setup time. The solver optimizes the batch size and sequence of the product batches entering the system. In [13], a real-time job shop scheduling problem is considered with restricted renewable and non-renewable resources and works' due dates. A simulation model is used to assess the idling of processing equipment and the efficiency of the workshop in terms of throughput. The GA solver is used to allocate the renewable and non-renewable resources to the works. Advantages of the study are consideration of both types of resources and the integration of the scheduling phase into a cloud manufacturing system.

Subcontracted renewable resources are considered in [14]. The authors optimize the schedule cost by allocation of own and subcontracted resources taking into account a time-cost contradiction. The method is discussed in detail in Section 5.

We consider the scheduling problem optimizing not only own renewable resources but also subcontracted ones. The problem's objective function is cost minimization of the engaged subcontracted resources. The problem restrictions include the time frame of the earliest and latest work start date and a limited amount of the non-renewable resources available.

The goal of the present paper is to develop and assess an imitation and heuristic (IH) method for the scheduling problem with subcontracted resources. We considered two scheduling theory methods of scheduling on parallel machines by V. S. Tanaev and Y. A. Mezentsev to be most suitable for solving the given problem. We also considered application of a commercial solver to the given problem. Use of simulation and heuristics together allows us to overcome the revealed disadvantages of the considered methods. The developed IH method was applied for schedule search in a project company.

3. Notations

3.1. Indices

We consider the following indices:

- g is an index of the renewable resource of the allocated competence, $g = \overline{1, Q_r}$;
- i is an index of the operation or order contained in the project p, $i = \overline{1, N_p}$;
- j is an index of the operation or order that cannot be executed at the same time with the order i because they are to be processed by a single machine g, $j = \overline{1, N_p}$;
- k is an index of the time interval specified by the time constraints in the V.C. Tanaev method, $k = \overline{1, \beta}$, $\beta \leq 2N_p - 1$;
- L is an index of the order that can be processed during E_k time interval in the V.C. Tanaev method, $L = \overline{1, n(k)}$;
- p is an index of the project contained in the portfolio, $p = \overline{1, P}$;
- r is an index of the renewable resource's competence, $r = \overline{1, R}$;
- st is an index of the dynamic programming stage of the Y.A. Mezentsev algorithm, $st = \overline{1, N_p}$;
- t is an index of the day, $t = \overline{\tau, T}$;
- v is an index of the non-renewable resource, $v = \overline{1, V}$;
- η is an index of the iteration stage of the IH algorithm, $\eta = \overline{1, \Psi}$.

3.2. Sets

We consider the following sets:

- $e_1 < e_2 < \ldots < e_{k+1}$ is a set of the variables τ_i^0 and dl_i values in the V.C. Tanaev method;

- $N_k = \{i_{k,1}, i_{k,2}, \ldots, i_{k,n(k)}\}$ is a set of orders that can be processed during E_k interval with index L in the V.C. Tanaev method;
- Pr_p is a set of operations in the project p;
- $Rown$ is a set of own renewable resources;
- Rsc is a set of subcontracted renewable resources;
- $Z(t,r)$ is a set of indices for operations performed at time $t \geq \tau$ using own or subcontracted renewable resource r;
- $Z(t,v)$ is a set of indices for operations performed at time $t \geq \tau$ using non-renewable resource v;
- $\Theta(t,r)$ is a set of indices for operations performed at time $t \geq \tau$ using own resource r;
- $Y(t,r)$ is a set of indices for operations performed at time $t \geq \tau$ using subcontracted renewable resource r.

3.3. Parameters

We identified the following parameters:

- $c_{p,i,r}$ is a duration of the time interval $M_{p,i,r}(\eta)$ in the IH algorithm;
- $d_i > 0$ is an operation duration in case of one project in the portfolio, i.e., $P = 1$;
- $dl_i > 0$ is a deadline for processing the order i; it can be calculated as: $dl_i = \tau_i^1 + d_i$;
- $d_{p,i} > 0$ is a duration of the operation i of the project p;
- $E_k = (e_k; e_{k+1}]$ is a time interval specified by time constraints in the V.C. Tanaev method, $E_k \subseteq (\tau_i^0, dl_i]$;
- $f_{g,st}(\tau_{st}^0, d_{st}, y_{g,st})$ is a completion time of the order i processed via the machine or renewable resource g at the stage st in the Y.A. Mezentsev algorithm;
- Ft is a total cost of the subcontracted renewable resources;
- H and K' are parameters of the Y.A. Mezentsev algorithm, they are intended to drop out some feasible solutions at each search stage to decrease the searching time;
- $Kres$ is a total cost of the used resources, own and subcontracted ones;
- Ksc is a predefined limit of the cost of subcontracted renewable resources;
- $M = \max_k n(k)$ is a quantity of machines or renewable resources, both own and subcontracted, required to process orders in the V.C. Tanaev method, $n(k) \leq M$;
- M' is the quantity of the needed subcontracted resources in the V.C. Tanaev method, $M' \leq (M - Q_r)$;
- ML is an arbitrarily assumed, sufficiently large constant in the Lingo method;
- N is the number of operations of the project portfolio;
- N_{st} is the current amount of the developed schedules at the stage st in the Y.A. Mezentsev algorithm;
- $n(k)$ is the quantity of all the orders constituting the N_k set in the V.C. Tanaev method;
- Q_r is the available quantity of the own renewable resource with competence r;
- $Q_{t,v}$ is the current quantity of each non-renewable resource v at time t;
- $q_{p,i,r} \geq 0$ is the required amount of renewable resources of allocated competence r to perform the operation i of the project p; $q_{p,i,r} = 0$ when the renewable resource r is not required to process the operation i of the project p;
- $q^-_{p,i,v} \leq 0$ and $q^+_{p,i,v} \geq 0$ are the amounts of non-renewable resource v consumed when the operation i of the project p starts and produced when the operation i ends, respectively; $q^-_{p,i,v} = 0$ when the non-renewable resource v is not required to start the operation i of the project p; the variable $q^+_{p,i,v} = 0$ when the non-renewable resource v is not produced as an outcome of the operation i of the project p;
- $SC_{p,t,r}$ is a cost of the subcontracted resource of allocated competence r engaged in the project p at the time t;
- $sr_{i,r}$ is the daily cost of performing the operation i using a unit quantity of the renewable resource r in case of one project in the portfolio, i.e., $P = 1$;
- $sr_{p,i,r} \geq 0$ is the daily cost of performing the operation i of the project p using a unit quantity of the renewable resource of allocated competence r;

- $sr'_{p,i,r} \geq 0$ is the total cost of performing the operation i of the project p using a subcontracted renewable resource r;
- Tdl is the global project's deadline that must not be exceeded, $Tdl = \max_i dl_i$;
- $U_{p,t,r}$ is the percentage utilization of the renewable resource r engaged in the project p at the time t.
- $U^0_{p,i,r}(\eta)$ is the average utilization of the renewable resource r for the operation i of the project p during the time interval $\delta_{p,i,r}(\eta)$ in the IH algorithm;
- $U^-_{p,i,r}(\eta)$ is the average utilization of the renewable resource r for the operation i of the project p during the time interval $\delta^-_{p,i,r}(\eta)$ in the IH algorithm;
- $U^+_{p,i,r}(\eta)$ is the average utilization of the renewable resource r for the operation i of the project p during the time interval $\delta^+_{p,i,r}(\eta)$ in the IH algorithm;
- $\Delta_k = e_{k+1} - e_k$ is the duration of the time interval E_k in the V.C. Tanaev method;
- $\delta_{p,i,r}(\eta)$ is the time interval where the utilization percentage of the renewable resource r for the operation i of the project p is equal to 100%, $U^0_{p,i,r}(\eta) = 100\%$, in the IH algorithm;
- $\delta^-_{p,i,r}(\eta)$ is the time interval $\delta_{p,i,r}(\eta)$ shifted to the left on the time axis in the IH algorithm;
- $\delta^+_{p,i,r}(\eta)$ is the time interval $\delta_{p,i,r}(\eta)$ shifted to the right on the time axis in the IH algorithm;
- $\lambda_{g,i}$ is the cost of processing the order i by the resource g;
- $\mu_g(t)$ is a function that shows the presence of subcontracted resources g at the time t, $g = \overline{Q_r, M}$;
- $\zeta_{p,i,r}$ is a threshold of the $sr'_{p,i,r}$ total cost of performing the operation i of the project p using a subcontracted renewable resource r.
- $\sigma(t)$ is a function of orders allocated to machines or renewable resources in the V.C. Tanaev method;
- $\tau^0_{p,i} \geq \tau$ and $\tau^1_{p,i} \geq \tau$ are the earliest and latest possible start times given for the operation i of the project p; $\tau^0_{p,i} = \tau^1_{p,i} = \tau$ if the operation i is only allowed to start at the time τ;
- τ^0_i and τ^1_i are orders' earliest and latest possible start times in case of one project in the portfolio, $P = 1$;
- $\hat{\tau}_{g,i}$ is the actual delay between start of the i-th order on the g-th machine upon the previous order completion in the Y.A. Mezentsev algorithm;
- $\varphi_{g,st}(\tau^0_i, d_i, y_{g,i})$, $i = \overline{1,st}$ is the completion time by the resource g for all the orders that exist at the stages from the first to the st-th in the Y.A. Mezentsev algorithm;
- $\varphi_{st}(\tau^0_i, d_i, y_{g,i})$, $i = \overline{1,st}$ is the minimal completion time for all the orders existing at the stages from the first to the st-th in the Y.A. Mezentsev algorithm;
- ω is the cost of processing the order per time unit in conventional units;
- Ψ is the predefined maximum number of the IH algorithm steps.

3.4. Decision Variables

The decision variables may vary regarding the different scheduling problem statements. We identify the following decision variables connected with the problem:

- $S_{k,g}(t)$ are integer variables of a sequence of orders to perform by each renewable resource g at each time interval E_k for the scheduling problem given in Section 5.2;
- $x_{p,i} \in \{\tau, \ldots, T\}$ are integer variables of the operation start times for the scheduling problem given in Section 4;
- $x_i \in \{\tau, \ldots, T\}$ are integer variables of the operation start times given in Section 5.3 in case of one project in the portfolio, i.e., $P = 1$;
- $y_{g,i}$ are Boolean variables for allocating the order i to the machine g for the problems given in Sections 5.1–5.3. The variable assumes a value of 1 if the order i is to be

executed by the machine g; otherwise, it equals 0. In Section 5.3, we assume that the machine g can be an own or subcontracted one;
- $\varepsilon_{i,j}$ are binary variables for the scheduling problem given in Section 5.3. The variable indicates orders that cannot be executed at the same time because they are to be processed by a single machine g. The variable $\varepsilon_{i,j}$ equals 1 if the order i is to be completed before the order j; otherwise, it equals 0; $(i, j) \in Pr_p$.

4. Scheduling Problem Statement

The scheduling problem statement is given by the authors in [21]. Below are the key points. The notations are presented in Section 3.

We assume that a set of operations in each project p appears in order of increasing of the operation's cost $sr_{p,i,r}$; $sr_{p,i,r} = 0$ if the subcontracted resource r is not required by the operation i of the project p.

We denote the set of indices for the operations performed at the time $t \geq \tau$ utilizing the renewable resource r as follows:

$$Z(t,r) = \{i \in [1, N_p] \; x_{p,i} \leq t < x_{p,i} + d_{p,i} \; \& \; q_{p,i,r} \neq 0 \; \& \; p \in [1, P]\} \quad (1)$$

We denote the set of indices for the operations performed at the time $t \geq \tau$ utilizing the non-renewable resource v as follows:

$$Z(t,v) = \left\{ i \in [1, N_p] \; x_{p,i} \leq t < x_{p,i} + d_{p,i} \; \& \; q^{-}_{p,i,v} \neq 0 \; \& \; p \in [1, P] \right\}. \quad (2)$$

We denote the set of indices for the operations performed at the time $t \geq \tau$ utilizing the own renewable resource r within the available amount Q_r:

$$\Theta(t,r) = \{i \in Z(t,r) \; \sum_{p \in [1,P]} \sum_i q_{p,i,r} \leq Q_r\}. \quad (3)$$

The set of indices for the operations performed at the time $t \geq \tau$ utilizing the subcontract resource r is defined as follows:

$$Y(t,r) = Z(t,r) \setminus \Theta(t,r) = \{i \in [1, N_p] \; i \in Z(t,r) \; \& \; i \notin \Theta(t,r)\}. \quad (4)$$

The percentage utilization of the resource r engaged in the project p at the time t is defined as follows:

$$U_{p,t,r} = \begin{cases} \left(\sum_{i \in \Theta(t,r)} q_{p,i,r} / Q_r \right) \cdot 100\%, & i \in \Theta(t,r), \\ 100\%, & i \in Y(t,r). \end{cases} \quad (5)$$

The cost of the subcontract resource of an allocated competence r involved in the project p at the time t is defined by the formula:

$$SC_{p,t,r} = \sum_{i \in Y(t,r)} sr_{p,i,r} \cdot q_{p,i,r}. \quad (6)$$

The current volume of the non-renewable resource v at the time t is defined as:

$$Q_{t,v} = Q_{\tau,v} + \sum_{\alpha=\tau}^{t} \sum_{p=1}^{P} \sum_{\substack{i \in Pr_p \\ \wedge \alpha = x_{p,i}}} q^{-}_{p,i,v} + \sum_{\alpha=\tau}^{t} \sum_{p=1}^{P} \sum_{\substack{i \in Pr_p \\ \wedge \alpha = x_{p,i} + d_{p,i}}} q^{+}_{p,i,v}. \quad (7)$$

The scheduling problem can be formalized as follows:

$$Ft = \sum_{p=1}^{P} \sum_{t=\tau}^{T} \sum_{r=1}^{R} SC_{p,t,r} \to min, \quad (8)$$

$$\sum_{p=1}^{P} \sum_{\substack{i \in Z(t,v) \\ \wedge t = x_{p,i}}} \left| q^{-}_{p,i,v} \right| \leq Q_{t,v}, \; \forall \, t = \overline{\tau, T}, \; \forall \, v = \overline{1, V}, \quad (9)$$

$$\tau^0_{p,i} \leq x_{p,i} \leq \tau^1_{p,i}, \ \forall\, p = \overline{1,P},\ \forall\, i = \overline{1,N_p}. \tag{10}$$

The objective function (8) minimizes the total cost of the subcontracted resources in the case of exceeding the availability of the own ones. The constraint (9) ensures availability of the required amount of non-renewable resources at the time of the operation start. The constraint (10) imposes a time frame on the start dates of the operations.

5. Scheduling Methods Application

We consider the schedule optimization problem for the parallel system having identical machines in the presence of delays processing the orders. The problem is one of the closest problems studied by the scheduling theory. According to [15], the problem is formalized as follows: $Pm/r_j/C_{max}$. For the problem, a parametric algorithm of dynamic programming with an alternative dropout option was proposed by Y. A. Mezentsev et al. [4].

Another closest problem studied by the scheduling theory is a parallel system schedule optimization using identical machines in the presence of orders' due dates. The problem is formalized as follows: $Pm/brkdwn, r_j/\sum \omega_j U_j$. As a solution to the problem, a schedule construction algorithm using the given due dates is proposed by V. S. Tanaev et al. [5].

We also considered Lingo commercial solver's application to the scheduling problem according to the scheme given in [14]. Here, the problem is modelled as a mixed binary linear program that minimizes the project cost, including subcontracted and own resources cost.

We applied the considered algorithms to solve a trial small-scale scheduling problem using two own renewable resources $Q_1 = 2$ of the same competence $R = 1$, and one project $P = 1$ with the number of the operations $N_1 = 7$. Seven orders, or operations, are fed at a different time into a parallel system with two identical machines, or renewable resources. Table 1 contains the initial information about the operation's duration and time frame between order processing's earliest possible start and latest finish. We assume the operations are ordered by the earliest start time.

Table 1. Initial information of the trial scheduling problem.

Order Number, i	Order Duration, d_i	Earliest Start Time, τ^0_i	Deadline, dl_i
1	2	0	3
2	3	0	4
3	2	1	4
4	4	2	7
5	3	3	7
6	2	5	8
7	4	5	10

The progress of solving the problem is given below.

5.1. Method Developed by Y. A. Mezentsev

We applied a parametric dynamic programming algorithm with the optional exclusion of the found alternatives to the considered scheduling problem. The notations are given in Section 3. The decision variables are Boolean variables $y_{g,i}$ allocating the order i to the machine g.

The dynamic algorithm variables can be calculated as given in the study [4]:

$$f_{g,st}\left(\tau^0_{st}, d_{st}, y_{g,st}\right) = \max\left\{0, \left[\tau^0_{st} y_{g,st} - \varphi_{g,st-1}\left(\tau^0_{st-1}, d_i, y_{g,i}\right)\right]\right\} + d_{st} y_{g,st}, \tag{11}$$

$$\varphi_{g,st}\left(\tau^0_i, d_i, y_{g,i}\right) = \left\{f_{g,st}\left(\tau^0_{st}, d_{st}, y_{g,st}\right) + \varphi_{g,st-1}\left(\tau^0_i, d_i, y_{g,i}\right)\right\}, i = \overline{1, st-1}, \tag{12}$$

$$\varphi_{st}\left(\tau^0_i, d_i, y_{g,i}\right) = \max_g \left\{\varphi_{g,st}\left(\tau^0_{st}, d_i, y_{g,i}\right)\right\}, i = \overline{1, st}. \tag{13}$$

The parametric algorithm includes the following stages:

1. Input of the initial data (τ_i^0, d_i, dl_i), $i = \overline{1, N_p}$ and H, K', Q_r, and N_p parameters. Set $\varphi_{g,0}(\tau_0^0, d_i, y_{g,i}) = 0$, $st = 0$;
2. $st = st + 1$;
3. If $st > N_p$ then go to Point 7;
4. Generate all the feasible schedules and calculate $f_{g,st}(\tau_{st}^0, d_{st}, y_{g,st})$ and the schedule length $\varphi_{g,st}(\tau_i^0, d_i, y_{g,i})$;
5. Check the number of the generated schedules N_{st}. If $N_{st} \leq K'$ then go to Point 2; otherwise, go to Point 6;
6. Discard Q_r^{H-1} out of the schedules generated at Point 4 with the maximum schedule length $\varphi_{g,st}(\tau_i^0, d_i, y_{g,i})$. Go to Point 2;
7. Choose the schedules with the minimum makespan. Determine the calendar plan by reverse dynamic programming.

We consider application of the parametric algorithm to the trial scheduling problem given in Table 1. Examples of the calculated algorithm characteristics are given in Tables 2 and 3 for the first and last algorithm stages. We set $H = 2$, $K' = 2^2 = 4$.

Table 2. Results of the first algorithm stage completion.

Stage 1	$x_{1,1}$	$x_{2,1}$	$f_{g,1}$	$\varphi_{g,1} = f_{g,1}$, $\varphi_1 = \max_g \{\varphi_{g,1}\}$
Order 1	1	0	(max{0,0-0} + 2,0) = (2,0)	$\varphi_1 = \max\{2,0\} = 2$
	0	1	(0,max{0,0-0} + 2) = (0,2)	$\varphi_1 = \max\{0,2\} = 2$

Table 3 contains the dark filled cells connected to the schedules providing the minimum local makespan on the given stage.

Table 4 contains the four schedules identified as the solutions of the given small-scale scheduling problem.

Since the machines are identical, schedule 1 is equal to schedule 4 and schedule 2 is equal to schedule 3. Schedules 1 and 2 differ by two last orders allocated to the opposite machines. All of the schedules reveal deadline violation on orders 6 and 7.

The obtained schedules are shown in Figure 1 in the form of a Gantt chart.

Table 3. Results of the last algorithm stage completion.

Stage 7	$x_{1,1}$	$x_{2,1}$	$x_{1,2}$	$x_{2,2}$	$x_{1,3}$	$x_{2,3}$	$x_{1,4}$	$x_{2,4}$	$x_{1,5}$	$x_{2,5}$	$x_{1,6}$	$x_{2,6}$	$x_{1,7}$	$x_{2,7}$	$f_{g,7}$	$\varphi_{g,7} = \{f_{g,7} + \varphi_{g,6}\}$, $\varphi_7 = \max_g\{\varphi_{g,7}\}$
Order 1+	1	0	0	1	1	0	0	1	1	0	0	1	1	0	$(\max\{0,5\text{-}8\} + 4,0) = (4,0)$	$\max[2+3+0+0+3+0+4, 0+0+3+4+0+2+0] = 12$
Order 2+	1	0	0	1	1	0	0	1	1	0	0	1	0	1	$(0,\max\{0,5\text{-}9\} + 4) = (0,4)$	$\max[2+3+0+0+3+0+0, 0+0+3+4+0+2+4] = 13$
Order 3+	1	0	0	1	1	0	0	1	1	0	1	0	1	0	$(\max\{0,5\text{-}9\} + 4,0) = (4,0)$	$\max[2+0+2+0+3+2+4, 0+3+0+4+0+0+0] = 13$
Order 4+	1	0	0	1	1	0	0	1	1	0	1	0	0	1	$(0,\max\{0,5\text{-}7\} + 4) = (0,4)$	$\max[2+0+2+0+3+2+0, 0+3+0+4+0+0+4] = 11$
Order 5+	1	0	0	1	1	1	1	0	1	0	0	1	1	0	$(\max\{0,5\text{-}7\} + 4,0) = (4,0)$	$\max[2+0+2+0+3+0+4, 0+3+0+4+0+2+0] = 11$
Order 6+	0	1	1	0	0	1	1	0	0	1	0	1	0	1	$(0,\max\{0,5\text{-}9\} + 4) = (0,4)$	$\max[2+0+2+0+3+0+0, 0+3+0+4+0+2+4] = 13$
Order 7	0	1	1	0	0	1	1	0	0	1	1	0	1	0	$(\max\{0,5\text{-}9\} + 4,0) = (4,0)$	$\max[0+3+0+4+0+2+4, 2+0+2+0+3+0+0] = 13$
	0	1	1	0	0	1	1	0	0	1	1	0	0	1	$(0,\max\{0,5\text{-}7\} + 4) = (0,4)$	$\max[0+3+0+4+0+2+0, 2+0+2+0+3+0+4] = 11$
	0	1	1	0	0	1	1	0	1	0	0	1	1	0	$(\max\{0,5\text{-}7\} + 4,0) = (4,0)$	$\max[0+3+0+4+0+0+4, 2+0+2+0+3+2+0] = 11$
	0	1	1	0	0	1	1	0	1	0	0	1	0	1	$(0,\max\{0,5\text{-}9\} + 4) = (0,4)$	$\max[0+3+0+4+0+0+0, 2+0+2+0+3+2+4] = 13$
	0	1	1	0	0	1	1	0	1	0	1	0	1	0	$(\max\{0,5\text{-}9\} + 4,0) = (4,0)$	$\max[0+0+3+4+0+2+4, 2+3+0+0+3+0+0] = 13$
	0	1	1	0	0	1	1	0	1	0	1	0	0	1	$(0,\max\{0,5\text{-}9\} + 4) = (0,4)$	$\max[2+3+0+0+3+0+0, 0+0+3+4+0+2+4] = 13$

Table 4. Schedules with minimal makespan.

Schedule	$x_{1,1}$	$x_{2,1}$	$x_{1,2}$	$x_{2,2}$	$x_{1,3}$	$x_{2,3}$	$x_{1,4}$	$x_{2,4}$	$x_{1,5}$	$x_{2,5}$	$x_{1,6}$	$x_{2,6}$	$x_{1,7}$	$x_{2,7}$	φ_7
1	1	0	0	1	1	0	0	1	1	0	1	0	0	1	11
2	1	0	0	1	1	0	0	1	1	0	0	1	1	0	11
3	0	1	1	0	0	1	1	0	0	1	1	0	0	1	11
4	0	1	1	0	0	1	1	0	0	1	0	1	1	0	11

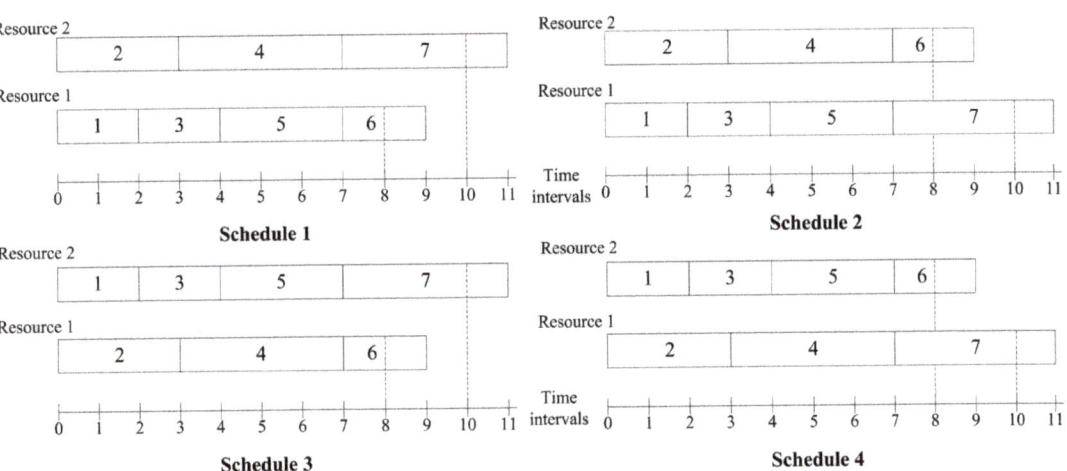

Figure 1. Schedules received by the Y. A. Mezentsev method with the minimum makespan.

As we can see, the schedules found by the Y. A. Mezentsev method for a parallel system with identical machines and orders' earliest start time do not support the orders' due date. Subcontracted resources are not considered by the method. Thus, $Ft = 0$ for the schedule found by Y. A. Mezentsev.

5.2. Method Developed by V. C. Tanaev

We considered an algorithm implemented by V. C. Tanaev. The notations used are given in Section 3.

The decision variables are schedules $S_{k,g}(t)$ for each resource g at each time interval E_k. These are integer type variables.

Two cases can occur: (1) $M \leq Q_r$ if the existing machines are enough to process all the orders within time intervals E_k; (2) $M > Q_r$ if the existing machines are not enough to process all the orders at the E_k time intervals and it is necessary to attract an amount $M' \leq (M - Q_r)$ of subcontracted machines.

An algorithm is given in [5] and contains the following stages. We modified the algorithm by adding the stages 4 to 7 to reduce the number of interruptions:

1. $k = 0$;
2. The next index of the time interval: $k = k + 1$. If $k > \beta$ then go to Point 9; otherwise, perform the following actions. Form a set of orders $N_k = \{i_{k,1}, i_{k,2}, \ldots, i_{k,n(k)}\}$, which can be processed at the E_k interval. The orders are sorted by descending the cost of the processing per time unit;
3. If $\min_{i \in N_k}\{d_i\} < \Delta_k$ then go to Point 8; otherwise, go to Point 4;
4. Start to browse a set of machines with given competence: $g = 0$;
5. The next index of the machine: $g = g + 1$. If $(g > M)$ OR $(N_k = \varnothing)$, then go to Point 2; otherwise, start to browse the set of orders from the beginning N_k: $L = 0$;

6. The next index of the order: $L = L + 1$. If $L > n(k)$, then go to Point 5; otherwise, go to Point 7;
7. Assign the machine g to the order $i_{k,L} \in N_k$ and form the schedule $S_{k,g}(t)$ at the E_k time interval:
   a. If $(k = 1)$ OR $(k \neq 1$ AND the order $i_{k,L}$ at the previous time interval E_{k-1} has not been assigned to the machine $g' \neq g)$ OR $(k \neq 1$ AND the order $i_{k,L}$ at the previous time interval E_{k-1} has been assigned to the machine $g' \neq g$, $g' > Q_r$ AND $n(k) \leq Q_r$), then assign the machine g to the order $i_{k,L}$:
      i. $d_i = d_i - \Delta_k$, $S_{k,g}(t) = i_{k,L}$, where $t = (e_k; e_{k+1}]$;
      ii. Eliminate the order $i_{k,L}$ from the set N_k. Go to Point 5;
   b. If $(k \neq 1$ AND the order $i_{k,L}$ at the previous time interval E_{k-1} has been assigned to the machine $g' \neq g$, $g' \leq Q_r$) OR $(k \neq 1$ AND the order $i_{k,L}$ at the previous time interval E_{k-1} has been assigned to the machine $g' \neq g$ and $g' > Q_r$ AND $n(k) > Q_r$), then go to Point 6;
8. Calculate the processing duration of all the orders constituting the N_k set according to the formula: $d_i' \begin{cases} d_i, d_i < \Delta_k \\ \Delta_k, d_i \geq \Delta_k \end{cases}$, $i \in N_k$. Apply a packing algorithm intended to assign the N_k set orders having d_i' durations to the set M machines for the E_k time interval. At this stage, we consider the following statements are true: $d_i' \leq \Delta_k, i \in N_k$ and $\sum_{i \in N_k} d_i' \leq M\Delta_k$:
   a. Calculate at the time interval $(e_k; e_k + M\Delta_k]$ a function $\sigma(t)$, where:
   $\sigma(t) = i_{k,1}$ at the time interval $(e_k; e_k + d_{i,k,1}]$;
   $\sigma(t) = i_{k,a}$ at the time interval $(e_k + \sum_{\alpha=1}^{A-1} d_{i,k,\alpha}; e_k + \sum_{\alpha=1}^{A} d_{i,k,\alpha}], A = \overline{2, n(k)}$;
   if $\sum_{i \in N_k} d_i < M\Delta_k$, then $\sigma(t) = 0$ at the time interval $(e_k + \sum_{i \in N_k} d_i; e_k + M\Delta_k]$;
   b. Form the schedule $S_{k,g}(t) = \sigma(t + (g-1)\Delta_k)$, $g = \overline{1, M}$;
   c. $d_i = d_i - d_i'$, $i \in N_k$. Go to Point 2;
9. The end of the algorithm.

We used the algorithm to solve the scheduling problem given in Table 1. We sorted the set N_p orders according to the information given in Table 1.

We allocated time marks $\{e_{\beta+1}\}, \beta + 1 = 9$ to form a set of time intervals $\{E_k\}$, where $k = \overline{1,8}$ (Figure 2).

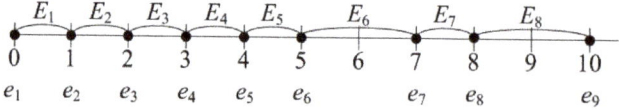

Figure 2. Time intervals of orders' processing for the trial problem.

Results of the method's execution upon the trial problem are given in Table 5.

Since $\max_k n(k) = 4$, the number of machines is defined as four. However, the maximum number of the machines used simultaneously turned out to be three as two orders were assigned to the machine r_1 at the time interval E_6 during schedule searching. Hence, we set $M = 3$, $Q_r = 2$, and $M' = 1$.

The schedule formed by the V. S. Tanaev method is shown in Figure 3.

Table 5. Schedule search with orders' due date by the V. S. Tanaev method.

Interval E_k and Its Length Δ_k	Set N_k	Schedule $S_{k,g}, g = \overline{1,M}$	Duration d_i, $i \in N_k$	Function $\sigma(t)$ Where $\min_{i \in N_k}\{d_i\} < \Delta_k$
$E_1 = (0;1]$ $\Delta_1 = 1$	$N_1 = \{i_1, i_2\}$	$S_{1,1}(t) = i_1$ $S_{1,2}(t) = i_2$ $S_{1,3}(t) = 0$	$t_1 = 2 - 1 = 1$ $t_2 = 3 - 1 = 2$	
$E_2 = (1;2]$ $\Delta_2 = 1$	$N_2 = \{i_1, i_2, i_3\}$	$S_{2,1}(t) = i_1$ $S_{2,2}(t) = i_2$ $S_{2,3}(t) = i_3$	$t_1 = 1 - 1 = 0$ $t_2 = 2 - 1 = 1$ $t_3 = 2 - 1 = 1$	
$E_3 = (4;5]$ $\Delta_3 = 1$	$N_3 = \{i_2, i_3, i_4\}$	$S_{3,1}(t) = i_4$ $S_{3,2}(t) = i_2$ $S_{3,3}(t) = i_3$	$t_4 = 4 - 1 = 3$ $t_2 = 1 - 1 = 0$ $t_3 = 1 - 1 = 0$	
$E_4 = (3;4]$ $\Delta_4 = 1$	$N_4 = \{i_4, i_5\}$	$S_{4,1}(t) = i_4$ $S_{4,2}(t) = i_5$ $S_{4,3}(t) = 0$	$t_4 = 3 - 1 = 2$ $t_5 = 3 - 1 = 2$	
$E_5 = (4;5]$ $\Delta_5 = 1$	$N_5 = \{i_4, i_5\}$	$S_{5,1}(t) = i_4$ $S_{5,2}(t) = i_5$ $S_{5,3}(t) = 0$	$t_4 = 2 - 1 = 1$ $t_5 = 2 - 1 = 1$	
$E_6 = (5;7]$ $\Delta_6 = 2$	$N_6 = \{i_4, i_5, i_6, i_7\}$	$S_{6,1}(t) = \begin{cases} i_4 \text{ if } t \in (5;6] \\ i_5 \text{ if } t \in (6;7] \end{cases}$ $S_{6,2}(t) = i_6$ $S_{6,3}(t) = i_7$	$t_4 = 1 - 1 = 0$ $t_5 = 1 - 1 = 0$ $t_6 = 2 - 2 = 0$ $t_7 = 4 - 2 = 2$	$\sigma(t) = \begin{cases} i_4 \text{ if } t \in (5;6] \\ i_5 \text{ if } t \in (6;7] \\ i_6 \text{ if } t \in (7;9] \\ i_7 \text{ if } t \in (9;11] \end{cases}$
$E_7 = (7;8]$ $\Delta_7 = 1$	$N_7 = \{i_7\}$	$S_{7,1}(t) = i_7$ $S_{7,2}(t) = 0$ $S_{7,3}(t) = 0$	$t_7 = 2 - 1 = 1$	
$E_8 = (8;10]$ $\Delta_8 = 2$	$N_8 = \{i_7\}$	$S_{8,1}(t) = \begin{cases} i_7 \text{ if } t \in (8;9] \\ 0 \text{ if } t \in (9;10] \end{cases}$ $S_{8,2}(t) = 0$ $S_{8,3}(t) = 0$	$t_7 = 1 - 1 = 0$	$\sigma(t) = \begin{cases} i_7 \text{ if } t \in (8;9] \\ 0 \text{ if } t \in (9;10] \end{cases}$

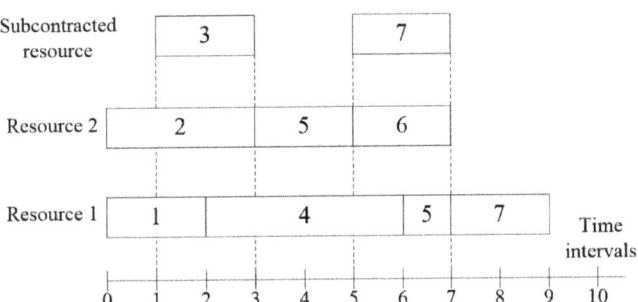

Figure 3. Schedule provided by the V. S. Tanaev method.

We introduce a function $\mu_g(t)$ for subcontracted resources $g = \overline{Q_r, M}$ and define it as follows: $\mu_g(t) = \begin{cases} 1, \text{ if } S_{k,g} \neq 0, t \in E_k \\ 0, \text{ else} \end{cases}$. We calculate the subcontracted resources cost according to the formula:

$$Ft = \omega \sum_{t=1}^{e_{\beta+1}} \sum_{g=Q_r}^{M} \mu_g(t) \qquad (14)$$

For the schedule found by the V. S. Tanaev method, the cost of the subcontracted resources is $Ft = 4\omega$.

The schedule found is not optimal according to the objective function (8). An example of a more effective schedule satisfying all the requirements (9)–(10) is shown in Figure 4; the subcontracted resources cost is $Ft = 2w$.

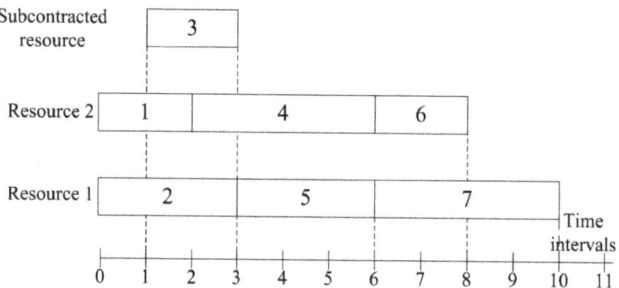

Figure 4. Schedule with the most effective resource allocation.

Without loss of generality, let us identify $w = 100$. Then, $Ft = 400$ for the schedule found by the V. C. Tanaev method.

5.3. Method Based on the Commercial Solver Application

We applied a method based on a Lingo solver [22] to the trial scheduling problem. We reduced our mathematical model into a model suitable for linear programming [14].

The decision variables are the following:

- x_i are integer variables indicating the start times of the orders, $i \in Pr_p$;
- $y_{g,i}$ are binary variables indicating what renewable resource g, either own or subcontracted, is allocated to a particular order i;
- $\varepsilon_{i,j}$ are binary variables indicating orders that cannot be executed at the same time because they are to be processed by a single resource g.

The mathematical model of this problem applied to a project is described as follows:

$$Kres = \sum_{g \in Rown \cup Rsc} \sum_{i \in Pr_p} \lambda_{g,i} \cdot y_{g,i} \to \min, \quad (15)$$

$$\sum_{g \in Rown \cup Rsc} y_{g,i} = 1, \ \forall i \in Pr_p, \quad (16)$$

$$x_i + d_i \leq x_j + ML \cdot (1 - \varepsilon_{i,j}) + ML \cdot (2 - y_{g,i} - y_{g,j}), \ \forall (i,j) \in Pr_p, \ \forall g \in Rown \cup Rsc, \quad (17)$$

$$x_j + d_j \leq x_i + ML \cdot \varepsilon_{i,j} + ML \cdot (2 - y_{g,i} - y_{g,j}), \ \forall (i,j) \in Pr_p, \ \forall g \in Rown \cup Rsc, \quad (18)$$

$$x_{Np} + d_{Np} \leq Tdl, \quad (19)$$

$$Ft = \sum_{g \in Rsc} \sum_{i \in Pr_p} \lambda_{g,i} \cdot y_{g,i} \leq Ksc, \quad (20)$$

$$x_i \geq \tau, \ \forall i \in Pr_p, \quad (21)$$

$$y_{g,i} \in \{0,1\}, \ \forall g \in Rown \cup Rsc, \ \forall i \in Pr_p, \quad (22)$$

$$\varepsilon_{i,j} \in \{0,1\}, \ \forall (i,j) \in Pr_p. \quad (23)$$

For the problem given in Table 1, the following variable values are set:

- The number of projects is $P = 1$;
- The cardinality of the $Rown$ set is $|Rown| = 2$;
- The cardinality of the Rsc set is $|Rsc| = 7$, where 7 is the number of orders that can be processed simultaneously;
- The cardinality of the Pr_p set is $|Pr_p| = 7$;

- $Ksc = 400$ that is limited by the maximum value of the subcontracted resources cost found by the V. S. Tanaev and Y. A. Mezentzev methods;
- $ML = 10000$;
- $Tdl = 10$ days according to the dl_i values from Table 1;
- $\lambda_{g,i} = \begin{cases} 1, & if\ g \in Rown \\ 100, & if\ g \in Rsc \end{cases} \forall i \in Pr_p$; we assume that the operation cost should be much lower when processed by the own resource comparing to the subcontracted.

It is defined for this problem that limited non-renewable resources are not considered while they are being introduced in the Formulation (8)–(10).

As a result of Lingo execution upon the linear programming problem (15)–(23), a set of alternative optimal schedules was formed providing the same minimum total used resources cost $Kres = 7$. Two optimal schedules are presented in Figure 5.

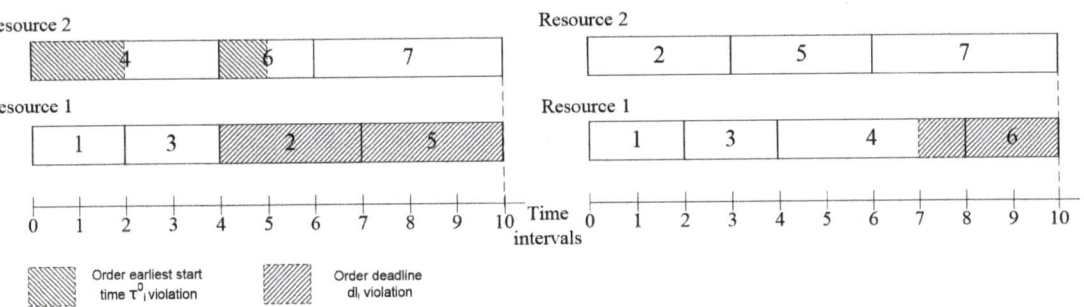

Figure 5. Two optimal schedules provided by the commercial solver.

As we can see, the commercial solver was able to reach subcontracting cost $Ft = 0$ for the schedules found, but there are violations of the deadline and the orders' earliest start time.

Thus, application of the commercial solver method leads to violation of the restrictions (9) and (10) of the problem considered.

6. Imitation and Heuristic Method

A key concept of the imitation and heuristic algorithm is integration of processes' imitation and some heuristic rules to improve the initial schedule. The IH method algorithm is based on an application of a multi-agent resource conversion process (MRCP) model [23]. The MRCP model is intended to describe discrete processes converting input non-renewable resources into output ones using renewable resources, or machines, throughout a given time interval.

An agent of the MRCP model is a decision maker model having formalized knowledge about resources' allocation using production rules. The MRCP model also includes a logistics agent. The logistics agent controls the current value and lifetime of the non-renewable resources and ensures fulfillment of the restriction (9) by launching the purchase or production process of the non-renewable resource required in case its current volume is decreased to a critical value or the resource's lifetime is exceeded.

The IH algorithm is a cycle with alternating stages of imitation and application of heuristic rules to improve the schedule. During the imitation stage, the schedule is fed to the MRCP model input, and the model evaluates the subcontracted resources cost according to the Formula (8). During the heuristic stage, the algorithm shifts the start days for the operations, where the subcontracted resources cost exceeds the given threshold. During shifting, the restriction (10) is ensured. The algorithm stops either in the absence of exceeding the threshold of the subcontracted resources cost, or in case a certain number of cycles is reached.

Let us consider the algorithm of the IH method used the MRCP model. The notations are given in Section 3.

The algorithm variables are calculated using the following formulas:

- $\delta_{p,i,r}(\eta) = [a;b]$ is a time interval where the utilization percentage of the renewable resource r for the operation i is equal to 100%, $U^0_{p,i,r}(\eta) = 100\%$;
- $c_{p,i,r} = b - a$ is the duration of the time interval $\delta_{p,i,r}(\eta)$;
- $\delta^-_{p,i,r}(\eta) = [(a - (\eta+1)\cdot c_{p,i,r}); (a - \eta \cdot c_{p,i,r})]$ is the time interval $\delta_{p,i,r}(\eta)$ shifted on the $(\eta+1)\cdot c_{p,i,r}$ days to the left on the time axis and satisfying the request: $U^-_{p,i,r}(\eta) \neq U^0_{p,i,r}(\eta)$;
- $U^-_{p,i,r}(\eta)$ is the average utilization of the renewable resource r for the operation i of the project p during the time interval $\delta^-_{p,i,r}(\eta)$:

$$U^-_{p,i,r}(\eta) = \sum_{t=a-(\eta+1)\cdot c_{p,i,r}}^{a-\eta\cdot c_{p,i,r}} U_{p,t,r}/c_{p,i,r}; \qquad (24)$$

- $\delta^+_{p,i,r}(\eta) = [(b + \eta \cdot c_{p,i,r}); (b + (\eta+1)\cdot c_{p,i,r})]$ is the time interval $\delta_{p,i,r}(\eta)$ shifted on the $(\eta+1)\cdot c_{p,i,r}$ days to the right on the time axis satisfying the request: $U^+_{p,i,r}(\eta) \neq\neq U^0_{p,i,r}(\eta)$;
- $U^+_{p,i,r}(\eta)$ is the average utilization of the renewable resource r for the operation i of the project p during the time interval $\delta^+_{p,i,r}(\eta)$:

$$U^+_{p,i,r}(\eta) = \sum_{t=b+\eta\cdot c(p,i,r)}^{b+(\eta+1)\cdot c(p,i,r)} U_{p,t,r}/c_{p,i,r}. \qquad (25)$$

The algorithm of the IH method includes the following stages:

1. Conduct experiments with the MRCP model, the input to which are the start dates of operations $x_{p,i}$. Form output model parameters: for each moment t and each project p, define a set of operations $Op_i \in Pr_p$ with indexes $i \in Y(t,r)$ using the subcontract renewable resource r to define the subcontracted resources cost $sr'_{p,i,r} = sr_{p,i,r} \cdot q_{p,i,r}$ for each operation; define the utilization $U_{p,t,r}$ for each renewable resource r used for the project p; define a function of the current resource volume $Q_{t,v}$ dependence on the time t for each non-renewable resource v. Set $p = 1$;
2. For the project p, define the operations with index $i = 1,..,N'$, $N' \le N_p$, where the subcontracted resources cost $sr'_{p,i,r}$ exceeds the given critical value $\xi_{p,i,r}$. If $N' \neq 0$, then set $i = 1, \eta = 1$ and go to Point 3; otherwise, go to Point 6;
3. Define the competence of the resource r for the operation i of the project p. Highlight a time interval $[x_{p,i}; x_{p,i} + d_{p,i}]$ with a duration $c_{p,i,r}$;
4. If the interval $\delta_{p,i,r}(\eta)$ exists, then go to Point 7; otherwise, go to Point 5;
5. If $(i+1) \le N'$, then set $i = i + 1$ and go to Point 3; otherwise, go to Point 6;
6. If $(p+1) \le P$ then set $p = p + 1$ and go to Point 2; otherwise, go to Point 14;
7. Calculate the utilizations $U^-_{p,i,r}(\eta)$ and $U^+_{p,i,r}(\eta)$ for the intervals $\delta^-_{p,i,r}(\eta)$ and $\delta^+_{p,i,r}(\eta)$;
8. If $U^-_{p,i,r}(\eta) < U^+_{p,i,r}(\eta)$, then go to Point 9; otherwise, go to Point 10;
9. If $\tau^0_{p,i} \le (x_{p,i} - c_{p,i,r}\cdot(\eta+1))$ –constraint (10) check– then shift the operation's start as follows:

    $x_{p,i} = x_{p,i} - c_{p,i,r}\cdot(\eta+1)$ –and go to Point 1; otherwise, go to Point 5;

10. If $U^-_{p,i,r}(\eta) > U^+_{p,i,r}(\eta)$, then go to Point 11; otherwise, go to Point 12;
11. If $(x_{p,i} + c_{p,i,r}\cdot(\eta+1)) \le \tau^1_{p,i}$, then shift the operation's start as follows:

    $x_{p,i} = x_{p,i} + c_{p,i,r}\cdot(\eta+1)$ –and go to Point 1; otherwise, go to Point 5;

12. If $U^-_{p,i,r}(\eta) = U^+_{p,i,r}(\eta)$ and $U^-_{p,i,r}(\eta) \neq U^0_{p,i,r}(\eta)$, then go to Point 9; otherwise, go to Point 13;
13. If $(\eta + 1) \leq \Psi$, then set $\eta = \eta + 1$ and go to Point 7; otherwise, go to Point 5;
14. The end of the algorithm.

We assess the number of iterations of the IH algorithm.

The number Ψ of search steps is a parameter of the algorithm. For each step, the algorithm sequentially bypasses all operations of the project portfolio to identify bottlenecks and eliminate them.

The number of operations of the project portfolio is calculated as follows: $N = \sum_{p=1}^{P} N_p$. Thus, the complexity of the proposed IH algorithm linearly depends on the number of operations in the project portfolio according to the formula: $\Psi \cdot N$. The quality of the schedule found is defined based on the value of the function (8) when restrictions (9)–(10) are fulfilled.

7. Case Study Results

The modified IH scheduling method was implemented in the BPsim software package including the BPsim simulation system and BPsim decision support system [24] with a common database.

The BPsim simulation system is used to develop and simulate the MRCP model of the processes under investigation. The system supports graphical MRCP notation, where a user can identify the types of nodes, i.e., operations and agents, logical links between them, and list the available non-renewable and renewable resources.

The following parameters can be assigned to each operation: the duration, start condition, amount of non-renewable and renewable resources required to complete the operation, and amount of produced non-renewable resources. To each agent, the behavior rules can be described in a form of if-then rules and assigned. The agents can affect the amount of resources and operations' start conditions.

The BPsim decision support system includes decision search diagrams based on the same resources used to build the simulation model. Development of the diagrams is based on the UML sequence diagrams [25] and Transact-SQL database management language [26]. The decision search diagrams are used to provide a visual comparison for multiple alternative decisions utilizing implemented user rules. The diagrams were applied to implement the algorithm of the IH scheduling method.

A case study was used to assess the subcontracted resources cost depending on the current schedule of 10 projects with 35 operations in a project company. The company has its own renewable resources, i.e., a staff of eight people with different skills (competencies). The following competencies of the staff were defined: documentation design (three people), carrying out an installation work (four people), and material supply (one person). The non-renewable resources of the company are the construction objects. The operation duration varied from 6 days to 90 days depending on the type of operation. The scheduling time interval was 430 days.

Following the proposed IH scheduling method, a simulation model was developed in the BPsim simulation system. The model's inputs are the labor costs and the agreed start dates for each operation, as well as the schedule. Assessment of the subcontracted resources cost was produced by the model.

For the initial schedule, the model outcome is presented in Figure 6a. For the schedule provided by the IH method, the model outcome is presented in Figure 6b.

The figures contain the percentage utilization of the three own resources with different competence marked with the red, blue, and green lines. When the blue and red resources utilization reaches the 100% level, it means the subcontracted resources with the matched competence are used during this time. The green resource utilization does not exceed 30% in both figures, so subcontracted resources with the same competence are not required.

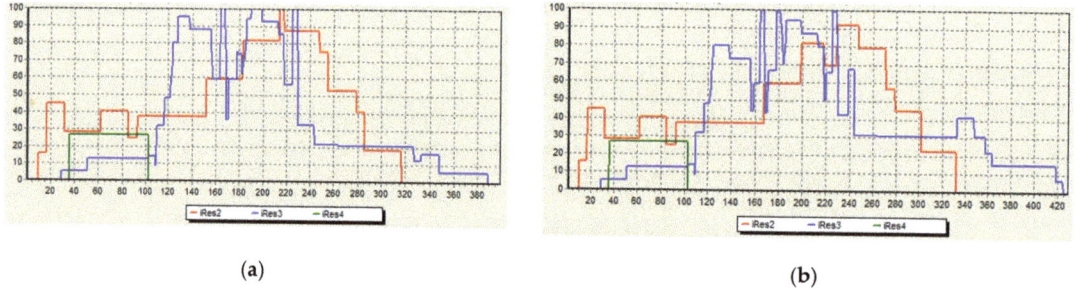

(a) (b)

Figure 6. Percentage utilization U of own resources iRes performing the project's portfolio depending on time: (**a**) Initial schedule; (**b**) Schedule provided by the IH method. Here, iRes2 is a resource with the documentation design competence, iRes3 is a resource with the competence to carry out installation works, and iRes4 is a resource with material supply competence.

The initial and the final schedules coincide up to 110 days due to the low resource utilization at the initial schedule during these days. Therefore, the IH algorithm has not shifted the operations' start day during the first 110 days in the final schedule.

Table 6 contains allocation of the subcontracted resources' cost in rubles and man per day (m/d) on the project's operations for the initial schedule.

Table 6. Subcontracted resources' cost allocation for the initial schedule.

Operation Name	Project 1	Project 3	Project 4	Project 5	Project 9
Development of technical specification in man per day (m/d)	3.9 m/d			1.68 m/d	
Installation and commissioning work (telemechanics) in m/d			2.45 m/d		
Installation and commissioning work (energy accounting) in m/d		11.54 m/d			8.53 m/d
Installation and commissioning work (telecommunications) in m/d			13.88 m/d		
Overall subcontract cost of the project in man per day	3.9 m/d	11.54 m/d	16.33 m/d	1.68 m/d	8.53 m/d
Overall subcontract cost of the project in rubles	10 249	95 990	65 998	6 746	70 953
Overall subcontract cost of the portfolio in m/d			41.98 man per day		
Overall subcontract cost of the portfolio in rubles			249,936 rubles		

For the initial schedule, the subcontracted resources are required for the projects 1, 3, 4, 5, and 9. The IH method allowed us to highlight several operations with a subcontracted resources cost exceeding the critical value equal to 60,000 rubles. The ones are installation works for projects 3 and 9 (95,990 and 70,953 rubles accordingly), and all operations for project 4 (65,998 rubles).

Table 7 contains allocation of the subcontracted resources' cost in rubles and m/d on the project's operations for the schedule provided by the IH method.

The IH method is proposed resource allocation and shifting of the operations' start date within a given timeframe. Although, the makespan of the found schedule was increased compared with the initial schedule, all operations and project deadlines were met.

It should be noted that the objective function considered is related to minimization of the subcontracted resources' cost while meeting restrictions. As a result, the subcontracted resources' cost of the schedule found was reduced compared with the initial schedule in terms of man per day from 42 m/d to 9 m/d and in terms of rubles from 249,936 rubles to 53,453 rubles for half a year, i.e., more than four times for both outcomes.

Table 7. Subcontracted resources' cost allocation for the schedule provided by the IH method.

Operation Name	Project 1	Project 3	Project 4	Project 5	Project 9
Development of technical specification in man per day (m/d)				1.68 m/d	
Installation and commissioning work (energy accounting) in m/d					4.04 m/d
Installation and commissioning work (telecommunications) in m/d			2 m/d		
Customer training in m/d	0.36 m/d	0.46 m/d			
Overall subcontract cost of the project in man per day	0.36 m/d	0.46 m/d	2 m/d	1.68 m/d	4.04 m/d
Overall subcontract cost of the project in rubles	946	3826	8330	6746	33,605
Overall subcontract cost of the portfolio in m/d			8.54 man per day		
Overall subcontract cost of the portfolio in rubles			53,453 rubles		

8. Discussions

The results of the scheduling methods' comparison are shown in Table 8.

Table 8. The results of the scheduling method comparison.

Method	Subcontracted Resources Accounting	Subcontracted Cost Optimization, (8)	Non-Renewable Resources Accounting, (9)	Deadlines Accounting (10)	Earliest Start Time Accounting (10)	Multi-Objective Optimization
Y. A. Mezentsev method [4]	No	No	No	No	Yes	No
V. S. Tanaev method [5]	Yes	No	No	Yes	No	No
Lingo based method [14]	Yes	Yes	No	No	No	No
Agent-based method [7]	Yes	No	Yes	Yes	Yes	Yes
GA-based method [9]	No	No	Yes	No	No	No
OptQuest based method [11]	No	No	No	Yes	No	Yes
Plant Simulation method [13]	No	No	Yes	Yes	No	Yes
IH method	Yes	Yes	Yes	Yes	Yes	No

As we can see from Table 8, all the methods have disadvantages when solving the project scheduling problem considered. The problem has the following features: search for a solution with the minimum subcontracted resources cost, large dimension of the search space, presence of the operations start time interval, and allocation of renewable and non-renewable resources with a restricted non-renewable resources lifetime. These features restrain the application of the scheduling theory methods considered. The Y. A. Mezentsev and V. S. Tanaev methods solve the scheduling problem while minimizing the makespan by tight packing of operations in compliance with part of the restrictions. However, the issues of attracting and optimizing the subcontracted resources cost are not given due attention. The Lingo-based method deals with subcontracting cost optimization, but the schedule found does not satisfy any constraint. The agent-based method lacks subcontracted resource optimization but considers all the restrictions, including the non-renewable resources one. The GA-based method and the commercial solver-based methods, OptQuest and GA of the Plant Simulation, do not consider the subcontracted resources and orders' earliest start times while the GA-based method accounts for restricted non-renewable resources.

Application of simulation multi-agent modeling with heuristic rules of the resource's allocation allows us to consider all features of the scheduling problem except multi-objective optimization. The HI method optimizes the subcontracted resources cost taking into account restricted renewable and non-renewable resources and orders' earliest start times. At the same time, a project duration may increase wherein the orders deadlines are met.

For the scheduling problem with 35 operations, eight renewable resources and one type of non-renewable resource, the running time of the HI algorithm is estimated at 6 min. This time is comparable to the running time of the methods based on agents, OptQuest, and Plant Simulation application. The HI algorithm running time is much more than the one of the methods proposed by Y. A. Mezentsev and V. S. Tanaev as well as the Lingo- and GA-based ones, estimated at tens of seconds. It should be noted that the search time for the simulation-based methods is always more than the search time for the methods based on the optimization algorithms. When simulating, a time for conducting one experiment varies from a few seconds to several minutes depending on the model dimension and simulation tool used. In case of solving the optimization problem, it is necessary to search for solutions in a search space while the alternative solution is estimated by conducting an experiment with the model; therefore, the total search time increases to tens or even hundreds of minutes. Nevertheless, simulation-based methods have the advantage of being able to determine the objective function and the constraints required without a reduction to a specific mathematical model used in optimization algorithms.

We also applied the HI algorithm to the scheduling problem with a large data set. A construction holding was considered with 302 building operations, 119 renewable resources or construction machinery, and 161 types of non-renewable resources or construction material. The detailed holding description is given in [27]. For the problem with a large data set, the running time of the HI algorithm is estimated at one hour and 38 min. The given duration of the algorithm operation is rather long compared to the obtained one for the small dimension problem but is acceptable for decision making that is not in real time.

9. Conclusions

The scheduling problem was formulated with the objective function of minimization of the subcontracted resources cost, presence of restrictions on non-renewable resources, and operations' earliest start time and due dates. Analysis of the different scheduling methods based on the scheduling theory, optimization algorithms, and agent-based simulation was conducted. The analysis revealed factors preventing application of the methods to the problem under consideration. The factors include lack of a search for the optimal allocation of renewable resources on operations in terms of minimizing the subcontracted resources' cost and lack of accounting for limited non-renewable resources. The given factors indicate the relevance of the development of the hybrid scheduling method based on integration of simulation and heuristic modeling.

The IH method algorithm was developed based on the multi-agent simulation model of the resource's allocation with operations performing and heuristic rules of the operations start days shifting. The IH method is used to search the schedule that meets the time and resources restrictions and has a minimum subcontracted resources cost.

A case study was conducted to assess the subcontracted resources cost for the real scheduling problem of a project company. Application of the IH method allowed us to compose a schedule that reduced the company's waste on subcontracted resources by more than four-fold for half a year compared with the schedule provided by decision makers based on their knowledge and experience.

Comparison of the IH method and the other scheduling heuristic methods was performed. The conditions were identified, under which the new IH method is more effective than the other ones. The conditions include a focus on optimizing the project portfolio cost with a fixed portfolio duration. In future, the authors plan to refine the IH method for solving the multicriteria problem of finding a schedule that is optimal in terms of the makespan and the renewable resources cost.

Author Contributions: Conceptualization, A.A. and K.A.; Methodology, A.A.; Software, K.A. and O.A.; Validation, O.A.; Formal Analysis, K.A.; Investigation, A.A. and O.A.; Writing—Original Draft Preparation, A.A.; Writing—Review and Editing, A.A.; Visualization, O.A.; Supervision, K.A.; Project

Administration, O.A.; Funding Acquisition, K.A. All authors have read and agreed to the published version of the manuscript.

Funding: This research was funded by Act 211 Government of the Russian Federation, contract no. 02.A03.21.0006.

Institutional Review Board Statement: Not applicable.

Informed Consent Statement: Not applicable.

Data Availability Statement: Not applicable.

Conflicts of Interest: The authors declare no conflict of interest.

References

1. Clark, C.E. The PERT model for distribution of an activity time. *Oper. Res.* **1962**, *10*, 405–406. [CrossRef]
2. Kannan, R. Graphical evaluation and review technique (GERT): The panorama in the computation and visualization of network-based project management. *Adv. Secur. Comput. Internet Serv. Appl.* **2014**, *9*, 165–179. [CrossRef]
3. Gimadi, E.K.; Goncharov, E.N.; Mishin, D.V. On some implementations of solving the resource constrained project scheduling problems. *Yugosl. J. Oper. Res.* **2019**, *29*, 31–42. [CrossRef]
4. Mezentsev, Y.A.; Estraykh, I.V.; Chubko, N.Y. Implementation of an efficient parametric algorithm for optimal scheduling on parallel machines with release dates. *J. Phys. Conf. Ser. Inf. Technol. Bus. Ind.* **2019**, *1333*, 022002. [CrossRef]
5. Tanaev, V.S.; Shkurba, V.V. *Introduction to the Scheduling Theory*; Science: Moscow, Russia, 1975.
6. Guizzi, G.; Vespoli, S.; Grassi, A.; Carmela Santillo, L. Simulation-based performance assessment of a new job-shop dispatching rule for the semi-heterarchical industry 4.0 architecture. In Proceedings of the 2020 Winter Simulation Conference (WSC), Orlando, FL, USA, 14–18 December 2020; pp. 1664–1675. [CrossRef]
7. Skobelev, P.; Zhilyaev, A.; Larukhin, V.; Grachev, S.; Simonova, E. Ontology-based open multi-agent systems for adaptive resource management. In Proceedings of the 12th International Conference on Agents and Artificial Intelligence, Valletta, Malta, 22–24 February 2020; pp. 127–135. [CrossRef]
8. Goncharov, E.; Leonov, V. Genetic algorithm for the resource-constrained project scheduling problem. *Autom. Remote Control* **2017**, *78*, 1101–1114. [CrossRef]
9. Xie, L.; Chen, Y.; Chang, R. Scheduling optimization of prefabricated construction projects by genetic algorithm. *Appl. Sci.* **2021**, *11*, 5531. [CrossRef]
10. Guizzi, G.; Revetria, R.; Vanacore, G.; Vespoli, S. On the open job-shop scheduling problem: A decentralized multi-agent approach for the manufacturing system performance optimization. In Proceedings of the 12th CIRP Conference on Intelligent Computation in Manufacturing Engineering, Gulf of Naples, Italy, 18–20 July 2018; pp. 192–197. [CrossRef]
11. Alvandi, S. Energy efficiency improvement through optimal batch sizing in job shop. *Mod. Appl. Sci.* **2020**, *14*, 6–19. [CrossRef]
12. Ištoković, D.; Perinić, M.; Doboviček, S.; Bazina, T. Simulation framework for determining the order and size of the product batches in the flow shop: A case study. *Adv. Prod. Eng. Manag.* **2019**, *14*, 166–176. [CrossRef]
13. Yu, H.; Han, S.; Yang, D.; Wang, Z.; Feng, W. Job shop scheduling based on digital twin technology: A survey and an intelligent platform. *Hindawi Complex.* **2021**, 1–12. [CrossRef]
14. Biruk, S.; Jaśkowski, P.; Czarnigowska, A. Minimizing project cost by integrating subcontractor selection decisions with scheduling. *IOP Conf. Ser. Mater. Sci. Eng.* **2017**, *245*, 072007. [CrossRef]
15. Pinedo, M.L. *Scheduling: Theory, Algorithms, and Systems Development*; Springer: Berlin/Heidelberg, Germany, 2008. [CrossRef]
16. Wooldridge, M.; Jennings, N. Intelligent agent: Theory and practice. *Knowl. Eng. Rev.* **1995**, *10*, 115–152. [CrossRef]
17. Goldberg, D.; Holland, J.H. Genetic algorithms and machine learning. *Mach. Learn.* **1988**, *3*, 95–99. [CrossRef]
18. OptQuest, the Official Web Site. Available online: https://www.opttek.com/products/optquest/ (accessed on 19 August 2021).
19. Anylogic Simulation Software, the Official Web Site. Available online: https://www.anylogic.com/ (accessed on 19 August 2021).
20. Optimize Production Logistics & Material Flow. Siemens Tecnomatix Plant Simulation, the Official Web Site. Available online: https://www.plm.automation.siemens.com/global/en/products/tecnomatix/logistics-material-flow-simulation.html (accessed on 19 August 2021).
21. Antonova, A.S.; Aksyonov, K.A. Analysis of the methods for accounting the renewable and non-renewable resources in scheduling. In Proceedings of the 7th ITTCS International Young Scientists Conference on Information Technology, Telecommunications and Control Systems, Innopolis, Russia, 17–18 December 2020. [CrossRef]
22. LINGO 19.0—Optimization Modeling Software for Linear, Nonlinear, and Integer Programming. Available online: https://www.lindo.com/index.php/products/lingo-and-optimization-modeling (accessed on 19 August 2021).
23. Aksyonov, K.; Bykov, E.; Aksyonova, O.; Goncharova, N.; Nevolina, A. The architecture of the multi-agent resource conversion processes. In Proceedings of the UKSim 11th European Modelling Symposium on Mathematical Modelling and Computer Simulation, Manchester, UK, 20–21 November 2017; pp. 61–64. [CrossRef]

24. Aksyonov, K.; Aksyonova, O.; Antonova, A.; Aksyonova, E.; Ziomkovskaya, P. Development of cloud-based microservices to decision support system. In Proceedings of the International Conference on Open Source Systems, OSS 2020, Innopolis, Russia, 12–14 May 2020; pp. 87–97. [CrossRef]
25. The official UML Web Site. Available online: http://www.uml.org (accessed on 19 August 2021).
26. SQL Server language reference. Available online: https://docs.microsoft.com/en-us/previous-versions/sql/sql-server-2005/ms189826(v=sql.90) (accessed on 19 August 2021).
27. Aksyonov, K.A.; Bykov, E.A.; Aksyonova, O.P.; Wang, K. Application of BPsim.DSS system for decision support in a construction corporation. *Appl. Mech. Mater.* **2013**, *256–259*, 2886–2889. [CrossRef]

Article

Non-Sequential Linear Construction Project Scheduling Model for Minimizing Idle Equipment Using Constraint Programming (CP)

Shu-Shun Liu [1,*], Agung Budiwirawan [2,3] and Muhammad Faizal Ardhiansyah Arifin [2,3]

1. Department of Civil and Construction Engineering, National Yunlin University of Science & Technology, Yunlin 640, Taiwan
2. Graduate School of Engineering Science and Technology, National Yunlin University of Science & Technology, Yunlin 640, Taiwan; d10210034@yuntech.org.tw (A.B.); d10610211@gemail.yuntech.edu.tw (M.F.A.A.)
3. Department of Civil Engineering, Universitas Negeri Semarang, Semarang 50229, Indonesia
* Correspondence: liuss@yuntech.edu.tw; Tel.: +886-908-073-815

Abstract: Over the last several decades, the scheduling of linear construction projects (LCPs) has been explored extensively by experts. The linear scheduling method (LSM), which focuses on work rate and work continuity, has the advantage of tackling LCPs' scheduling problems. The traditional LSM uses work continuity to monitor resource allocation continuity on the premise that activities with the same type of work use the same crew. However, some LCPs require a combination of different types of equipment to comprise the crew. Sometimes, parts of different crews require the same types of equipment, and sometimes, the same crew requires different equipment configurations. This causes the pattern of work continuity to be different from the pattern of resource allocation continuity. Therefore, we propose an optimization model of the LSM to minimize idle equipment on a non-sequential linear construction project—i.e., a road network maintenance project. This model is intended to minimize the number of idle equipment and their idle time to achieve more efficient scheduling for linear construction projects. This model offers novel details of resource allocation continuity assessment by taking into account equipment combination and configuration (ECC). Therefore, the scheduling concept used by the proposed model is named the linear scheduling model with ECC (LSM–ECC). The model was developed using constraint programming (CP), as CP has good performance and robustness in the optimization field. The model was implemented to a representation of a road network maintenance project and has satisfactory results.

Keywords: linear project; linear scheduling method; equipment idleness; constraint programming; equipment combination and configuration

Citation: Liu, S.-S.; Budiwirawan, A.; Arifin, M.F.A. Non-Sequential Linear Construction Project Scheduling Model for Minimizing Idle Equipment Using Constraint Programming (CP). *Mathematics* **2021**, *9*, 2492. https://doi.org/10.3390/math9192492

Academic Editors: Frank Werner and Armin Fügenschuh

Received: 18 August 2021
Accepted: 28 September 2021
Published: 5 October 2021

Publisher's Note: MDPI stays neutral with regard to jurisdictional claims in published maps and institutional affiliations.

Copyright: © 2021 by the authors. Licensee MDPI, Basel, Switzerland. This article is an open access article distributed under the terms and conditions of the Creative Commons Attribution (CC BY) license (https://creativecommons.org/licenses/by/4.0/).

1. Introduction

1.1. Background

A construction project consists of a set of activities related to each other to achieve the project's objectives [1]. Achieving the project's objectives in the life cycle of the construction project often faces many management problems. One of the prevalent problems is resource availability. Resources are often limited and expensive. Thus, resource management plays an essential factor in project scheduling [2]. Managing resources—equipment, materials, and crews—efficiently in satisfying all project activities becomes an inevitable aspect for construction managers in controlling the project schedule to achieve the project's goals.

The efficiency of resource management is not only just ensuring resources in satisfying all the activities, but also optimizing the resource usages. Meanwhile, ensuring project running on schedule with limited resources makes resource management more difficult. Therefore, a detailed and thorough resource scheduling that aligns with the activity schedule is a crucial necessity in construction project scheduling [3,4].

A linear construction project (LCP) is a challenge for a project manager to schedule efficiently. This type of project shares the resources in different spaces either in a sequential or parallel manner. The manager needs to allocate enough resources for maintaining work continuity in various locations of a project, minimizing idle resources, and finishing the project within contract duration [5,6]. Sometimes a conflict of resource usage occurs during the execution of LCPs because the movement of activities' resources is not well planned upon limited shared space [3,7]. Consequently, the schedule developed for LCPs not only accounts for precedence constraints of activities but also considers the space and time constraints of the movement of the activity resource [8]. The critical path method (CPM) emphasizes scheduling based on precedence relationships between activities [9] and is less able to provide monitoring of work continuity for LCPs. On the other hand, the linear scheduling method (LSM) can provide good monitoring of work continuity, so that the LSM is more suitable for scheduling linear construction projects.

One way to achieve the desired scheduling efficiency is to minimize idle resources. Idle resources are defined as resources that have been mobilized but not utilized. Idle resources are unproductive resources, but these idle resources still incur costs, e.g., maintenance costs, rent, and depreciation. By minimizing the amount and time of idle resources, it will minimize unproductive costs which ultimately results in more efficient scheduling.

A road network maintenance project consists of the same type of road maintenance being applied to several road sections. This characteristic includes the road network maintenance project as an LCP. Therefore, an LSM is suitable to be applied as a scheduling method. However, road network maintenance projects have different characteristics compared with regular LCPs to be handled by traditional LSM.

Traditional LSM uses crews as resources used to complete works. The efficiency of linear construction project scheduling is seen from the continuity of work by assuming that the same type of work uses the same crew, while different types of work use different crews. This cannot be applied to road network maintenance projects, which use a combination of heavy equipment as resources to complete their activities. Using the term crew from traditional LSM, the crew used to complete an activity is a set of equipment consisting of several types of equipment as sub-crew and divided into main equipment and supporting equipment. Sometimes, the supporting equipment needed to complete one type of activity is also needed as support equipment to complete another type of activity. Therefore, the profile of work continuity becomes different from the profile of equipment allocation continuity. Based on this condition, we propose a concept of resource allocation dividing the crew into smaller units—sub-crew—which is called equipment combination and configuration (ECC). A more in-depth explanation of ECC is in Section 3.1. Based on the linear scheduling method with ECC concept (LSM–ECC), the authors propose an optimization model for linear construction project scheduling by minimizing idle resources at the sub-crew level, i.e., equipment.

From the point of view of the execution sequence of the repetitive units, traditional LSM requires engineers to define the execution sequence of the repetitive units of an LCP. For sequential/serial linear construction projects, engineers do not have much trouble deciding which repetitive units to begin with. However, the sequence of maintenance work on each road segment of a road network maintenance project is flexible because usually, each road segment has almost the same accessibility. Therefore, engineers have to think about which road segment to work on first to achieve an efficient schedule. To solve this problem, an optimization model that can provide engineers with suggestions for the execution sequence of each road segment to obtain efficient scheduling is needed, even suggestions for non-sequential/parallel execution if possible.

1.2. Objectives

Based on the aforementioned conditions, the authors propose an optimization model to minimize idle equipment for non-sequential linear construction projects. This model is expected to monitor the continuity of resource allocation at the sub-crew level, i.e., equipment, and minimize the equipment idleness.

This study aims to develop an optimization model of a non-sequential linear construction project to minimize idle equipment using constraint programming (CP). This optimization model is expected to be able to: (1) monitor the mobilization and allocation of equipment; (2) minimize idle equipment in quantity and time; (3) provide the engineer with suggestions on the work sequence of the project's repetitive unit.

Constraint programming (CP) was used to build this model because CP has several advantages, namely: CP defines a model built using objective functions and a set of constraints without having to define procedures and calculation steps; and logical constraints in CP are easy to define with the help of an optimization programming language (OPL) compared to ordinary mathematical models.

1.3. Paper Structure

The remainder of this paper is structured as follows: Section 2 discusses some previous studies related to scheduling theory, non-sequential linear scheduling, as well as resource allocation and resource-leveling. Section 3 presents the material and method conducted in this study. Section 4 discusses the result of the proposed model applied in three different scenarios. Lastly, Section 5 presents the general conclusion and the opportunity for future research related to this study.

2. Literature Review

There are two main categories of construction projects, namely repetitive and non-repetitive construction projects. A repetitive project consists of multiple repetitive units and requires timely movement of construction resources from one unit to the next unit to repeat the same activities [5]. Repetitive projects can also be classified as typical and non-typical. Activities in a typical repetitive project have the same crew productivity rates that are repeated on different repetitive units. Most of the real construction projects are adjacent to non-typical repetitive projects, which have different repeated productivity rates on different repetitive units [10]. Breaking the continuity of the same activities between repetitive units creates work gaps that cause idle resources and bring additional costs [11]. Hence, maintaining the work continuity in repetitive projects then will ensure constant usage of construction resources and minimizing equipment idle time [10].

Repetitive projects can be divided into linear and non-linear projects according to the linear geometric pattern [12]. In terms of linear projects, this type has repetitive units—a sequence of construction activities [13]. Linear construction projects include characteristics as a series of linear repetitive activities, such as railways, highways, pipelines, and tunnels, while high-rise buildings with typical floors and typical housing projects are considered non-linear repetitive projects [12].

One of the methods that is often used in scheduling is the network-based scheduling method. Critical path method (CPM) and project evaluation and review technique (PERT) are examples of network-based methods [14]. The network-based scheduling method focuses on scheduling based on the precedence relationship between activities. Because this method has a strong definition of the precedence relationship between activities, the schedule of activity—i.e., start time, finish time, etc.—could be calculated easily, and this method is suitable for automatic schedule calculation, such as using computer programs. However, this method does not show the work rate and work continuity of activities. The resource-driven scheduling method focuses on work performance and continuity. Line of balance (LOB) and the linear scheduling method (LSM) are examples of resource-driven scheduling methods.

In construction projects, resources are divided into two main categories, namely renewable resources and non-renewable resources. A renewable resource is a resource that can be repeatedly utilized without replenishment, and a non-renewable resource is a one-time consumable resource and the usage of this resource cannot be repeated [15]. This study is focused on managing renewable resources, especially construction equipment at a project level. Based on these resource management problems and the crucial resource management role, this study attempts to overcome these challenges from the perspective of the repetitive linear construction project.

The main reason this study utilized LSM is that the network-based scheduling method has great drawbacks in the application of linear construction projects since the network planning methods is difficult in ensuring the work continuity of a linear construction project and leads to a greater risk of idle time of the renewable resource [16]. In the network planning method, more repetitive activities will lead the network growth and make scheduling visualization intricate [17]. Additionally, network schedules are only able to provide a one-dimensional graph in terms of their informational content, which solely shows how sequentially connection activities occur upon a time [18].

LSM depicts the construction schedule of a linear construction project by a rectangular coordinate according to the characteristics of a linear construction project with the horizontal and vertical axes representing the spatial position and schedule of a project, respectively [19]. The two-dimensional coordinate system in LSM broadens the scope of information that can be communicated including the key elements inside the system, for instance, activities, rate of activities, and buffer between activities are employed for illustrating the project schedule then the LSM diagram will be formed [14].

According to the spatial location of the activity, a linear-type activity can be categorized into two types, namely full-span and partial-span linear activity [20]. The major characteristic of a linear type activity in LSM is a rate concept that denotes the spatial progress of the linear activity in unit time, and this concept becomes a differentiation feature compare to the critical path method (CPM) [21]. The slope of a linear-type activity in the LSM diagram indicates the rate of that activity. The slope of linear-type activity can be varied in proportion, depending on the resource usage of the activity. Thus, the accuracy in developing a linear schedule is extremely dependent on the capacity production of the activity resources [22].

The distance between two activities in the horizontal and vertical directions in the LSM diagram is, respectively, named as the distance and time buffer [23]. The buffer in the LSM concept depends on the technical constraints, managerial policy, or other conditions. Furthermore, the minimum/maximum time buffer is defined as the minimum/maximum time between two activities; similarly, the minimum/maximum distance buffer is defined as the minimum/maximum distance between two activities. Commonly the minimum buffer is easy to fulfill; however, in some cases, a maximum buffer needs to interrupt the activity or adjusting its productivity [14].

LSM has a similar concept for controlling activity path (CAP), such as a critical path in a network-based scheduling method. The float activities in the network scheduling method exist in the non-critical path after the critical path is calculated. A similar concept in LSM called the rate float for the float of non-controlling activities or non-controlling segments of activities in the schedule created by LSM appeared while the CAP has been established [2,24]. The number of possible changes in the production rate for a non-controlling linear activity can be specified by the rate float before the non-controlling linear activity becomes a controlling activity. In other words, the rate float is also defined as "the difference between the planned production rate of an activity and the lowest possible production rate without interfering in the buffer" [24]. By shifting non-controlling activities on the available rate float, the LSM model can adjust resource allocation and minimize resource fluctuations to obtain the resource leveling model without changing the original duration schedule [19,25].

Among those aforementioned examples of linear construction projects and challenges of resource management in linear construction projects, this study is focused on scheduling equipment in road network maintenance projects. There are only a few studies that discuss road network maintenance at the project level with detailed renewable resource scheduling. Mizutani et al. [26] proposed an optimal solution for pavement repair that considers work zone policies at the highway network level. Huang and Lin [27] proposed an arc routing problem approach to solve construction machinery schedules for road maintenance. Aarabi and Batta [28] proposed scheduling for pothole repair using a vehicle routing problem without focusing on machine management scheduling. Research focused on using the linear scheduling method (LSM) for highway construction projects has been performed in prior studies [5,20,29–34]. However, none of those studies have discussed the detail of equipment scheduling, considering equipment idleness in the concept of equipment combination and configuration, as well as mobilization and demobilization of the machinery.

Two leading concepts are associated with how to manage project resources, namely resource allocation and resource leveling [2,35]. The concept of resource allocation is to reschedule the project activity to efficiently manage the limited resources by allowing to exceed project duration planning as minimum as possible [36]. While the concept of resource leveling is to make the resource usage curve during the construction project as flat as possible so it can assist in avoiding short-term peaks and troughs, reduce resource costs and management costs, as well as avert needless losses by keeping the original project duration [36–38]. Accordingly, both of these concepts deal with two dissimilar resource sub-problems that can solely be utilized to a project one after the other rather than simultaneously.

Referring to the basic concept of resource leveling, in smoothing the histogram of resource profile to be as flat as possible, it brings an exact deployment of resources within a project that can minimize renewable resource cost [39]. However, considering the complex mixture of activity relationships in the scheduling, the objective function on resource leveling can be nonlinear and makes the graph that represents the resource profiles become extremely discontinuous. Thus, a small change in resource consumption on activity may create a vast change in the resource profile. According to these advantages of the resource leveling concept and the challenge to solve resource leveling problems, some previous studies have applied various approaches to solving resource leveling problems on the network schedules or linear schedules. Using the definition of buffer and the concept of rate float in the LSM schedule, some previous studies have attempted to resolve the problem of resource leveling in construction projects. Lucko [39] utilized a singularity function in LSM to optimize the resource leveling profile while considering the resource rate changes. Tang et al. [38] present linear scheduling of a railway construction project to level the resource profile for optimum resource usage. Tang et al. [19], proposed a two-stage scheduling system model and algorithm for linear construction project resource leveling to automatically generate a linear schedule including the resource leveling; this study performs according to the example data of highway construction project conducted by Matilla and Abraham [24]. Su and Lucko [17] proposed a combination of LSM and LOB to optimize multiple crew scheduling within and between repetitive activities with singularity functions. By employing new constraints in LSM, namely total resource constraint, resource utilization constraint, and construction mileage constraints, Wang et al. [40] strove to maximize the space–time flexibility of construction activities and optimize the LCP's resource-leveling. Esfahan et al. [21] considered equipment congestion in road construction proposed a space–time float concept for optimizing the resource scheduling of an LCP. Damci et al. [2], in the framework of LOB, present multi-resource leveling optimization with the principle of optimum crew size and natural rhythm. Ipsilandis [11] adopted the resource leveling concept to minimize project duration or to minimize resource work breaks in linear repetitive projects.

Sometimes a resource allocation problem can be called a resource-constrained project scheduling problem (RCPSP), since the main concept of the resource allocation model is developed to solving resource conflicts by rescheduling activities while minimizing the additional project duration [41]. The main consideration of traditional RCPSP is how to deal with a set of n activities needed to schedule and to minimize a project's completion time and meet two main constraints: (1) the precedence constraint, and (2) the limited availability of resources [13]. RCPSP and its variants have been extensively investigated by researchers during the last several decades, since the pioneering work of experts [6–8] about mathematical programming formulations of scheduling problems.

Traditional RCPSP uses given and normally constant resource allocations throughout each activity. Different from the traditional RCPSP, Fündeling and Trautmann [42] proposed a model in which resource requirements and resource allocations must be determined. This resulted in a different "work profile", which was not limited to a rectangular shape as the traditional RCPSP has been, and "work content", which was defined as the total amount of resource required to finish an activity. For more general uses, "resource profile" and "resource requirement" are used instead of "work profile" and "work content", respectively, since resources are not restricted only to human resources.

Recent computer technology has opened many opportunities in solving large-scale and difficult mathematical models efficiently. Taking advantage of this condition, new mathematical models for RCPSP have been formulated and extensively compared by experts [25–27]. Naber and Kolisch [43] proposed four discrete-time model formulations of a resource-constrained project scheduling problem with flexible resource profiles (FRCPSP) and compared the model efficiency in terms of solution quality and computational time. These models used decision variables based on previous research [26,28–31] to achieve the shortest project completion time. Leu and Hwang [44] consider using resource sharing in repetitive precast production proposed optimization schedule based on the LOB method. Liu and Wang [45] proposed a resource allocation optimization model in an LCP by employed constraint programming (CP). Zhang et al. [46], in the framework of the line of balance (LOB) method, focus on the learning effect to minimize total resource usage while satisfied all of the demands of work continuity and the target deadline of every activity. Hyari and El-Rayes [5] simultaneously minimize project duration and maximize crew work continuity in bridge construction utilizing the LSM method by considering typical and non-typical repetitive activities. Kong and Dou [47] solve resource-constrained project scheduling problems under multiple time constraints that include a duration constraint of activity, temporal constraint, and resource calendar constraint.

By the nature of resource leveling characteristics, this type of resource management concept is more relevant to linear scheduling. Nevertheless, some previous research has already attempted to combine the resource-leveling concept and resource allocation concept in one single scheduling optimization model. For example, Hegazy employed a genetic algorithm (GA) technique [36], Jun and El-Rayes [41] developed a multi-objective optimization model based on a GA module, Koulinas and Anagnostopoulos utilized bi-objective models [35], Francis Siu et al. applied an integer linear programming technique [48], and Khanzadi et al. [49] utilized a colliding body optimization (CBO) algorithm and charged system search (CSS) technique. Tang et al. [50] solve scheduling optimization problems in transportation-type linear construction projects using a constraint programming (CP) technique.

Total project duration in this study after the optimization process should be the same as with the original duration or can be shorter than the original duration. Thus, this characteristic makes this study adopt the properties of resource-leveling concepts. Moreover, minimizing equipment idleness has a similar concept to make resource usage histogram as flat as possible. However, on the other side, the objective of minimizing equipment resource idleness must also consider the available number of the items of equipment, where this condition has similar properties with resource allocation concepts. Therefore, this study simultaneously utilized both the concept of resource leveling and resource allocation to solve equipment management problems on a road network maintenance project.

This study applied a combination of different types of equipment in one single work crew to serve several activities. Therefore, the resource allocation problem in this study will also adopt the resource sharing concept to maximize the available resource, because, in construction projects, the concept of resource sharing is suitable when meeting the condition of resource shortage [51]. The concept of resource sharing means to work with greater efficiency or produce extra benefits by using finite resources [52].

Although those previous studies perform a combination concept of resource leveling and resource allocation, none of those previous studies consider the work zone safety during the highway network maintenance by minimizing the lag time between activities and utilizing flexible resource profile in the framework of non-sequential LSM to solve resource idleness problems. Furthermore, none of those studies were able to provide a single group crew (equipment fleet) that combine two different types of equipment, called main equipment and supporting equipment. The supporting equipment will be applied as resource sharing to serve two activities. These will be the key techniques of this study in solving management renewables resource problems on a highway maintenance project.

Table 1 denotes the most relevant research prior to the proposed model.

Table 1. List of the most relevant researches prior to this proposed model.

Research	Objective	Resource	Activity Duration	Method
[2]	Resource leveling	Multiple types of resource.	Production rate and duration are based on the resource which requires longest time.	Genetic algorithm (GA)
[44]	Multi objective (production duration, resource amount, minimum makespan)	Multiple types of resources.	Activity duration is based resource allocation.	Genetic algorithm (GA)
[38]	Resource leveling	The type resource is implicitly represented by the type of work.	Duration is based on resource's production rate.	Constraint programming (CP)
[40]	Resource leveling	Multiple types of resources used by an activity.	Duration is based on resource's production rate.	Quantum-behaved particle swarm optimization (QPSO)
[50]	Resource leveling	The type resource is implicitly represented by the type of work.	Duration is based on resource's production rate.	Constraint programming (CP)
Proposed model	Resource idleness minimization	Multiple types of resources used by an activity.	Activity duration is based resource allocation and equipment combination and configuration (ECC).	Constraint programming (CP)

LSM, work continuity, and resource allocation have been studied extensively by experts in recent decades. However, as far as the author knows, to obtain the efficiency of resource allocation as an objective function, previous studies have always used the concept of resource leveling, which minimizes the deviation of resource allocation from a certain reference line or minimizes daily resource allocation changes. In contrast, the proposed model does not use the traditional resource leveling concept. Alternatively, to achieve resource allocation efficiency, the proposed model minimizes the deviation of resource allocation with mobilized resources, as indicated in Figure 1. In addition, this model also introduces the concept of equipment combination and configuration (ECC), which, to some degree, is almost the same as the concept of shareable resources. The concept of ECC will be discussed in the ECC section.

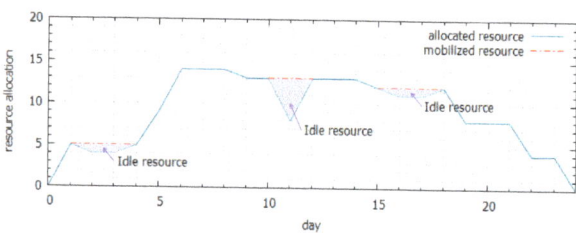

Figure 1. Idle equipment definition.

3. Materials and Methods

3.1. Model Concept

The optimization model of linear scheduling for minimizing idle equipment was developed based on a road-network maintenance project, hereinafter referred to as the project. A representation of a road-network maintenance project—the object of this study—consists of five repetitive units (road segments). Each road segment consists of asphalt stripping activity, asphalt resurfacing activity, and road marking activity (Figure 2).

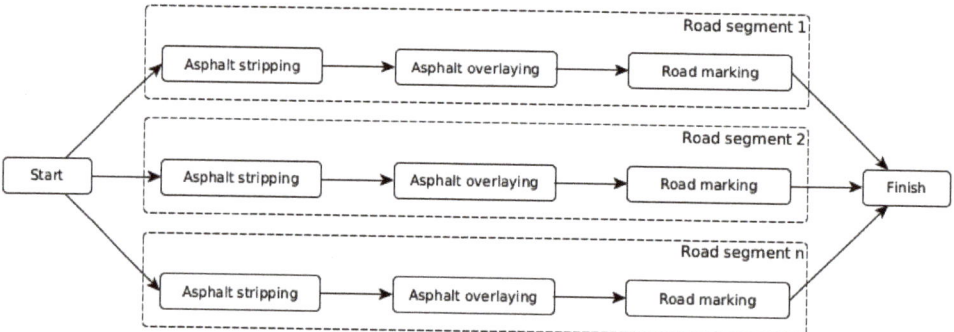

Figure 2. Road-network maintenance project.

3.1.1. Project Characteristics

A road-network maintenance project relies on heavy equipment to finish its activities. For example, an asphalt stripping activity uses an asphalt-milling machine (AM) to grind and strip the existing asphalt layer. Thus, in this optimization model, the schedule of activities is represented as the schedule of the main equipment used by related activity, and vice versa. Table 2 denotes the main equipment used by each activity type in the project.

Table 2. Main equipment used by activities.

Activity Name	Main Equipment
Asphalt stripping	Asphalt milling machine (AM)
Asphalt resurfacing	Asphalt finisher machine (AF)
Road marking	Road marking machine (ME)

3.1.2. Equipment Combination and Configuration (ECC)

Traditional LSM uses the term 'crew' to describe the resources used to do work. Therefore, in traditional LSM, the continuity of resource allocation can be represented as a continuity of work, and vice versa. However, at the project site, the crew consists of several types of interdependent resources, hereinafter referred to as sub-crew, so that the continuity of one type of work is not always the same as the allocation continuity of the related sub-crew.

In the road network maintenance project, a crew needed to finish a work consists of several sub-crews—i.e., equipment. The main equipment requires supporting equipment to work properly. Asphalt stripping activity requires AM as the main item of equipment to peel off the existing road surface. The material resulting from the peeling of the road surface must be disposed of using dump trucks (DT). Because the capacity and duty cycle between AM and DT are different, each unit of AM requires several units of DTs; in this case, for example, one unit of AM requires five units of DTs. Such an arrangement is proposed by this study as equipment combination and configuration (ECC). Table 3 denotes the ECC applied to this project.

Table 3. Equipment combination and configuration (ECC) for each crew.

Activity Name	Crew	Main Equipment	Supporting Equipment
Asphalt stripping	1	Asphalt milling machine (AM)	5 Dump trucks (DT)
Asphalt resurfacing	2	Asphalt finisher machine (AF)	4 Dump trucks (DT), 2 pneumatic rollers (PR)
Road marking	3	Road marking machine (ME)	-

ECC creates a different situation compared to traditional LSM, which uses work continuity to assess scheduling efficiency. Figures 3 and 4 depict the ECC schema and the resource allocation monitoring scheme, respectively.

Dump trucks serve as supporting equipment for Crew 1 for asphalt stripping activity and also as supporting equipment for Crew 2 for asphalt overlaying activity (Figure 3). From a traditional LSM perspective, asphalt stripping activity and asphalt overlaying activity use different resources, namely Crew 1 and Crew 2, so that the continuity of the allocation of Crew 1 and Crew 2 is in line with the continuity of asphalt stripping activity and asphalt overlaying activity, respectively. However, from the ECC perspective, dump trucks are allocated to asphalt stripping activity and asphalt overlaying activity, so that the continuity of dump truck allocation follows the scheduling of these two types of activities (Figure 4).

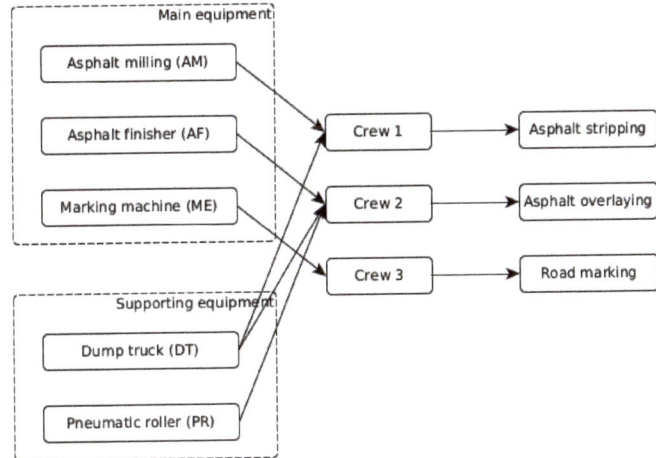

Figure 3. Equipment combination and configuration (ECC) schema.

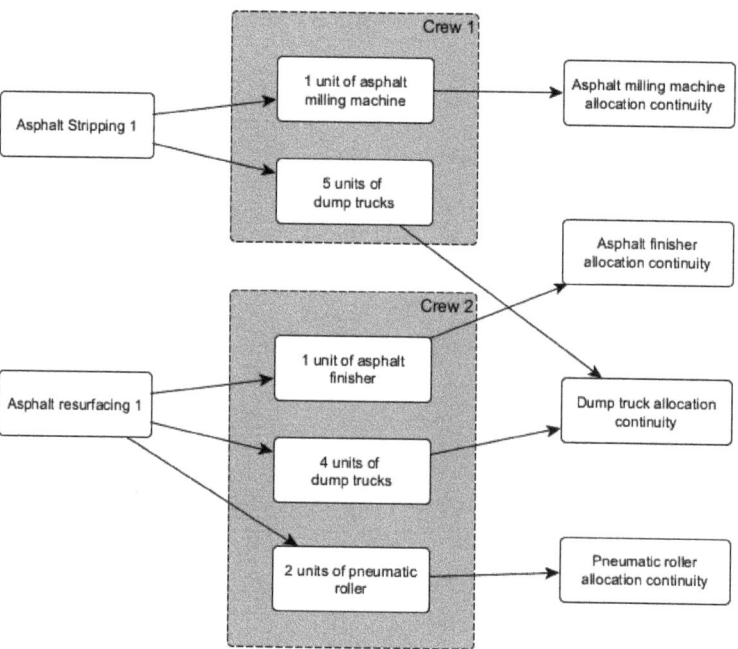

Figure 4. Equipment allocation continuity schema.

3.1.3. Model Features

This optimization model proposed two features to improve the practicality and efficiency of equipment allocation. Those features are (1) resource dependency between main equipment and supporting equipment, and (2) flexible resource profile.

Resource dependency in this model is defined as the dependency between main equipment and supporting equipment. The traditional linear scheduling method (LSM) applies the term crew as a resource of activities. One type of crew is allocated to a particular type of activity and shared among the same type of activities. The proposed model divides the crew into equipment. In this model, the crew consists of some main equipment and supporting equipment. Therefore, shared resources in the proposed model are not at the crew level but the equipment level instead. This opens possibilities of shared equipment between different types of work. Figure 5 shows crew allocation to activities in a traditional way and the addition of the resource dependency concept. Besides the main equipment used to finish an activity, this model also considers supporting equipment, which is important in maintaining the work performance of the main equipment. Asphalt stripping activity needs an asphalt milling machine (AM) as the main equipment. However, an asphalt milling machine needs dump trucks (DT) to take the product of the asphalt stripping process to the dumping area. The number of dump trucks supporting the asphalt milling machine is important to match both types of equipment's work rates. Besides the asphalt milling machine, the asphalt finisher machine—the main equipment of asphalt overlaying activity—also needs dump trucks to support its work. This model also tackles the condition where some main items of equipment of several activities need the support of one type of supporting equipment. This condition became a challenge especially when there is a shortage of resource availability.

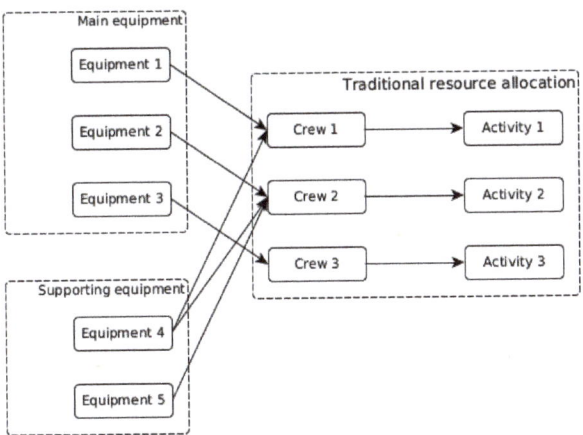

Figure 5. Resource allocation in traditional LSM and proposed resource dependency concept.

Each road segment of this project consists of three serial activities, which are (1) asphalt stripping, (2) asphalt resurfacing, and (3) road marking. Asphalt stripping starts the works on every segment. Asphalt milling machines (AM) act as the main equipment for this activity. The asphalt milling machine is supported by dump trucks (DT) to collect and take the product of the asphalt milling machine to the dumping area. In this model, some dump trucks (DT) are assigned to one asphalt milling machine (AM) to match the work rate between the asphalt milling machine and the dump trucks to achieve the most efficient work rate. The unmatched work rate of either the main equipment or the supporting equipment would result in lower overall work performance. Asphalt resurfacing is the succeeding activity of the asphalt stripping activity. Asphalt finishers (AF), pneumatic rollers (PR), and dump trucks (DT) are used in this asphalt resurfacing activity. Asphalt finishers are the main equipment used to resurface the road segment supported by several pneumatic rollers and dump trucks. Pneumatic rollers compact the new pavement surface, and dump trucks supply hot-mixed asphalt from an asphalt mixing plant to the asphalt finisher. Several pneumatic rollers and dump trucks are assigned to one asphalt finisher to match the work rate among the equipment to achieve the most efficient work performance. Road marking is the last activity to execute on each road segment. Road marking is the succeeding activity of asphalt resurfacing activity. Road marking machines (ME) are assigned to this road marking work as the main equipment. Figure 3 shows the schema of main and supporting equipment allocation for this model.

Flexible resource profiles are the second feature of the proposed model as an addition to the linear scheduling method. In the traditional linear scheduling method, the resource allocation profile commonly has a constant rate; thus, it makes the shape of a bar. As mentioned above, for road safety reasons, this model omits time floating buffer, which causes a disadvantage in minimizing idle equipment. To deal with this problem, this model implements flexible shapes of resource allocation profiles to provide a different allocation of equipment during the transition between the same type of activities on different road segments. However, it is not an easy task to move equipment between activities or between repetitive units. Thus, this model limits the shape of the resource profile only as a bar or trapezoid shape.

3.1.4. Model Objective

The objective of this model is to minimize equipment idleness caused by valleys of equipment allocation profile (Figure 1). Idle equipment is not calculated by the difference between maximum available equipment and allocated equipment at a time, but the difference between mobilized equipment and allocated equipment at a time. The proposed model achieves the objective by eliminating these valleys.

3.2. Model Formulation

The optimization model for minimizing idle equipment of road network maintenance project was developed following the model development workflow (Figure 6) and using constraint programming engine of IBM ILOG CPLEX Optimization Studio version 12.10 and OPL modeling language.

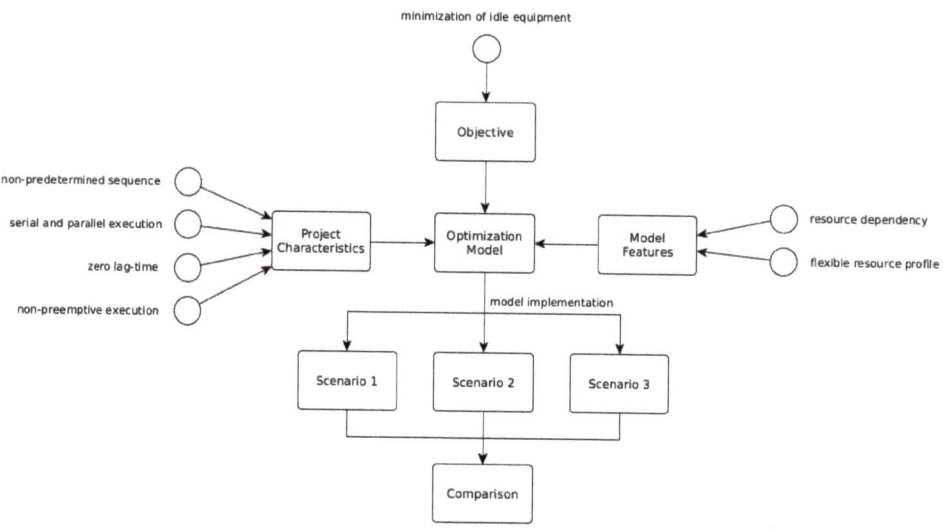

Figure 6. Optimization model development workflow.

This model binds activities to the allocation of their main equipment. Therefore, the proposed scheduling model schedules activities as the main equipment scheduling. The precedence relationships between activities are implemented as precedence relationships between main items of equipment.

This model consists of five parts, which are: (1) input data; (2) decision variable; (3) decision expression; (4) objective function; and (5) constraints.

3.2.1. Input Data

Input data are parameters set before the optimization process started. These data are not changed during the optimization process. Input data are expressed by input variables and structured using indices. Tables 4 and 5 denote indices and input variables used by the proposed model, respectively.

Table 4. Indices used by variables of the proposed model.

Index	Description	Range
i, m	Index of equipment type	1 ... number of equipment types
j	Index of a road segment	1 ... number of road segments
k	Index of time (day)	1 ... contract period

Table 5. Input variables used by the proposed model.

Variable	Description
u	Number of equipment types
r	Number of road segments
d	Contract period
pr_{jim}	Precedence relationship between main items of equipment. This variable holds the precedence relationship data between predecessor m and successor i in road segment j. It contains values of zero and one. For example, $pr_{123} = 1$ means on the road segment 3, main equipment 1 is the predecessor of main equipment 2.
PEr_{ij}	The number of main items of equipment required to finish an activity. This variable holds the number of main items of equipment i required on the road segment j. $PEr_{12} = 4$ means 4 unit-day of main equipment type 1 is required to finish the related activity on road segment 2. If one unit of main equipment type 1 is allocated, the activity will be finished in four days, or if two units of main equipment 1 are allocated, the activity is finished in two days, etc.
R_{jim}	The ratio between supporting equipment and main equipment. This variable describes the relationship between supporting equipment i and main equipment m on the road segment j. $R_{241} = 5$ means that, on the road segment 2, equipment 4 is the supporting equipment of equipment 1, with a ratio of 5:1.
Av_i	The number of available items of equipment. This variable holds the number of available items of equipment i.

3.2.2. Decision Variables

Decision variables are the objects of the optimization process. The optimization algorithm searches the optimized values of the decision variables to achieve the objective function. The values of these variables are changed during the optimization process until they reach the considered optimum values based on the objective function.

PE_{ijk} is used as the decision variables of this model. PE_{ijk} is the number of allocated main items of equipment, i, on the road segment j on day k. $PE_{123} = 2$ means 2 units of main equipment type 1 are allocated on road segment 2 on day 3.

3.2.3. Decision Expressions

CPLEX constraint programming and OPL programming language allow the user to use decision expressions. Decision expressions are mathematical expressions for variables that are not the target of the optimization algorithm. For example, a mathematical expression $b = a + 2$, a is the decision variable; therefore, the value of a is based on the optimization algorithm. However, b is not the target of the optimization algorithm. The value of b only follows the value of a; thus, to define the value of b, a decision expression is used.

E_{ijk} is the allocation of equipment i of road segment j on day k. Equipment allocation is the total allocation of the equipment as main equipment and supporting equipment; thus, it is the sum of the allocation number as main equipment and the allocation number of supported main equipment multiplied by the ratio of supporting equipment to main equipment (Equation (1)).

$$E_{ijk} = ME_{ijk} + \sum_m SMratio_{jim} \times ME_{mjk} \qquad (1)$$

$Ereq_{ij}$ is the required allocation of equipment i on the road segment j. The required equipment allocation is the sum of the allocation as main equipment and as supporting equipment. This variable acts as a control variable whether the allocation of particular equipment has reached the required allocation or not (Equation (2)).

$$Ereq_{ij} = MEreq_{ij} + \sum_m SMratio_{jim} \times MEreq_{mj} \qquad (2)$$

Equation (3) defines variable Ed_{ik} as the total allocation of equipment i on day k of all road segments. This variable is used in a constraint to limit the equipment allocation to the available equipment on one day.

$$Ed_{ik} = \sum_j E_{ijk} \qquad (3)$$

Er_{ij} is the total allocation of equipment i of road segment j. This expression (Equation (4)) defines this variable, which is used to track the total allocation of particular items of equipment on a road segment. Combined with variable $Ereq_{ij}$, this variable is used to check if the allocation of a particular item of equipment has met the requirement.

$$Er_{ij} = \sum_k E_{ijk} \qquad (4)$$

Equation (5) defines variable $Etot_{ijk}$, which tracks the total allocation of equipment i of road segment j from day 1 to day k. This variable is used to track whether an activity has finished or not.

$$MEtot_{ijk} = \sum_{n=1}^{k} ME_{ijn} \qquad (5)$$

Variable F_{ijk}, defined by Equation (6), identifies whether equipment i of road segment j has finished on day k. It has the value of zero and one. The value of zero means the equipment i of road segment j has not been completely allocated on day k. The value of one means it has been allocated completely.

$$F_{ijk} = \begin{cases} 1 & \Leftrightarrow MEtot_{ij,k-1} = MEreq_{ij} \\ 0 & \Leftrightarrow MEtot_{ij,k-1} \neq MEreq_{ij} \end{cases} \qquad (6)$$

Equation (7) defines variable $precPass_{jimk}$, which is the indicator if equipment m as the predecessor of equipment i of road segment j has finished on day k. If $precPass_{jimk}$ has the value of zero, it means equipment m of road segment j is not considered finished. Equipment m is considered finished if equipment m is not the predecessor of equipment i, or equipment m has been allocated as required.

$$precPass_{jimk} = \begin{cases} 1 & \Leftrightarrow \left(prec_{jim} = 0\right) \vee \left(prec_{jim} = 1 \wedge F_{mjk} = 1\right) \\ 0 & \Leftrightarrow \neg\left[\left(prec_{jim} = 0\right) \vee \left(prec_{jim} = 1 \wedge F_{mjk} = 1\right)\right] \end{cases} \qquad (7)$$

Equations (8) and (9) define variables $Ebmax_{ik}$ and $Eamax_{ik}$, which are the maximum daily allocation of equipment i from day 1 to day k and the maximum daily allocation of equipment i from day k to the last day, respectively. These variables are used as the benchmark for idle equipment calculation.

$$Ebmax_{ik} = \max_{m \in 1..k} Ed_{im} \qquad (8)$$

$$Eamax_{ik} = \max_{m \in k..d} Ed_{im} \qquad (9)$$

Variables eb_{ik} and ea_{ik}, defined by Equations (10) and (11), are the number of idle items of equipment based on maximum daily allocation benchmark before day k and after day k, respectively. The value of these variables would be assessed by Equation (12) to determine the value of idle equipment. Variable e_{ik}, defined by Equation (12), is the number of idle items of equipment on day k.

$$eb_{ik} = \begin{cases} Ebmax_{i,k-1} - Ed_{ik} & \Leftrightarrow Ed_{ik} < Ebmax_{i,k-1} \\ 0 & \Leftrightarrow Ed_{ik} \geq Ebmax_{i,k-1} \end{cases} \qquad (10)$$

$$ea_{ik} = \begin{cases} Eamax_{i,k+1} - Ed_{ik} & \Leftrightarrow Ed_{ik} < Eamax_{i,k+1} \\ 0 & \Leftrightarrow Ed_{ik} \geq Eamax_{i,k+1} \end{cases} \qquad (11)$$

$$e_{ik} = \begin{cases} eb_{ik} \Leftrightarrow eb_{ik} \leq ea_{ik} \wedge (eb_{ik} > 0 \wedge ea_{ik} > 0) \\ ea_{ik} \Leftrightarrow eb_{ik} > ea_{ik} \wedge (eb_{ik} > 0 \wedge ea_{ik} > 0) \\ 0 \Leftrightarrow \neg(eb_{ik} > 0 \text{ and } ea_{ik} > 0) \end{cases} \quad (12)$$

Variables ET_{ik} and T_{ik}, defined by Equations (13) and (14), are variables for identifying whether equipment i and any equipment is allocated on day k. Variable T_k is also used to count the duration of the project.

$$ET_{ik} = \begin{cases} 1 \Leftrightarrow Ed_{ik} > 0 \\ 0 \Leftrightarrow Ed_{ik} \leq 0 \end{cases} \quad (13)$$

$$T_k = \begin{cases} 1 \Leftrightarrow \sum_i ET_{ik} > 0 \\ 0 \Leftrightarrow \sum_i ET_{ik} \leq 0 \end{cases} \quad (14)$$

3.2.4. Objective Function

The objective of this optimization model is to minimize idle equipment of a road-network maintenance project. Variable e_{ik}, which is defined by Equation (12), is used to define the objective function of this model. The objective function (Equation (15)) is the sum of variable e_{ik} for all items of equipment i and day k.

$$\sum_x \sum_k e_{ik} \quad (15)$$

3.2.5. Constraints

Equations (16)–(24) are constraints regulating the precedence relationship between main items of equipment. These equations are expressed using the syntax OPL programming language used by IBM ILOG CPLEX CP engine. These constraints are defined using conditional assessment; thus, OPL implication syntax is used for these definitions.

Equations (16)–(19) regulate the precedence relationship for main equipment, which do not have other items of equipment as their predecessors. These constraints are started by a conditional assessment of whether a particular item of equipment has other items of equipment as its predecessor, shown by the expression of $\sum_m prec_{jim} = 0$. If the assessment results in a true value, it means equipment i does not have any predecessors. Equation (16) defines that, if equipment i of road segment j does not have any predecessors, then equipment i could be allocated on day k.

$$\sum_m prec_{jim} = 0 \Rightarrow E_{ijk} \geq 0 \quad (16)$$

Constraints defined by Equations (17) and (18) state that if equipment i of road segment j does not have any predecessors allocated on the previous days, and it has not reached the number required allocation, it needs to be allocated on day k.

$$\sum_m prec_{jim} = 0 \wedge E_{ij,k-1} > 0 \wedge Etot_{ij,k-1} < Ereq_{ij} \Rightarrow E_{ijk} \geq 0 \quad (17)$$

$$\sum_m prec_{jim} = 0 \wedge E_{ij,k-1} > 0 \wedge Etot_{ij,k-1} < Ereq_{ij} \Rightarrow E_{ijk} \leq Ereq_{ij} - Etot_{ij,k-1} \quad (18)$$

Equation (19) defines that, if equipment i of road segment j has reached the number of required allocations, it is not allowed to be allocated anymore.

$$\sum_m prec_{jim} = 0 \wedge Etot_{ij,k-1} = Ereq_{ij} \Rightarrow E_{ijk} = 0 \quad (19)$$

Equations (20)–(24) regulate the precedence relationship for the main items of equipment, which have other items of equipment as their predecessors. These constraints are started by a conditional assessment of whether particular items of equipment have other items of equipment as their predecessors, shown by the expression of $\sum_m prec_{jim} > 0$. If the assessment results in a true value, it means the items of equipment i have some predecessors. Equation (20) defines that, if equipment i of road segment j does not have any predecessors, then equipment i could be allocated on day k.

Equation (20) defines that, if the items of equipment i of road segment j have predecessors, they are not allowed to be allocated on day 1.

$$\sum_m prec_{jim} > 0 \Rightarrow E_{ij1} = 0 \tag{20}$$

If item of equipment i of road segment j on day k has not had all its predecessors considered finished, it is not allowed to be allocated. This condition is defined by Equation (21). The expression $\sum_m precPass_{jimk} < u$ shows that the number of items of equipment considered finished on day k is smaller than the registered equipment; thus, it means some of the predecessors are still not finished.

$$\sum_m prec_{jim} > 0 \wedge \sum_m precPass_{jimk} < u \Rightarrow E_{ijk} = 0 \tag{21}$$

The expression $\sum_m precPass_{jimk} = u$ of Equations (22) and (23) shows that all registered equipment is considered finished on day k; thus, it means all predecessors of equipment i have been finished. This condition allows for equipment i to be allocated on day k.

$$\sum_m prec_{jim} > 0 \wedge \sum_m precPass_{jimk} = u \wedge Etot_{ij,k-1} < Ereq_{ij} \Rightarrow E_{ijk} > 0 \tag{22}$$

$$\sum_m prec_{jim} > 0 \wedge \sum_m precPass_{jimk} = u \wedge Etot_{ij,k-1} < Ereq_{ij} \Rightarrow E_{ijk} \leq Ereq_{ij} - Etot_{ij,k-1} \tag{23}$$

Equation (24) defines that, if item of equipment i of road segment j on day $k-1$ has been allocated as required, equipment i is not allowed to be allocated anymore.

$$\sum_m prec_{jim} > 0 \wedge \sum_m precPass_{jimk} = u \wedge Etot_{ij,k-1} = Ereq_{ij} \Rightarrow E_{ijk} = 0 \tag{24}$$

Constraints expressed by Equations (25) and (26) limit the maximum daily allocation of each item of equipment to the number of available items of equipment and reach the required allocation on each road segment, respectively.

$$Ed_{ik} \leq Eavail_i \tag{25}$$

$$Er_{ij} = Ereq_{ij} \tag{26}$$

The constraint defined by Equation (27) makes sure that the project is started on day 1.

$$T_1 = 1 \tag{27}$$

3.2.6. Model Implementation Scenario

The road-network maintenance project scheduling—termed as the scheduling problem for the rest of this paper—was solved by three scenarios. The first scenario solves the scheduling problem by traditional LSM. This scenario acts as a benchmark for comparison to other scenarios. The second scenario solves the scheduling problem using the proposed model with the limitation of rectangular resource profiles. The third scenario solves the scheduling problem using the proposed model utilizing flexible resource profiles. Table 6 shows the comparison of the scenarios.

Table 6. Comparison of scenarios.

No.	Aspect	Scenario 1	Scenario 2	Scenario 3
1	Solving method	Traditional LSM	Proposed model	Proposed model
2	Execution order	predetermined	Not predetermined	Not predetermined
3	Execution sequence	Serial	Parallel	Parallel
4	Resource profile	Rectangular	Rectangular	Trapezoidal
5	Resource	Crew	Main and supporting equipment	Main and supporting equipment

The scheduling problem solved has similar data for each scenario, except precedence relationship between road segments and type of resource used by activities. Tables 5 and 6 show resource requirements and resource availability for the project scheduling problem, respectively. The precedence relationship of Scenario 1 is different from that of the other two scenarios. Since it has a predetermined execution sequence, Scenario 1 has an additional precedence relationship of asphalt milling machine allocation between road segments. It is set that road segment 1 is executed the first time and road segment 5 is executed the last. Figure 7 shows the precedence relationship used by Scenario 1. Scenario 1 also has a different type of resource. It does not have equipment dependency; thus, it uses the more traditional sharing, that is, 'crew' level resource sharing. In this scenario, the crew is the main equipment of the particular activity without dependency on supporting equipment—as shown in Figure 8. Scenarios 2 and 3 use precedence relationship and equipment allocation schema, as shown Figures 1 and 3, respectively.

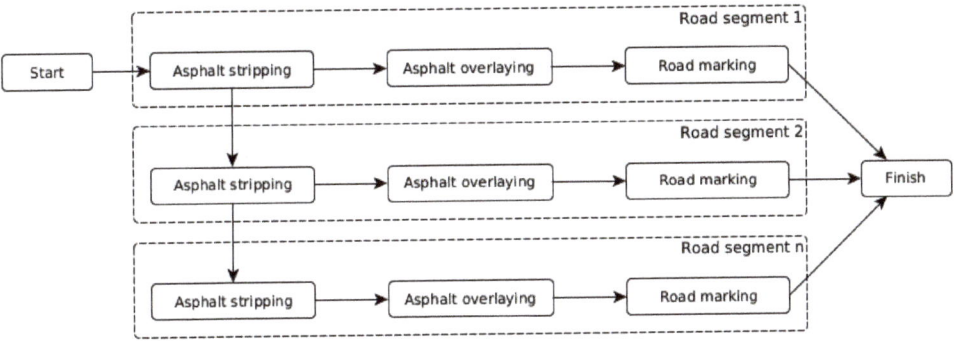

Figure 7. Precedence relationship of scenario 1.

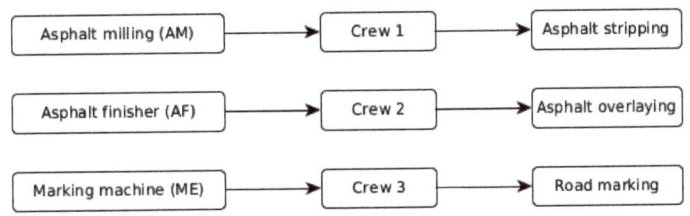

Figure 8. Equipment allocation schema for Scenario 1.

4. Result and Discussion

Road network maintenance project scheduling problem was solved in three scenarios. Scenario 1 used the traditional linear scheduling method (LSM) to solve the scheduling problem. Repetitive units—road segment works—were executed sequentially using a predetermined sequence. Scheduling began with the execution of road segment 1 and ended with road segment 5. When an activity is completed on a road segment, the crew will carry out the same type of activity on the next road segment without having to wait for the entire sequence of activities on the previous road segment to complete. Rectangular or uniform resource profiles were applied in this scenario. This scenario served as a comparison reference for the other two scenarios, i.e., Scenario 2 and Scenario 3.

Scenario 2 and Scenario 3 applied the proposed model for minimizing idle equipment. Road segments' execution does not have to be sequential. In addition to sequential execution, road segments could be carried out in parallel as long as the required equipment is still available. If road segments are carried out sequentially, the order in which the work is undertaken is not predetermined. This optimization model would determine the sequence of road segments' execution to achieve the model's objectives. Scenario 2 and Scenario 3 utilized rectangular and flexible resource profiles to solve the scheduling model, respectively. These scenarios—Scenario 2 and Scenario 3—were compared to Scenario 1 to show the advantages offered by the proposed model. The result of Scenario 2 would also be further compared to the result of Scenario 3 to show the difference caused by flexible resource profiles.

The model was run on the Intel Xeon Silver 4112 platform with 16 GB of RAM. The time needed to complete one scenario was about 20 min.

4.1. Scenario 1: Traditional LSM

The traditional linear scheduling method (LSM) was used to solve the scheduling problem as Scenario 1. Tables 7 and 8 show resource requirements and resource availability of this scenario, respectively. The road segment execution had a predetermined sequence, starting from road segment 1 and ending by road segment 5—as shown by Figure 7, and followed by a serial execution order. Resource allocation for this scenario was limited to a rectangular resource profile and did not implement the relationship between main equipment and supporting equipment. The resource allocation schema used by this scenario is shown in Figure 8. While most of the characteristics of this scenario follow traditional LSM, this scenario limits the schedule not to have a lag time between activities in one road segment, and once an activity is started, it has to be continuously executed until the activity is finished.

Table 7. Resource requirements.

Road No.	Pavement Stripping		Pavement Overlaying			Road Marking
	AM (Unit-Day)	DT (Unit-Day)	AF (Unit-Day)	DT (Unit-Day)	PR (Unit-Day)	ME (Unit-Day)
1	2	10	5	20	10	2
2	5	25	13	52	26	5
3	3	15	7	28	14	3
4	2	10	6	24	12	2
5	3	15	7	28	14	3

Table 8. Available resource.

No.	Equipment Type	Amount	
1	Asphalt stripping equipment (AM)	3	unit
2	Asphalt finishing equipment (AF)	3	unit
3	Marking equipment (ME)	1	unit
4	Dump truck (DT)	14	unit
5	Pneumatic roller (PR)	6	unit

The road segment execution follows a serial sequence. When asphalt stripping activity in one segment finished, it was followed by its succeeding activity—asphalt spreading activity. At the same time, asphalt stripping activity on the next road segment was allowed to start. Asphalt stripping activity of the next road segment could start when it met other constraints, such as to ensure that it could be followed by its succeeding activities without lag time. Schedule and equipment allocation is shown in Figure 9.

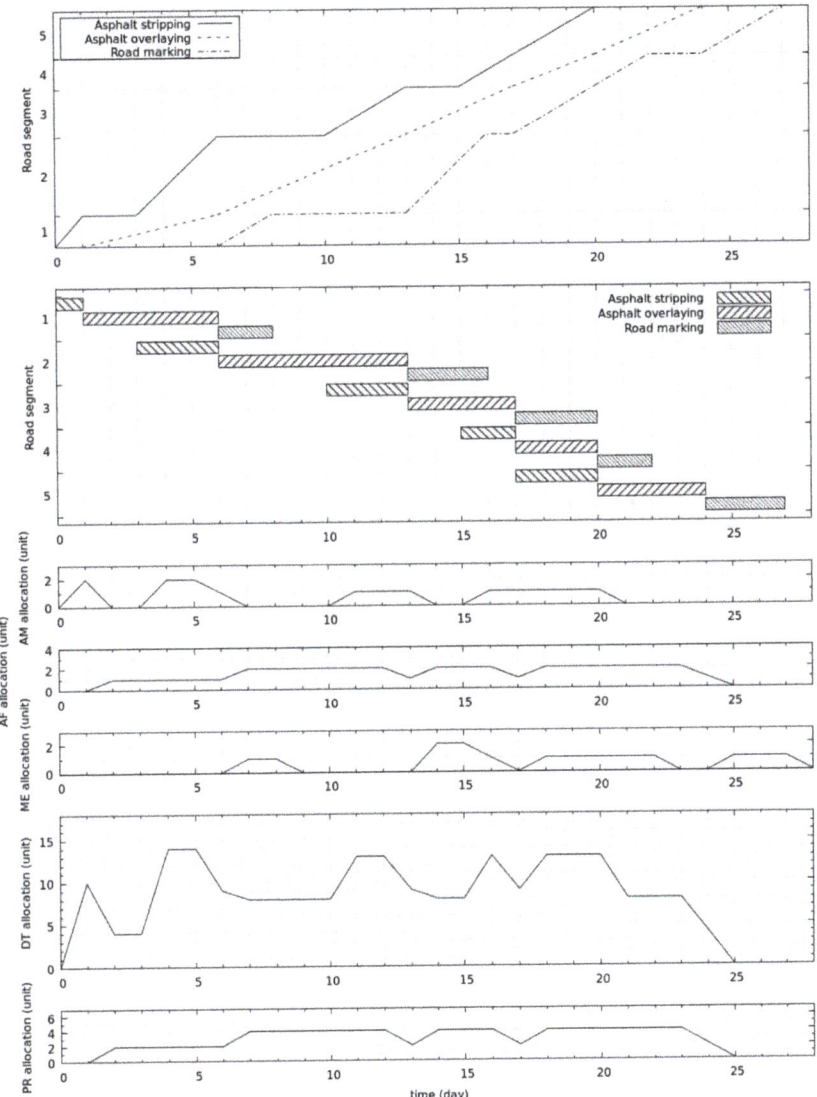

Figure 9. Schedule and equipment allocation of Scenario 1.

The schedule resulted in Scenario 1 having a project execution duration of 28 days and 87 units of idle equipment. Idle equipment resulted from this scenario is caused by the lack of resource availability and time buffer. The number of resources available in this scenario did not have enough amount to apply different work rates to the activities. Combined with the lack of time buffer, this scenario could not shorten or lengthen each activity's duration and it could not offset the activity's execution timing.

4.2. Scenario 2: Proposed Model with the Rectangular Resource Profile

Scenario 2 used the proposed model to solve the scheduling problem with a limitation of rectangular resource profile. This scenario has the same resource requirement and resource availability as Scenario 1—shown in Tables 7 and 8.

This scenario opens the possibilities of executing repetitive units by serial or parallel sequence. All supporting equipment availability are set to a high enough amount that these supporting items of equipment can support the main items of equipment that need them, even in parallel execution.

The result of the linear schedule of this scenario is shown in Figure 10. Asphalt stripping and asphalt spreading activities were executed in parallel at some span of duration, while road marking activities were executed in serial sequence. Figure 10 also shows the equipment allocation schedule. Some idle equipment happened at the total allocation of asphalt milling machine, road marking machine, and dump truck.

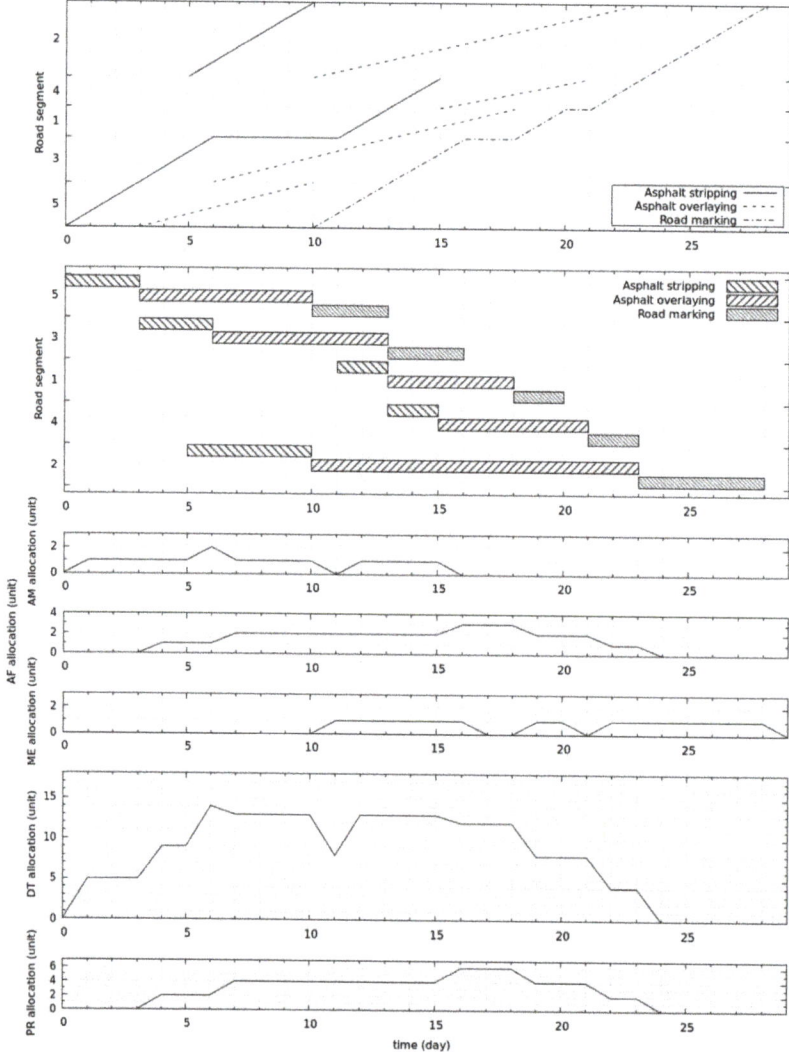

Figure 10. Schedule and equipment allocation of Scenario 2.

The optimization model results project execution duration of 29 days and nine units of idle equipment. The ability to execute activities in parallel applied in this scenario allowed the model to flatten the resource allocation profile and reduced a significant amount of idle equipment compared to the traditional LSM. However, this model is limited to a rectangular resource profile; thus, it could not apply a gradual increase in resource allocation at the beginning and a gradual decrease in resource allocation at the end of an activity's execution.

4.3. Scenario 3: Proposed Model with the Flexible Resource Profile

The proposed model with a flexible resource profile was applied to the scheduling problem as Scenario 3. This scenario opens the alternatives of executing repetitive units by serial or parallel sequence and allocating main equipment using flexible resource profiles to minimize idle equipment. Tables 7 and 8 show resource requirements and resource availability of this scenario, respectively. All supporting equipment availability is set to a high enough amount that these supporting items of equipment can support the main items of equipment that need them, even in parallel execution.

Figure 11 shows the schedule and equipment allocation of Scenario 3. Asphalt stripping and asphalt spreading activities were executed in parallel at some span of duration, while road marking activities were executed in serial sequence. Since this scenario allowed the implementation of flexible resource profiles, the main equipment allocations of each road segment were shaped into trapezoids. Thus, the model could apply a gradual increase and gradual decrease in resource allocation at the beginning and the end of the activity's execution duration.

The optimization model results project execution duration of 25 days and nine units of idle equipment. The ability to execute repetitive units—road segments—in fully or partially parallel execution and the implementation of flexible resource profile could decrease a significant amount of idle equipment compared to the traditional LSM and result in a shorter duration compared to the second scenario.

4.4. Comparison

This model was intended to minimize idle equipment in a non-sequential linear construction project, i.e., a road-network maintenance project. This model proposed several improvements compared to previous models, which are: (1) non-predetermined road segment execution sequence; (2) serial and parallel execution alternatives; (3) dependency between main equipment and supporting equipment; and (4) application of flexible resource profile. Thus, this model was implemented into three scenarios to prove that the model could handle the proposed improvements. The first scenario uses traditional LSM to solve the scheduling problem and acts as the basis for comparison to the other scenarios. The second scenario applies the proposed model with the limitation of rectangular resource profiles and has abundant supporting equipment availability. The third scenario implements the proposed model with flexible resource profiles. The second and third scenario was intended to show the advantage of the proposed model compared to the traditional LSM. Table 9 shows the comparison of the optimization results among implemented scenarios.

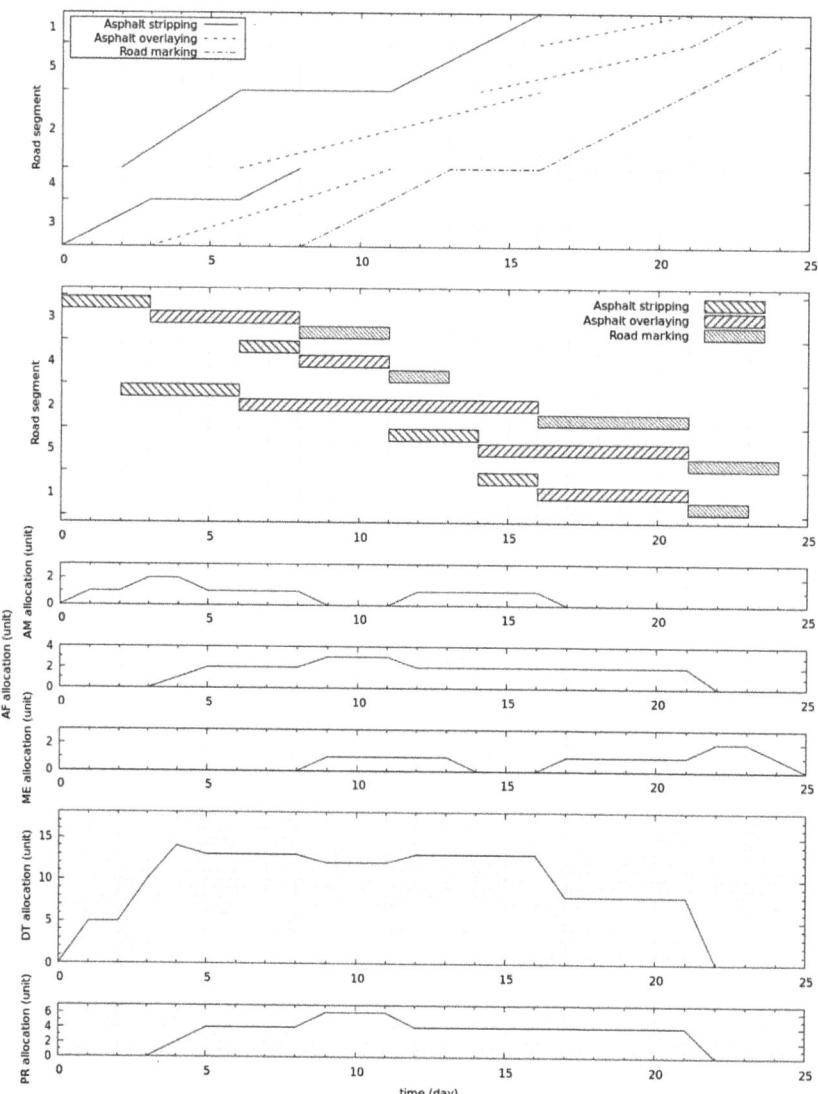

Figure 11. Schedule and equipment allocation of Scenario 3.

Table 9. Comparison of the optimization results.

No.	Aspect	Scenario 1	Scenario 2	Scenario 3
1	Idle equipment	87 unit-day	9 unit-day	9 unit-day
2	Duration	28 days	29 days	25 days
3	Execution order	predetermined	Not predetermined	Not predetermined
4	Execution sequence	Serial	Parallel	Parallel
5	Resource profile	Rectangle	Rectangle	Trapezoid
6	AM utilization	2	2	2
7	AF utilization	2	3	3
8	ME utilization	2	1	2

Table 9. *Cont.*

No.	Aspect	Scenario 1	Scenario 2	Scenario 3
9	DT utilization	14	14	24
10	PR utilization	4	6	6

The proposed model proved to be able to implement the proposed improvement mentioned above. Scenario 1, implementing traditional LSM, assigned repetitive units—road segments—by serial predetermined execution order. The equipment allocation schedule (Figure 9) shows some idle equipment happened because there is not any possibility to float a particular activity to meet work continuity.

The result of Scenario 2 and Scenario 3 shows that the proposed model was able to implement parallel execution for the project's repetitive units as long as it met the constraint of resource availability. In addition to the ability of parallel execution, Scenario 3 also shows the advantage of flexible resource profile to rectangular resource profile. The flexible resource profile allowed the implementation of a gradual increase and gradual decrease in resource profile at the beginning and the end of an activity execution duration, respectively. Scenario 2 and Scenario 3 have nine unit-days of idle equipment compared to 87 unit-days of idle equipment in Scenario 1. In addition to the minimization of idle equipment, Scenario 3 also has 25 days of project duration, which is shorter than in other scenarios.

5. Conclusions

The proposed model has succeeded in achieving its objectives from several points of view, namely project characteristics, model features, and model objectives. From the point of view of project characteristics, this model successfully accommodates the characteristics of a road network maintenance project, which are: (1) non-predetermined sequence of repetitive unit's execution, (2) serial or parallel execution of repetitive units, (3) zero lag time between activities inside a repetitive unit, and (4) non-preemptive execution of each activity (shown by the Gantt charts of Figures 10 and 11).

The proposed model succeeded to implement the features offered by this model. Resource dependency between main equipment and supporting equipment opens a new sight of resource allocation continuity. The crew consists of main equipment and supporting equipment and the number of equipment assigned to the crew. Compared to traditional LSM, which considers resource sharing at the crew level, this model shares resources at the equipment level. Thus, resource allocation continuity is observed by the allocation of a particular type of equipment, whether that particular equipment is assigned to the same type or a different type of crew.

The next feature of the proposed model is the implementation of flexible resource profiles. This model proved to obtain shorter project duration by the implementation of flexible resource profiles compared to the more traditional resource profiles—rectangular resource profiles.

From an optimization objective point of view, the proposed model has succeeded to minimize idle equipment of a non-sequential linear construction project by using constraint programming. This model calculated idle equipment based on the difference between actual mobilized equipment and the actual allocated equipment at a particular time. This calculation is more realistic than the usage of maximum mobilized equipment or possible available PR equipment as the reference value of the idle equipment calculation.

From several points of view mentioned above, it could be concluded that compared to the traditional linear scheduling method, this model has achieved several advantages, which are: (1) this model could schedule linear construction project without predetermined execution order; (2) this model could execute repetitive units in a serial or parallel way; (3) this model included a dependency between main equipment and supporting equipment, compared to traditional LSM which consider this resource as crew; (4) this model presented the alternative of using flexible resource profile; and (5) this model could minimize idle equipment, and thus this model could deliver equipment allocation continuity.

This model has several branches as future works. From the concept of resource allocation, this model applied the concept of dependency between main equipment and supporting equipment, which are renewable resources. This concept of dependency between renewable resources could be further explored to the concept of dependency between unrenewable resources and the procurement schedule of renewable and unrenewable redsources.

This research is applied to a case of a road network maintenance project. To further ensure that this resource allocation optimization model can be applied to general cases of road network maintenance projects, it is necessary to apply this model to other similar projects as further research.

Author Contributions: Conceptualization, S.-S.L., A.B. and M.F.A.A.; formal analysis, S.-S.L., A.B. and M.F.A.A.; investigation, S.-S.L., A.B. and M.F.A.A.; methodology, S.-S.L.; writing—original draft preparation, A.B. and M.F.A.A.; writing—review and editing, S.-S.L.; visualization, A.B. and M.F.A.A.; supervision, S.-S.L. All authors have read and agreed to the published version of the manuscript.

Funding: This research received no external funding.

Institutional Review Board Statement: Not applicable.

Informed Consent Statement: Not applicable.

Data Availability Statement: Not applicable.

Conflicts of Interest: The authors declare no conflict of interest.

References

1. Pinedo, M.L. *Scheduling: Theory, Algorithms, and Systems*, 4th ed.; Springer: New York, NY, USA, 2012; Volume 4.
2. Damci, A.; Arditi, D.; Polat, G. Multiresource leveling in line-of-balance scheduling. *J. Constr. Eng. Manag.* **2013**, *139*, 1108–1116. [CrossRef]
3. Tao, S.; Wu, C.; Sheng, Z.; Wang, X. Space-time repetitive project scheduling considering location and congestion. *J. Comput. Civ. Eng.* **2018**, *32*. [CrossRef]
4. Wu, C.; Wang, X.; Lin, J. Optimizations in project scheduling: A state-of-art survey. In *Optimization and Control Methods in Industrial Engineering and Construction*; Springer: Berlin/Heidelberg, Germany, 2014; pp. 161–177.
5. Hyari, K.; El-Rayes, K. Optimal planning and scheduling for repetitive construction projects. *J. Manag. Eng.* **2006**, *22*, 11–19. [CrossRef]
6. Zhang, L.; Dai, G.; Zou, X.; Qi, J. Robustness-based multi-objective optimization for repetitive projects under work continuity uncertainty. *Eng. Constr. Archit. Manag.* **2020**. [CrossRef]
7. Esfahan, N.R.; Razavi, S. Uncertainty-aware linear schedule optimization: A space-time constraint-satisfaction approach. *J. Constr. Eng. Manag.* **2017**, *143*. [CrossRef]
8. Moselhi, O.; Hassanein, A. Optimized scheduling of linear projects. *J. Constr. Eng. Manag.* **2003**, *129*, 664–673. [CrossRef]
9. Adeli, H.; Karim, A. *Construction Scheduling, Cost Optimization and Management*; CRC Press: Boca Raton, FL, USA, 2001; ISBN 0429076770.
10. Bakry, I.; Moselhi, O.; Zayed, T. Optimized acceleration of repetitive construction projects. *Autom. Constr.* **2014**, *39*, 145–151. [CrossRef]
11. Ipsilandis, P.G. Multiobjective linear programming model for scheduling linear repetitive projects. *J. Constr. Eng. Manag.* **2007**, *133*, 417–424. [CrossRef]
12. Roghabadi, M.A.; Moselhi, O. Optimized crew selection for scheduling of repetitive projects. *Eng. Constr. Archit. Manag.* **2020**, *28*, 1517–1540. [CrossRef]
13. Brucker, P.; Drexl, A.; Möhring, R.; Neumann, K.; Pesch, E. Resource-constrained project scheduling: Notation, classification, models, and methods. *Eur. J. Oper. Res.* **1999**, *112*, 3–41. [CrossRef]
14. Lucko, G.; Araújo, L.G.; Cates, G.R. Slip chart–inspired project schedule diagramming: Origins, buffers, and extension to linear schedules. *J. Constr. Eng. Manag.* **2016**, *142*, 4015101. [CrossRef]
15. Liu, S.S.; Budiwirawan, A.; Arifin, M.F.A.; Chen, W.T.; Huang, Y.H. Optimization model for the pavement pothole repair problem considering consumable resources. *Symmetry* **2021**, *13*, 364. [CrossRef]
16. Katsuragawa, C.M.; Lucko, G.; Isaac, S.; Su, Y. Fuzzy linear and repetitive scheduling for construction projects. *J. Constr. Eng. Manag.* **2021**, *147*. [CrossRef]
17. Su, Y.; Lucko, G. Linear scheduling with multiple crews based on line-of-balance and productivity scheduling method with singularity functions. *Autom. Constr.* **2016**, *70*, 38–50. [CrossRef]

18. Lucko, G.; Gattei, G. Line-of-balance against linear scheduling: Critical comparison. *Proc. Inst. Civ. Eng. Manag. Procure. Law* **2016**, *169*, 26–44. [CrossRef]
19. Tang, Y.; Liu, R.; Sun, Q. Two-stage scheduling model for resource leveling of linear projects. *J. Constr. Eng. Manag.* **2014**, *140*, 4014022. [CrossRef]
20. Harmelink, D.J.; Rowings, J. Linear scheduling model: Development of controlling activity path. *J. Constr. Eng. Manag.* **1998**, *124*, 263–268. [CrossRef]
21. Roofigari-Esfahan, N.; Paez, A.; N. Razavi, S. Location-aware scheduling and control of linear projects: Introducing space-time float prisms. *J. Constr. Eng. Manag.* **2015**, *141*, 06014008. [CrossRef]
22. Duffy, G.A.; Oberlender, G.D.; Seok Jeong, D.H. Linear scheduling model with varying production rates. *J. Constr. Eng. Manag.* **2011**, *137*, 574–582. [CrossRef]
23. Harmelink, D.J. Linear scheduling model: Float characteristics. *J. Constr. Eng. Manag.* **2001**, *127*, 255–260. [CrossRef]
24. Mattila, K.; Abraham, D. Resource leveling of linear schedules using integer linear programming. *J. Constr. Eng. Manag.* **1998**, *124*, 232–244. [CrossRef]
25. El-Rayes, K.; Jun, D.H. Optimizing resource leveling in construction projects. *J. Constr. Eng. Manag.* **2009**, *135*, 1172–1180. [CrossRef]
26. Mizutani, D.; Nakazato, Y.; Lee, J. Network-level synchronized pavement repair and work zone policies: Optimal solution and rule-based approximation. *Transp. Res. Part C Emerg. Technol.* **2020**, *120*, 102797. [CrossRef]
27. Huang, S.-H.; Lin, P.-C. Multi-treatment capacitated arc routing of construction machinery in Taiwan's smooth road project. *Autom. Constr.* **2012**, *21*, 210–218. [CrossRef]
28. Aarabi, F.; Batta, R. Scheduling spatially distributed jobs with degradation: Application to pothole repair. *Socioecon. Plann. Sci.* **2020**, *72*, 100904. [CrossRef]
29. Georgy, M.E. Evolutionary resource scheduler for linear projects. *Autom. Constr.* **2008**, *17*, 573–583. [CrossRef]
30. Tang, Y.; Liu, R.; Wang, F.; Sun, Q.; Kandil, A.A. Scheduling optimization of linear schedule with constraint programming. *Comput. Civ. Infrastruct. Eng.* **2018**, *33*, 124–151. [CrossRef]
31. Hojjat, A.; Samanwoy, G.-D. Mesoscopic-wavelet freeway work zone flow and congestion feature extraction model. *J. Transp. Eng.* **2004**, *130*, 94–103. [CrossRef]
32. Adeli, H.; Karim, A. Scheduling/cost optimization and neural dynamics model for construction. *J. Constr. Eng. Manag.* **1997**, *123*, 450–458. [CrossRef]
33. Xiaomo, J.; Hojjat, A. Freeway work zone traffic delay and cost optimization model. *J. Transp. Eng.* **2003**, *129*, 230–241. [CrossRef]
34. Shayanfar, E.; Schonfeld, P. Selecting and scheduling interrelated road projects with uncertain demand. *Transp. A Transp. Sci.* **2019**, *15*, 1712–1733. [CrossRef]
35. Koulinas, G.K.; Anagnostopoulos, K.P. Construction resource allocation and leveling using a threshold accepting–based hyper-heuristic algorithm. *J. Constr. Eng. Manag.* **2012**, *138*, 854–863. [CrossRef]
36. Hegazy, T. Optimization of resource allocation and leveling using genetic algorithms. *J. Constr. Eng. Manag.* **1999**, *125*, 167–175. [CrossRef]
37. Moselhi, O.; Lorterapong, P. least impact algorithm for resource allocation. *Can. J. Civ. Eng.* **1993**, *20*, 180–188. [CrossRef]
38. Tang, Y.; Liu, R.; Sun, Q. Schedule control model for linear projects based on linear scheduling method and constraint programming. *Autom. Constr.* **2014**, *37*, 22–37. [CrossRef]
39. Lucko, G. Integrating efficient resource optimization and linear schedule analysis with singularity functions. *J. Constr. Eng. Manag.* **2011**, *137*, 45–55. [CrossRef]
40. Wang, Z.; Hu, Z.; Tang, Y. Float-based resource leveling optimization of linear projects. *IEEE Access* **2020**, *8*, 176997–177020. [CrossRef]
41. Heon Jun, D.; El-Rayes, K. Multiobjective optimization of resource leveling and allocation during construction scheduling. *J. Constr. Eng. Manag.* **2011**, *137*, 1080–1088. [CrossRef]
42. Fündeling, C.-U.; Trautmann, N. A Priority-rule method for project scheduling with work-content constraints. *Eur. J. Oper. Res.* **2010**, *203*, 568–574. [CrossRef]
43. Naber, A.; Kolisch, R. MIP Models for resource-constrained project scheduling with flexible resource profiles. *Eur. J. Oper. Res.* **2014**, *239*, 335–348. [CrossRef]
44. Leu, S.-S.; Hwang, S.-T. Optimal repetitive scheduling model with shareable resource constraint. *J. Constr. Eng. Manag.* **2001**, *127*, 270–280. [CrossRef]
45. Liu, S.S.; Wang, C.J. Optimization model for resource assignment problems of linear construction projects. *Autom. Constr.* **2007**, *16*, 460–473. [CrossRef]
46. Zhang, L.; Zou, X.; Kan, Z. Improved strategy for resource allocation in repetitive projects considering the learning effect. *J. Constr. Eng. Manag.* **2014**, *140*, 4014053. [CrossRef]
47. Kong, F.; Dou, D. Resource-constrained project scheduling problem under multiple time constraints. *J. Constr. Eng. Manag.* **2021**, *147*. [CrossRef]
48. Siu, M.-F.F.; Lu, M.; AbouRizk, S. Resource supply-demand matching scheduling approach for construction workface planning. *J. Constr. Eng. Manag.* **2016**, *142*, 4015048. [CrossRef]

49. Khanzadi, M.; Kaveh, A.; Alipour, M.; Aghmiuni, H.K. Application of CBO and CSS for resource allocation and resource leveling problem. *Iran. J. Sci. Technol. Trans. Civ. Eng.* **2016**, *40*, 1–10. [CrossRef]
50. Tang, Y.; Sun, Q.; Liu, R.; Wang, F. Resource leveling based on line of balance and constraint programming. *Comput. Civ. Infrastruct. Eng.* **2018**, *33*, 864–884. [CrossRef]
51. Bendoly, E.; Perry-Smith, J.E.; Bachrach, D.G. The perception of difficulty in project-work planning and its impact on resource sharing. *J. Oper. Manag.* **2010**, *28*, 385–397. [CrossRef]
52. Xu, J.; Meng, J.; Zeng, Z.; Wu, S.; Shen, M. Resource sharing-based multiobjective multistage construction equipment allocation under fuzzy environment. *J. Constr. Eng. Manag.* **2013**, *139*, 161–173. [CrossRef]

MDPI
St. Alban-Anlage 66
4052 Basel
Switzerland
Tel. +41 61 683 77 34
Fax +41 61 302 89 18
www.mdpi.com

Mathematics Editorial Office
E-mail: mathematics@mdpi.com
www.mdpi.com/journal/mathematics

www.ingramcontent.com/pod-product-compliance
Lightning Source LLC
LaVergne TN
LVHW070735100526
838202LV00013B/1243